CZECHOSLOVAK ACADEMY OF SCIENCES

TRANSACTIONS

of the

EIGHTH PRAGUE CONFERENCE

on

INFORMATION THEORY, STATISTICAL DECISION FUNCTIONS, RANDOM PROCESSES

held at

Prague, from August 28 to September 1, 1978

VOLUME A

1978

D. REIDEL PUBLISHING COMPANY

DORDRECHT : HOLLAND/BOSTON : U.S.A.

The Library of Congress Cataloged the First Issue of this Title as Follows:

Conference on Information Theory, Statistical Decision Functions, Random Processes.

Transactions. 1st- conference; 1956-

Prague, Publishing House of the Czechoslovak Academy of Sciences.

v. diagrs. 25 cm. (Československá akademie věd. Sekce technická.

Studie a prameny, sv. 16

English, Russian, French, and German.

1. Information theory-Congresses. 2. Statistical decision. 3. Stochastic processes.

QA273.C743 519 58-42106

ISBN 978-94-009-9859-9 ISBN 978-94-009-9857-5 (eBook)
DOI 10.1007/ 978-94-009-9857-5

TRANSACTIONS include contributions of authors reprinted directly by a photographic method.
For this reason the authors are fully responsible for the correctness of their text.

Sold and distributed in the U.S.A., Canada and Mexico by D. Reidel Publishing Company, Inc., Lincoln Building, 160 Old Derby Street, Hingham, Mass. 02043, U.S.A.

Published by D. Reidel Publishing Company, P. O. Box 17, Dordrecht, Holland, in co-edition with ACADEMIA, Publishing House of the Czechoslovak Academy of Sciences, Prague

Softcover reprint of the hardcover 1st edition 1978

CONTENTS

PREFACE

During the past Prague Conferences, the Organizing Committee had to face with a great regret permanent and even increasing delays in publishing the Transactions. The delays seemed to be out of any our reach as long as the traditional manuscript delivery rules and traditional printing technique were supposed to be employed. This is why we decided, starting with this Conference, for a new method of papers collection and printing. We want to thank all participants of our conference, in particular those who contributed to the volumes A and B, for understanding this our step.

Our thanks are due to Academician Jaroslav Kožešník, the scientific editor of these volumes, and to the editorial board for reviewing all papers and fulfiling many printing management duties. We also thank Academia Publishing House for printing the whole volume in time.

ORGANIZING COMMITTEE

of the

EIGHTH PRAGUE CONFERENCE

on

INFORMATION THEORY,

STATISTICAL DECISION FUNCTIONS,

RANDOM PROCESSES

PROBLEMES LIES A LA DETERMINATION DES SPECTRES DE PUISSANCE EN THEORIE DES FONCTIONS ALEATOIRES

A. Blanc-Lapierre

Gif-sur-Yvette (France)

RESUME

La détermination expérimentale du spectre de puissance, au sens physique, d'une grandeur fluctuante pose des problèmes d'analyse spectrale et d'ergodisme. Ces deux types de problèmes sont examinés conjointement pour une grandeur représentée par une fonction aléatoire $X(t,\omega)$, pas nécessairement stationnaire, mais présentant des caractères de permanence suffisants pour que la puissance moyenne puisse être définie. On examine aussi le cas de spectres liés à certaines transformées non linéaires $F\{X(t,\omega)\}$ de $X(t,\omega)$. Enfin, on discute le rôle joué dans ce type de problèmes par les fonctions certaines admettant une répartition asymptotique des valeurs.

1 - INTRODUCTION.

La détermination expérimentale du spectre de puissance moyenne d'une grandeur fluctuante $X(t)$, "présentant des caractères suffisants de permanence" -précisément pour que la puissance moyenne existe !- mobilise des méthodes de filtrage (i) ou de corrélation (ii) que, sous forme abstraite, nous schématisons comme suit :

i) Filtrage

$$(1\text{-}1) \qquad \bar{F}_X(\nu) = \overline{\left|\mathfrak{F}_{-\infty,\nu}\{X(t)\}\right|^2} = \lim_{T\to\infty} \overline{\left|\mathfrak{F}_{-\infty,\nu}\{X(t)\}\right|^2}^{\,0\longrightarrow T}$$

où $\overline{[\quad]}^{\,a\longrightarrow b}$ $[a < b]$ représente la moyenne temporelle de $[\quad]$ sur $[a,b]$, $\mathfrak{F}_{-\infty,\nu}$ est le filtre de gain 1 sur $(-\infty,\nu)$ et zéro ailleurs, et $\bar{F}_X(\nu)$ la fonction de répartition du spectre de puissance moyenne de X.

ii) Corrélation

(1-2) $$\overline{C}_X(\tau) = \overline{X(t)X^*(t-\tau)} = \lim_{T\to\infty} {}^{0}\overline{X(t)X^*(t-\tau)}^{T}$$

Les modèles représentant des grandeurs fluctuantes doivent tenir compte de l'irreproductibilité essentielle de celles-ci, au cours du temps et entre expériences "macroscopiquement identiques". Relativement à (1-1) et (1-2), ces modèles doivent assurer l'existence des moyennes temporelles qui y figurent et faire qu'elles soient significatives c'est-à-dire reproductibles sur tout l'ensemble Ω des expériences ω macroscopiquement identiques. Dans les modèles aléatoires, on considère X comme une fonction aléatoire (f.a.) $X(t,\omega)$ définie sur l'ensemble Ω des épreuves ω . L'exigence de reproductibilité macroscopique pose le problème des propriétés ergodiques de X. On peut vouloir éviter tout caractère aléatoire et ne travailler que sur une seule fonction : celle qui est révélée à l'expérimentateur. Dans cette optique, on développe un modèle déterministe. Cela revient, en fait, à travailler sur l'une des trajectoire $X(t,\omega_o)$ de $X(t,\omega)$. $X(t,\omega_o)$ est, alors, traitée comme une fonction certaine à laquelle l'irreproductibilité au cours du temps impose des propriétés analytiques très particulières : les fonctions à utiliser sont, alors, des fonctions admettant une répartition asymptotique des valeurs. Ce sont les implications de ces considérations pour la détermination des spectres de puissance que nous analysons, d'abord pour les modèles aléatoires puis pour les modèles déterministes. Dès maintenant, notons que, par delà ses aspects techniques, l'étude de l'ergodisme n'est autre que le raccord entre ces deux types de modèles. Nous voulons étudier ces problèmes, non seulement pour le spectre de X(t) lui-même, mais aussi pour ceux d'un ensemble assez large de transformées non linéaires instantanées $F\{X(t)\}$ de X(t). En prenant F développable en série entière, cela conduit à l'étude des moyennes temporelles suivantes où k et ℓ sont des entiers ≥ 0 :

(1-3) $$\overline{\mathcal{F}_{-\infty,\nu}\{X^k(t)\}\left[\mathcal{F}_{-\infty,\nu}\{X^{\ell}(t)\}\right]^*} \quad (\alpha) \quad \text{ou} \quad \overline{X^k(t)\left[X^*(t-\tau)\right]^{\ell}} \quad (\beta)$$

2 - MODELES ALEATOIRES ET DETERMINATION DES SPECTRES DE PUISSANCE MOYENNE

2-1. HYPOTHESES SUR $X(t,\omega)$. a) Second ordre. X(t) est avant tout une f.a. de second ordre [(1) pp. 455, 493 et (2) pp. 342-470] , centrée , continue en m.q. et de covariance harmonisable, c'est-à-dire telle que :

(2-1) $$\Gamma_X[t,t'] = E\{X(t)X^*(t')\} = \int_{-\infty}^{+\infty}\int_{-\infty}^{+\infty} e^{2\pi i\,[\nu t-\nu' t']}\, d^2\gamma_X(\nu,\nu')$$

avec

(2-2) $$\int_{-\infty}^{+\infty}\int_{-\infty}^{+\infty} \left|d^2\gamma_X(\nu,\nu')\right| < C < +\infty$$

Dans ces conditions, X(t,ω) est, lui-même, harmonisable :

$$(2\text{-}3)\qquad X(t,\omega) = \int_{-\infty}^{+\infty} e^{2\pi i\nu t}\, dx(\nu,\omega) \quad (\alpha)\ ;\quad d^2\gamma_X(\nu,\nu') = E\{dx(\nu)dx^*(\nu')\} \quad (\beta)$$

Nous ne supposons pas, pour l'instant, X(t,ω) stationnaire de second ordre (c'est-à-dire à dx(ν,ω) orthogonaux). Cependant, l'hypothèse contenue dans (2-2) impose à X(t,ω) certains caractères de permanence : par exemple, quels que soient les filtres $\mathcal{F}_1$ et $\mathcal{F}_2$, de gains ≤ 1 en module, on a : $\left|E\{\mathcal{F}_1\{X(t_1)\}\ [\mathcal{F}_2\{X(t_2)\}]^*\}\right| < C$.

b) Moments d'ordre supérieur au second [hypothèses nécessaires pour les moments contenus dans (1-3)] : nous supposons que X(t,ω) est de la classe $\Phi(\infty)$ au sens de (2) [p. 366] c'est-à-dire tel que, $\forall$: K, $\mathcal{E}_1$, $\mathcal{E}_2$, ..., $\mathcal{E}_K$, l'intégrale :

$$(2\text{-}4)\qquad \int_{-\infty}^{+\infty}\cdots\int_{-\infty}^{+\infty} \left|E\{dx_{\mathcal{E}_1}(\nu_1)\ \ldots\ dx_{\mathcal{E}_K}(\nu_K)\}\right|$$

est finie [les $\mathcal{E}$ valent $\pm$ 1 et $a_{\mathcal{E}}$ vaut a pour $\mathcal{E} = +1$ et a^* pour $\mathcal{E} = -1$].

2-2. PROBLEMES INTRODUITS PAR LA MOYENNE TEMPORELLE

1° FILTRAGE ASSOCIE A LA MOYENNE TEMPORELLE. Soient Z(t,ω) la f.a. dont on aura à prendre la moyenne temporelle dans des expressions du type (1-3) [avec X = X(t,ω)] et z(ν,ω) la f.a. intervenant dans la représentation harmonique de Z selon (2-3α). L'opération ${}^{t}\overline{Z(t,\omega)}^{t+T}$ est un filtre linéaire qui, pour $T \longrightarrow \infty$, tend vers un filtre limite, $\mathcal{F}_{\nu=0}$, de gain $G(\nu) = 1$ pour $\nu = 0$ et $G(\nu) = 0$ pour $\nu \neq 0$. $\mathcal{F}_{\nu=0}$ isole, si elle existe, la raie $\nu = 0$. On a :

$$(2\text{-}5)\qquad \lim_{T\to\infty} \text{m.q. } {}^{0}\overline{Z(t,\omega)}^{T} = dz(0,\omega),\ \text{p.s.} = 0 \text{ si } d^2\gamma_Z(0,0) = 0$$

De même :

$$(2\text{-}6)\qquad \lim_{T\to\infty} \text{m.q. } {}^{0}\overline{\exp[-2\pi i\nu_0 t]\, Z(t,\omega)}^{T} = dz(\nu_0,\omega),\ \text{p.s.} = 0 \text{ si } d^2\gamma_Z(\nu_0,\nu_0) = 0$$

Les moyennes temporelles en moyenne quadratique de ce type isolent les "raies spectrales" de Z(t,ω). Leurs résultats ne peuvent être indépendants de ω que si ces raies ont une puissance moyenne nulle. Les raies spectrales constituent un obstacle à l'ergodisme.

2° ERGODISME - LA LOI FORTE DES GRANDS NOMBRES POUR Z(t,ω). L'ergodisme du modèle considéré implique que, Z(t,ω) -c'est-à-dire : X(t,ω) s'il s'agit de l'ergodisme du premier ordre ou l'une ou l'autre des expressions dont on prend la moyenne temporelle dans (1-3) pour l'ergodisme des moments d'ordre supérieur- soit tel que : (i) ${}^{0}\overline{Z(t,\omega)}^{T}$ ait une limite $\overline{Z(\omega)}$ presque sûre -p.s.- pour $T \longrightarrow \infty$ et que (ii) cette limite soit p.s. indépendante de ω . De façon très peu restrictive, il revient au même de dire que l'on doit avoir :

(2-7) $$\lim_{T \to \infty} \text{p.s.} \; {}^{0}\overline{Z(t,\omega)}^{T} = \overline{Z(t,\omega)} = \lim_{T \to \infty} {}^{0}\overline{E\{Z(t,\omega)\}}^{T} = \overline{Z}$$

ou encore :

(2-8) $$\lim_{T \to \infty} \text{p.s.} \; {}^{0}\overline{Z'(t,\omega)}^{T} = \overline{Z'(t,\omega)} = 0 \quad (\alpha) \quad \text{avec } Z' = Z - E\{Z\} \quad (\beta)$$

On peut donner des conditions portant uniquement sur le second ordre de $Z'(t,\omega)$, suffisantes pour assurer (2-8). Elles sont liées aux valeurs de la covariance $\Gamma_{Z'}(t,t')$ pour les grands $|t|$ et $|t'|$ -ou à celles des $d^2\gamma_{Z'}(\nu,\nu')$ près de $\nu = \nu' = 0$ c'est-à-dire aux propriétés des $dz'(\nu,\omega)$ près de $\nu = 0$. Donnons ici deux exemples de telles conditions suffisantes, renvoyant à l'annexe pour des développements plus détaillés.

Exemple 1 : Condition portant sur $\Gamma_{Z'}(t,t')$ [cf. Théorème II de l'annexe].

Théorème 1 - S'il existe deux constantes certaines M et α , $M > 0$ et $\alpha > 2$, telles que, pour les grandes valeurs de T, on ait :

(2-9) $$\frac{1}{T^2} \int_0^T \int_0^T \Gamma_{Z'}(t,t')\,dt\,dt' \leqslant \frac{M}{(\text{Log } T)^{\alpha}}$$

alors ${}^{0}\overline{Z'(t,\omega)}^{T} \xrightarrow{\text{p.s.}} \overline{Z'(t;\omega)} = 0$ pour $T \to \infty$.

Exemple 2 : Condition portant sur les $d^2\gamma_{Z'}(\nu,\nu')$ [cf. Théorème III de l'annexe].

Désignons par $\sigma_{Z'}[\nu]$ $[\nu > 0]$ l'intégrale de $|d^2\gamma_{Z'}(\nu_1,\nu_1')|$ sur $[|\nu_1| < \nu, |\nu_1'| < \nu]$.

Théorème 2 - Si, pour les faibles valeurs de $|\nu|$, on a :

(2-10) $$\sigma_{Z'}[\nu] \leqslant \left\{ 1/\left[|\text{Log } |\nu|| \right]^{2+\varphi} \right\} \qquad \varphi > 0$$

alors, ${}^{0}\overline{Z'(t,\omega)}^{T} \xrightarrow{\text{p.s.}} \overline{Z'(t,\omega)} = 0$ pour $T \to \infty$.

3° LES "MULTIPLICITES STATIONNAIRES" ET LEUR ROLE DANS LE PROBLEME CONSIDERE.

Les considérations de ce paragraphe concernent essentiellement l'ergodisme des moments d'ordre supérieur au premier. Prenons un exemple : soit l'expression ${}^{0}\overline{Y_1(t)Y_2^*(t-\tau)}^{T}$ où Y_1 et Y_2 sont les filtrées respectives de X(t) dans deux filtres $\mathcal{F}_1$ et $\mathcal{F}_2$ de gains uniformément bornés. Cette moyenne temporelle rentre, pour $T \to \infty$ et $k = \ell = 1$, dans le cadre de celles introduites en (1-3). Elle joue un rôle important dans la détermination expérimentale du spectre de puissance moyenne de X(t). D'autre part, les considérations que nous allons développer s'étendent aisément à des valeurs de $(k,\ell) > 1$. La fonction $Z(t,\omega)$ doit alors être prise égale à :

(2-11) $$Z(t,\omega) = Y_1(t)Y_2^*(t-\tau) = \underset{m.q.}{\int_{-\infty}^{+\infty}}\int_{-\infty}^{+\infty} e^{2\pi i[\nu-\nu']t}\, e^{2\pi i\nu\tau}\, G_1(\nu)G_2^*(\nu')\,dx(\nu)\,dx^*(\nu')$$

Nous conservons, naturellement, l'hypothèse (2-2) et nous utilisons (2-4) avec K = 4. En faisant tendre $T \longrightarrow \infty$, on filtre, dans ${}^{0}\overline{Y_1(t)Y_2^*(t-\tau)}^{T}$, la fréquence zéro pour $Z(t,\omega)$, c'est-à-dire que, dans le plan $\nu \times \nu'$, on ne conserve que les éléments $dx(\nu)dx^*(\nu')$ situés sur la première diagonale $\nu = \nu'$. On a donc :

(2-12) $$\lim_{T\longrightarrow\infty} \text{m.q.}\ {}^{0}\overline{Y_1(t)Y_2^*(t-\tau)}^{T} = \mathcal{L}(\tau,\omega) = \underset{m.q.}{\int_{-\infty}^{+\infty}} e^{2\pi i\nu\tau} G_1(\nu)G_2^*(\nu)\,|dx(\nu)|^2$$

Naturellement, nous désirons :

1) que la convergence m.q. intervenant dans la moyenne temporelle considérée ci-dessus soit, aussi, une convergence p.s.,

2) que la limite soit indépendante de ω : $\mathcal{L}(\tau,\omega) = E\{\mathcal{L}(\tau,\omega)\}$

Le fait d'avoir $\mathcal{L}(\tau,\omega) \overset{p.s.}{=} E\{\mathcal{L}(\tau,\omega)\}$ dépendra des propriétés des éléments aléatoires $dx(\nu)dx^*(\nu')$ sur $\nu - \nu' = 0$. La convergence presque sûre souhaitée sera, en outre, fonction des propriétés de ces mêmes éléments au voisinage de $\nu - \nu' = 0$ ou, plus précisément, des propriétés de $E\{dx(\nu_1)dx^*(\nu_1')dx^*(\nu_2)dx(\nu_2')\}$ simultanément près de $\nu_1 = \nu_1'$ et de $\nu_2 = \nu_2'$. Dans le cas stationnaire, auquel nous ne nous restreignons pas ici, ces questions sont discutées en détail dans (2) [p. 394 et suivantes]. Au total, on peut donner des conditions sur les $dx(\nu)dx^*(\nu')$ situés sur $\nu = \nu'$ et sur son voisinage pour assurer l'ergodisme du moment considéré sous la forme :

(2-13) $$\lim_{T\longrightarrow\infty} \text{p.s.}\ {}^{0}\overline{Y_1(t)Y_2^*(t-\tau)}^{T} = E\{\mathcal{L}(\tau)\} = \lim_{T\longrightarrow\infty} {}^{0}\overline{E\{Y_1(t)Y_2^*(t-\tau)\}}^{T}$$

Si on avait considéré la limite de l'expression ${}^{0}\overline{Y_1(t)Y_2(t-\tau)}^{T}$, il aurait fallu, dans ce qui précède, remplacer $\nu = \nu'$ par $\nu = -\nu'$. Plus généralement, la considération de ${}^{0}\overline{[Y_1(t)]^k[Y_2^*(t-\tau)]^{\ell}}^{T}$ -en poussant jusqu'à $K = k+\ell$ la condition (2-4)- aurait fait intervenir la multiplicité :

(2-14) $$\nu_1 + \nu_2 + \ldots \nu_k - \nu_{k+1} \ldots - \nu_{k+\ell} = 0$$

De façon générale, les propriétés ergodiques des moments d'ordres supérieurs du type de ceux que nous venons de considérer -avec toutes les extensions liées au choix des valeurs de k et de ℓ et au fait qu'on peut, pour chaque facteur Y, mettre Y ou Y^*- sont liées à celles des éléments aléatoires

(2-15) $$dx_{\varepsilon_1}(\nu_1)dx_{\varepsilon_2}(\nu_2) \ldots\ldots dx_{\varepsilon_K}(\nu_K)$$

sur les multiplicités (2-16) ci-dessous et sur leurs voisinages.

(2-16) $\varepsilon_1 \nu_1 + \varepsilon_2 \nu_2 + \ldots . \ \varepsilon_K \nu_K = 0$

Les multiplicités (2-16) sont dites "multiplicités stationnaires" car, si les moments de X sont stationnaires, les $E\{dx_{\varepsilon_1}(\nu_1) \ldots . \ dx_{\varepsilon_K}(\nu_K)\}$ ne peuvent être différents de zéro que sur ces multiplicités [cf. (2) p. 424].

3 - MODELES DETERMINISTES

3-1. NOTATIONS ET HYPOTHESES. INTRODUCTION DES FONCTIONS ADMETTANT UNE REPARTITION ASYMPTOTIQUE DES VALEURS. Soit $\mathcal{H}$ un espace vectoriel de fonctions H(t) [certaines !] , mesurables, à valeurs complexes [$H(t) \in C$ = plan complexe] . On suppose $\mathcal{H}$ invariant par translation et les H admettant une répartition asymptotique des valeurs (6) dans le sens suivant : $\forall \Phi$, continue et borné sur C, $\overline{\Phi[H_j(t)]}$ existe pour tout $H_j \in \mathcal{H}$. Il existe alors une mesure de probabilité $P(\omega)$, $\omega \in \Omega = C^{\mathcal{H}}$, telle que, $\forall \Phi$ et H_j on ait :

(3-1) $$\overline{\Phi[H_j(t)]} = \int_C \Phi(x) d\mu(x)$$

$\mu(x)$ correspondant à la projection de P sur $x = H_j$.

La mesure de probabilité ainsi introduite découle de la possibilité d'une pondération liée à la notion de temps relatif de réalisation [au sens de $\lim_{T \to \infty} \{(\Delta t)/T\}$]. L'existence de cette probabilité permet, dans le cadre strictement déterministe où nous nous plaçons maintenant, d'introduire des concepts de nature aléatoire.

Par ce procédé, détaillé en (7) et (8), on peut associer :
- à tout $H_j \in \mathcal{H}$, une variable aléatoire (v.a.) $h_j(\omega)$ $[\omega \in C^{\mathcal{H}}]$,
- à tout $H_j(t)$ et à l'ensemble de ses translatées $H_j(t+\lambda)$, $[-\infty < \lambda < +\infty]$, une fonction aléatoire stationnaire $h_j(\lambda,\omega)$.

Les lois de ces v.a. et de ces f.a. sont conformes à P et telles que, sous réserve d'existence, toute moyenne temporelle [côté H] correspond à une espérance mathématique égale [côté h] . On peut alors, éventuellement sous des hypothèses supplémentaires adéquates, étendre aux $H \in \mathcal{H}$ les résultats du calcul des probabilités relatif à l'indépendance, le caractère gaussien, le filtrage, l'analyse harmonique... etc. Nous reprenons rapidement, ci-dessous, les points essentiels relatifs à la fonction aléatoire stationnaire $h[\lambda,H;\omega]$ correspondant à H(t) et à ses translatées.

3-2. PROPRIETES DU SECOND ORDRE DES $h(\lambda,H;\omega)$

Les propriétés des f.a. stationnaires du second ordre s'appliquent aux $h(\lambda,H;\omega)$. Elles induisent des propriétés corrélatives pour les $H(t+\lambda)$. Il en est ainsi pour tout ce qui touche au filtrage linéaire, à la corrélation, aux spectres, à l'analyse harmonique [cf. (2) pp. 342-470 et (1) pp. 464-489] . On notera que la filtrée $\mathcal{F}\{h(\lambda,H;\omega)\}$ de la f.a. $h(\lambda,H;\omega)$ est la f.a. de λ $h[\lambda,\mathcal{F}[H];\omega]$ associée à la filtrée $\mathcal{F}[H(t)]$ de $H(t)$.

L'orthogonalité est définie par la nullité d'une espérance mathématique pour les h ou par celle de la moyenne temporelle correspondante pour les H. On distinguera entre l'orthogonalité de deux fonctions H et H' qui traduit la nullité de la moyenne temporelle de $H(t)H'^*(t)$ ou de $E\{h(0,H;\omega)h^*(0,H';\omega)\}$ et l'orthogonalité de ces deux fonctions et de leurs translatées qui traduit, pour τ quelconque, la nullité de la moyenne temporelle de $H(t+\tau)H'^*(t)$ ou de $E\{h(\tau,H;\omega)h^*(0,H';\omega)\}$.

Si, par exemple, on a :

(3-2) $$H(t) = \sum_{m.q.} A_n e^{2\pi i\nu_n t} \quad \text{et} \quad H'(t) = \sum_{m.q.} A'_n e^{2\pi i\nu'_n t}$$

l'orthogonalité de H et H' et de leurs translatées implique que les deux ensembles $\nu = \{\nu_n\}$ et $\nu' = \{\nu'_n\}$ n'aient pas d'élément commun.

En ce qui concerne les propriétés harmoniques, on introduit, pour $h(\lambda,H;\omega)$ la représentation suivante, classique pour les f.a. stationnaires de second ordre :

(3-3) $$h(\lambda,H;\omega) = \int_{-\infty\ (m.q.)}^{+\infty} e^{2\pi i\nu\lambda}\, d_\nu\, \eta(\nu,H;\omega)$$

où les $d_\nu\,\eta(\nu,H;\omega)$ sont orthogonaux. Cette orthogonalité à sa correspondance dans $\mathcal{H}$: si $\mathcal{F}_1$ et $\mathcal{F}_2$ sont des filtres disjoints $[G_1(\nu)G_2(\nu) \equiv 0]$, on a :

(3-4) $$\overline{\mathcal{F}_1[H(t)]\ .\ [\mathcal{F}_2[H(t+\tau)]]^*} = E\{h(\lambda,\mathcal{F}_1[H];\omega)\,.\,h^*(\lambda+\tau,\mathcal{F}_2[H];\omega)\}$$

$$= E\{\mathcal{F}_1\{h(\lambda,H;\omega)\}\,.\,[\mathcal{F}_2\{h(\lambda+\tau,H;\omega)\}]^*\}$$

$$= 0$$

Les passages à la limite intervenant dans la définition des filtres linéaires, ou dans celle de l'intégrale (3-3), sont entendus en moyenne quadratique : au sens d'espérances mathématiques dans (Ω,P) ou de moyennes temporelles, côté $\mathcal{H}$.

3-3. PROPRIETES D'ORDRES SUPERIEURS DES $h[\lambda,H;\omega]$

Ensemble $\mathcal{H}_{(\infty)}$. Par analogie avec la classe $\Phi(\infty)$ de la théorie des f.a. [cf. (2) p. 365 et p. 419], nous introduirons ici des ensembles $\mathcal{H}_{(\infty)}$. Un ensemble $\mathcal{H}$ sera dit du type $\mathcal{H}_{(\infty)}$, si, en plus des hypothèses de définition de $\mathcal{H}$, les H(t) satisfont, encore, aux suivantes.

Quels que soient : l'entier $K>0$, les $H_j(t)$ [distincts ou non], les $\varepsilon_j = \pm 1$ et les τ_j [$j = 1, 2, \ldots K$], on a :

a) $\overline{H_{\varepsilon_1}(t+\tau_1)\ldots H_{\varepsilon_K}(t+\tau_K)}$ existe et vaut :

(3-5) $$E_{(P)}\{h_{\varepsilon_1}(\tau_1,H_1;\omega)\ldots h_{\varepsilon_K}(\tau_K,H_K;\omega)\}$$

b)

(3-6) $$\int_{-\infty}^{+\infty}\cdots\int_{-\infty}^{+\infty}|E_{(P)}\{d_{\nu_1}\eta_{\varepsilon_1}(\nu_1,H_1;\omega)\ldots d_{\nu_K}\eta_{\varepsilon_K}(\nu_K,H_K;\omega)\}| < +\infty$$

Si $\mathcal{H}$ est un $\mathcal{H}_{[\infty]}$, on peut énoncer le résultat suivant :

Théorème 3. Pour tout ensemble $\{\nu_j,\varepsilon_j,H_j\}$ [$j=1,2,\ldots K$] ne satisfaisant pas à :

(3-7) $$\varepsilon_1\nu_1 + \varepsilon_2\nu_2 + \ldots + \varepsilon_K\nu_K = 0$$

on a :

(3-8) $$E_{(P)}\{d_{\nu_1}\eta_{\varepsilon_1}(\nu_1,H_1;\omega)\ldots d_{\nu_K}\eta_{\varepsilon_K}(\nu_K,H_K;\omega)\} = \overline{[\mathcal{F}d_{\nu_1}[H_1(t)]]_{\varepsilon_1}\ldots[\mathcal{F}d_{\nu_K}[H_K(t)]]_{\varepsilon_K}}$$
$$= 0$$

où $\mathcal{F}d_{\nu_j}$ est le filtre laissant passer la bande $[\nu_j,\nu_j+d_{\nu_j}]$.

Seuls peuvent différer de zéro les éléments du type (3-8) qui sont distribués sur les multiplicités stationnaires (3-7).

3-4. FONCTIONS H INDEPENDANTES AINSI QUE LEURS TRANSLATEES (même définition que dans le cas de l'orthogonalité). Si deux ensembles $H_j(t)$ [$j=1,2,\ldots$] et $H'_\ell(t)$ [$\ell=1,2,\ldots$] sont indépendants, l'un par rapport à l'autre, au sens des fonctions et de leurs translatées, les moyennes temporelles et les espérances mathématiques se scindent en deux facteurs et, $\forall$ les τ_j et les τ'_j, on a :

(3-9) $$\overline{H_{1_{\varepsilon_1}}(t+\tau_1)\dots H'_{1_{\varepsilon'_1}}(t+\tau'_1)\dots} = E_{(P)}\{h_{\varepsilon_1}(\tau_1,H_1;\omega)\dots h_{\varepsilon'_1}(\tau'_1,H'_1;\omega)\dots\}$$
$$= \overline{H_{1_{\varepsilon_1}}(t+\tau_1)\dots} \cdot \overline{H'_{1_{\varepsilon'_1}}(t+\tau'_1)\dots}$$
$$= E_{(P)}\{h_{\varepsilon_1}(\tau_1,H_1;\omega)\dots\}\ E_{(P)}\{h_{\varepsilon'_1}(\tau'_1,H'_1;\omega)\dots\}$$

Théorème 4. Pour que les deux fonctions H et H' introduites en (3-2) soient indépendantes ainsi que leurs translatées, il faut et il suffit que les deux ensembles $\mathcal{V}$ et $\mathcal{V}'$ définis ci-dessous n'aient d'autre élément commun que $\nu = 0$. $\mathcal{V}$ est constitué par les combinaisons linéaires d'un nombre fini d'éléments ν_n à coefficients $\gtreqless 0$. $\mathcal{V}'$ est constitué de la même façon du côté des ν'_n.

3-5. FONCTIONS H GAUSSIENNES. Pour simplifier, supposons les H réels. Soit $\mathcal{H}_1 \subset \mathcal{H}$. Nous dirons que tous les $H \in \mathcal{H}_1$ sont gaussiennes dans leur ensemble ainsi que leurs translatées, si, $\forall$ K et L, $\forall$ les $H_k \in \mathcal{H}_1$ (k=1,2,...K) et $\forall$ les λ_ℓ $[\ell=1,2,\dots L]$, les LK v.a. $h[\lambda_\ell,H_K;\omega]$ correspondant aux $H_k(t+\lambda_\ell)$ sont gaussiennes dans leur ensemble au sens de P. S'il en est ainsi, les fonctions aléatoires $h[\lambda,H_k;\omega]$ constituent un ensemble de f.a. de Laplace-Gauss. Toutes les propriétés de ces f.a. induisent alors des propriétés corrélatives pour les fonctions certaines $H \in \mathcal{H}_1$.

4 - RACCORD ENTRE MODELES ALEATOIRES ET MODELES DETERMINISTES - ERGODISME

Ce paragraphe établit un pont entre les deux points de vue correspondant respectivement aux modèles aléatoires (§ 2) et aux modèles déterministes (§ 3).

Nous sommes, au début du § 3, partis d'un ensemble de fonctions certaines $H \in \mathcal{H}$ essentiellement caractérisé par une invariance vis-à-vis des translations et par l'existence des répartitions asymptotiques. A cet ensemble, nous avons associé un ensemble de f.a. stationnaires $h[\lambda,H;\omega]$

(4-1) $$\mathcal{H} \longrightarrow h[\lambda,H;\omega]$$

On peut chercher à faire le cheminement inverse. Partons d'un ensemble $\mathcal{E}'$ de f.a. $X_j(t,\omega') \in \mathcal{E}'$, strictement stationnaires dans leur ensemble, définies sur les épreuves $\omega' \in \Omega'$. Considérons une épreuve particulière ω'_o. Les $X_j(t,\omega'_o)$ correspondants constituent-ils -ou non- un ensemble $\mathcal{H}$? Est-il possible que la réponse soit affirmative avec une probabilité 1 ? Suggérons, dans ses grandes lignes, la construction d'un exemple où il en est ainsi. Nous partons d'une f.a. à accroissements indépendants et stationnaires, soit $\mathcal{N}[\theta,\omega']$ et nous considérons les fonctions aléatoires :

(4-2) $$X_{R_j}(t,\omega') = \int_{-\infty}^{+\infty} R_j(t-\theta)\,d\mathcal{N}(\theta,\omega')$$

Par généralisation du raisonnement donné en (2), à la page 368, on peut montrer que, sous réserve d'astreindre R(t) à des conditions très larges de régularité, d'intégrabilité et de décroissance pour les grands $|t|$, il est possible d'imposer à l'ensemble des $X_j(t,\omega')$, d'une part, de faire partie d'un ensemble $\Phi(\infty)$ et, d'autre part, de posséder un ergodisme suffisant (conséquence de la décroissance des R pour $|t| \longrightarrow \infty$ et du fait que des conditions suffisantes du type (2-9) ou (2-10) peuvent être remplies) pour que la réponse à la question posée soit affirmative.

Soit, alors, P' la loi temporelle de l'ensemble des $X_j(t,\omega')$. Presque sûrement, on peut affirmer ce qui suit. Considérons les réalisations $X_j(t,\omega'_o)$ correspondant à une épreuve particulière ω'_o. Procédons sur cet ensemble comme nous l'avons fait sur $\mathcal{H}$ au paragraphe (3-1). Nous obtenons alors une loi asymptotique $P(\omega'_o)$. Cette loi n'est autre que P' (p.s.).

Il y a plus. On peut construire des ensembles $\mathcal{E}''$ de f.a. $X_j(t,\omega'') \in \mathcal{E}''$ non stationnaires mais ergodiques au sens du § 2-2, 2°, c'est-à-dire vérifiant les relations du type :

(4-3) $$\lim_{T\to\infty} \frac{1}{T}\int_0^T F[X_{j_1}(t+\tau_1)\ldots X_{j_M}(t+\tau_M)]\,dt$$
$$\overset{p.s.}{=} \lim_{T\to\infty} \frac{1}{T}\int_0^T E\{F[X_{j_1}(t+\tau_1)\ldots X_{j_M}(t+\tau_M)]\}\,dt$$

F étant une fonction de $[X_{j_1}(t_1),\ldots,X_{j_M}(t_M)]$ assez générale et la moyenne temporelle du second membre de (4-3) étant supposée exister. Alors, sur presque tous les ω'', on définira une loi asymptotique $P''(\omega'')$ qui sera indépendante de ω'', soit $P''(\omega'') = P$. Naturellement, P différera de la loi temporelle P" des $\{X_j(t,\omega'')\}$ d'où l'on est parti.

De toute évidence, P est stationnaire (par construction) alors que P" ne l'est pas. En particulier, seuls les éléments du type

(4-4) $$E\{dx_{j_1,\varepsilon_1}(\nu_1)\,dx_{j_2,\varepsilon_2}(\nu_2)\ldots\ldots\}$$

situés sur une "multiplicité stationnaire" du type (3-7) joueront un rôle dans P où ils auront, d'ailleurs, la même valeur que dans P".

De même, si P" présente le caractère markovien, il n'en sera pas de même, en général, pour P par suite d'une sorte d'effet de mélange produit par la moyenne temporelle.

De même, en général, le caractère gaussien se perd en passant de P'' à P.

5 - REMARQUE

Faisons une remarque sur un problème de transposition de fréquence. Soit $X(t,\omega)$ une f.a. réelle, gaussienne, centrée, stationnaire, à spectre uniforme sur $|\nu| < \Omega$ [densité f]. Considérons $Y(t,\omega) = X(t)\cos 2\pi\nu_0 t$ $[\nu_0 > \Omega]$. On peut noter ce qui suit :

5-1. Y(t,ω) n'est pas stationnaire et, dans sa représentation du type (2-3α), les $dy(\nu,\omega)$ ne sont pas orthogonaux. On peut cependant conserver un développement doublement orthogonal en utilisant comme fonctions de base les $\varphi(\nu,t) = \exp[2\pi i\nu t]$. $\cos 2\pi\nu_0 t$. On a, alors, évidemment :

$$Y(t) = \int_{-\Omega}^{+\Omega} \varphi[\nu,t]\, dx(\nu) \tag{5-1}$$

où les $dx(\nu)$ sont orthogonaux au sens de l'espérance mathématique, les $\varphi[\nu,t]$ l'étant au sens de la moyenne temporelle. Il s'agit là, sur un cas évident, de l'application du théorème de Karhunen - Loève, convenablement étendu à des fonctions possédant sur $[-\infty,+\infty]$ une puissance moyenne finie, non nulle.

5-2. L'expérimentateur va analyser Y(t,ω) en prenant des moyennes temporelles : il détruit la répartition gaussienne des valeurs. $Y(t,\omega)$ est gaussien mais il n'est pas stationnaire tandis que sa loi asymptotique n'est pas gaussienne mais est stationnarisée [spectre uniforme sur $\nu_0 - \Omega < |\nu| < \nu_0 + \Omega$ avec une densité (f/4)]. On peut se demander, si, par un filtrage linéaire convenable de $Z(t) = \mathcal{F}\{Y(t)\}$, on ne pourrait pas obtenir un Z stationnaire. Soit $G(\nu)$ le gain de $\mathcal{F}$. Z sera stationnaire si :

$$G(\nu_0+h)G(\nu_0-h) \equiv 0 \tag{5-2}$$

Mais, alors, on perd la moitié de la bande de $Y(t,\omega)$, ce qui peut être génant. On peut alors se contenter de sauver, pour la loi asymptotique, uniquement le caractère gaussien unidimensionnel. Dans sa thèse de Docteur-Ingénieur, M. Tahami (Université de Paris-Sud, Ecole Supérieure d'Electricité -1977-) montre que ce résultat restreint est atteint si on a :

$$\int_{-\Omega}^{+\Omega} G(\nu_0-h)G(\nu_0+h)\,dh = 0 \tag{5-3}$$

ANNEXE

LA LOI FORTE DES GRANDS NOMBRES POUR LES F.A. DE SECOND ORDRE

I-1. HYPOTHESES ET NOTATIONS

$Z(t,\omega)$: f.a. de second ordre(+), mesurable, continue en m.q. et de covariance harmonisable [cf. § (2-1) p. 2] . Soient, dans le cas général, $\Gamma_Z(t,t')$ et $\gamma_Z(\nu_1,\nu_2)$ les fonctions Γ et γ relatives à Z [cf. § (2-1)] et, dans le cas stationnaire, $C_Z(\tau)$ et $F_Z(\nu)$ la fonction de corrélation et le spectre de Z. On rappelle que l'on a en général :

(I-1) $$E\left\{\left|{}^{0}\overline{Z(t,\omega)}^{T}\right|^2\right\} = \frac{1}{T^2}\int_{-\infty}^{+\infty}\int_{-\infty}^{+\infty}\Gamma_Z(t_1,t_2)\,dt_1dt_2$$

$$= \int_{-\infty}^{+\infty}\int_{-\infty}^{+\infty} e^{\pi i[\nu_1-\nu_2]T}\,\frac{\sin\pi\nu_1 T}{\pi\nu_1 T}\cdot\frac{\sin\pi\nu_2 T}{\pi\nu_2 T}\;d^2\gamma_Z(\nu_1,\nu_2)$$

et, dans le cas stationnaire, $\mathfrak{R}$ désignant la partie réelle :

$$E\left\{\left|{}^{0}\overline{Z(t,\omega)}^{T}\right|^2\right\} = \frac{2}{T}\int_0^T \left(1-\frac{s}{T}\right)\mathfrak{R}\,C_Z(s)\,ds = \frac{2}{T^2}\int_0^T dt\int_0^t \mathfrak{R}\,C_Z(s)\,ds$$

(I-2) $$= \int_{-\infty}^{+\infty}\frac{\sin^2\pi\nu T}{\pi^2\nu^2T^2}\,dF_Z(\nu) = 2\,\mathfrak{R}\,\overline{\overline{C_Z}}(T)$$

$$\text{avec}\quad \overline{C_Z}(T) = \frac{1}{T}\int_0^T C_Z(t)\,dt \quad\text{et}\quad \overline{\overline{C_Z}}(T) = \frac{1}{T}\int_0^T \frac{s}{T}\,\overline{C_Z(s)}\,ds$$

(+) On ne suppose, dans cette annexe, que la stationnarité de second ordre et nullement la stationnarité stricte qui permettrait d'appliquer le théorème de Birkhoff.

I-2. CONDITIONS SUFFISANTES POUR LA LOI FORTE

1° M. Loève (1) [cf. p. 458] , dans des cas stationnaires et non stationnaires et A. Blanc-Lapierre et R. Brard (4) dans le cas stationnaire ont donné de telles conditions vers 1945. Toujours dans le cas stationnaire, ces conditions ont été améliorées par I.N. Verbitskaya (5) puis par A. Blanc-Lapierre et A. Tortrat (3). Nous allons ici, étendre les résultats de (3) au cas non stationnaire.

2° Résultats de base. Pour que ${}^{0}\overline{Z(t,\omega)}^{T_n} \longrightarrow 0$, p.s., si $T_n \longrightarrow \infty$, il suffit que la série de terme général $E\left\{\left|{}^{0}\overline{Z(t,\omega)}^{T}\right|^2\right\}$ converge. De cette remarque et du fait que $E\left\{|Z(t)|^2\right\}$ admet une borne uniforme, on déduit le théorème suivant, que Z soit ou non stationnaire, mais conforme aux hypothèses données pour X à propos des équations (2-1) et (2-2)

Théorème I - La convergence ${}^{0}\overline{Z(t,\omega)}^{T} \xrightarrow{p.s.} 0$ est assurée si, pour une suite $T_n \nearrow \infty$, on a :

$$\text{(I-3)} \qquad \sum E\left\{\left|{}^{0}\overline{Z(t,\omega)}^{T_n}\right|^2\right\} < +\infty \quad (\alpha) \quad \text{et} \sum\left[\frac{T_{n+1}}{T_n} - 1\right]^2 < +\infty \quad (\beta)$$

On notera que, p.s., Z(t) se comporte comme $Z(T_n)$ pour $T_n \nearrow \infty$ si $T_n \in$ (1-3β). Il en est notamment ainsi pour toute suite $T_n^{(0)} = n^{\lambda}$ $[\lambda > 0]$ ou même $T_n^{(1)} = \exp\{n^{\gamma}\}$ $[0 < \gamma < 1/2]$.

3° Cas non stationnaire. i) Condition sur Γ_Z. En utilisant la suite $T_n^{(1)}$, on obtient le résultat suivant :

Théorème II - S'il existe deux constantes certaines, M et α, $M > 0$ et $\alpha > 2$, telles que, pour les grandes valeurs de T, on ait :

$$\text{(I-4)} \qquad \frac{1}{T^2}\int_0^T\int_0^T \Gamma_{Z',Z'}(t,t')\,dt\,dt' < \frac{M}{(\text{Log } T)^{\alpha}} \qquad (+)$$

alors, ${}^{0}\overline{Z'(t,\omega)}^{T} \xrightarrow{p.s.} \overline{Z'(t,\omega)} = 0$ pour $T \longrightarrow \infty$.

(+) I.N. Verbitskaya (5) a obtenu, pour Z réel, la condition de même nature :

$$\sum_1^{\infty} n^2 \; \overline{C_Z(2^k)} < +\infty$$

et a pu en déduire la condition suffisante intégrale $\displaystyle\int_1^{\infty} [\text{Log}^2 x/x]\; \overline{C_Z(x)}\,dx < +\infty$.

ii) Conditions sur le spectre. Soient, respectivement, $Z_1(t,\omega)$ et $Z_2(t,\omega)$ les composantes de $Z(t,\omega)$ correspondant aux domaines de fréquences $|\nu| < \nu_T$ et $|\nu| \geq \nu_T$, où ν_T est une valeur positive que nous lierons à T. On a :

(I-5) $$ {}^{0}\overline{Z(t)}^{T} = {}^{0}\overline{Z_1(t)}^{T} + {}^{0}\overline{Z_2(t)}^{T} $$

avec

(I-6) $$ \begin{cases} {}^{0}\overline{Z_1(t)}^{T} = \displaystyle\int_{|\nu|<\nu_T} e^{\pi i \nu T} \frac{\sin \pi \nu T}{\pi \nu T}\, dz(\nu) \\ \text{et} \\ {}^{0}\overline{Z_2(t)}^{T} = \displaystyle\int_{|\nu|\geq\nu_T} e^{\pi i \nu T} \frac{\sin \pi \nu T}{\pi \nu T}\, dz(\nu) \end{cases} $$

Posons

(I-7) $$ \sigma[\nu_T] = \int_{-\nu_T}^{+\nu_T}\int_{-\nu_T}^{+\nu_T} |d^2\gamma_Z(\nu_1,\nu_2)| \quad \text{et} \quad C = \int_{-\infty}^{+\infty}\int_{-\infty}^{+\infty} |d^2\gamma_Z(\nu_1,\nu_2)| $$

On a :

(1-8) $$ E\left\{\left|{}^{0}\overline{Z_1(t)}^{T}\right|^2\right\} \leqslant \sigma[\nu_T] \quad (\alpha) \qquad E\left\{\left|{}^{0}\overline{Z_2(t)}^{T}\right|^2\right\} \leqslant \frac{C}{\pi^2\nu_T^2 T^2} \quad (\beta) $$

La convergence ${}^{0}\overline{Z(t,\omega)}^{T} \xrightarrow{p.s.} 0$ sera assurée s'il est possible de trouver une suite $T_n \in$ [I-3 (β)] pour laquelle les deux séries $\sum_1^{\infty} \sigma\{\nu_{T_n}\}$ et $\sum_1^{\infty} [1/\nu_{T_n}^2 T_n^2]$ convergent. Cette double convergence sera assurée s'il existe deux nombres >0, φ_1 et φ_2 tels que :

(I-9) $$ \frac{1}{\nu_n^2 T_n^2} < \frac{1}{n^{1+\varphi_1}} \quad (\alpha) \quad \text{et} \quad \sigma[\nu_{T_n}] < \frac{1}{n^{1+\varphi_2}} \quad (\beta) $$

En prenant, pour T_n, la suite $T_n^{(1)}$, on établit :

Théorème III - La convergence ${}^{0}\overline{Z(t,\omega)}^{T} \xrightarrow{p.s.} 0$ est assurée si, pour les faibles valeurs de $|\nu|$, on a :

(I-10) $$ \sigma[|\nu|] \Longleftarrow \left\{1 / \left[|\mathrm{Log}\,|\nu||\right]^{2+\varphi}\right\}, \quad \varphi > 0 $$

4° Cas particulier de la stationnarité de second ordre. En désignant par $F_Z(\nu)$ le spectre de la f.a. stationnaire Z(t), on a, alors : $\sigma[|\nu|] = F_Z[|\nu|] - F_Z[-|\nu| + 0]$. On retrouve alors, comme conséquence du théorème III des résultats déjà établis dans (3) :

- si, près de $\nu = 0$, le spectre $F_Z(\nu)$ admet une densité $f_Z(\nu)$ de la forme $O(1/|\nu|^{\mu})$ [avec, nécessairement $\mu < 1$], alors ${}^{0}\overline{Z(t,\omega)}^{T} \xrightarrow{p.s.} 0$.

- La même conclusion vaut encore pour une densité $O(1/\{|\nu| \, |\mathrm{Log}\, |\nu||^{\mu}\})$ avec $\mu > 3$.

Remarque relative au cas stationnaire. Appelons L l'événement constitué par la convergence de $\overline{{}^{0}Z(t,\omega)}^{T}$ vers zéro, si $T \longrightarrow \infty$. Soit l'ensemble Φ de tous les $\{F,C\}$ = {spectre, fonction de corrélation} et Z_F une f.a. de spectre F. Nous avons donné, ci-dessus, des conditions suffisantes sur F pour que Prob L = 1 pour tout X_F. D'autre part, on a donné des exemples de spectres F pour chacun desquels il existe au moins un X_F tel que Prob L $<$ 1 (3). Le problème de la définition de la séparatrice entre les deux sous-ensembles Φ_1 et Φ_2 $[\Phi_1 + \Phi_2 = \Phi]$ correspondant à ces deux situations reste ouvert.

REFERENCES

(2) Blanc-Lapierre A. et Fortet R. (1953) : Théorie des fonctions aléatoires. Masson Paris.

(3) Blanc-Lapierre A. et Tortrat A. (1968) : Comptes Rendus Académie des Sciences Paris, série A, 267, p. 740.

(4) Blanc-Lapierre A. et Brard R. (1945) : Comptes Rendus Académie des Sciences Paris, série A, 220, p. 134 et (1946) : Bull. Soc. Math. Fr. p. 102.

(7) Blanc-Lapierre A. et Lefèvre C. (1972) : Analyse harmonique généralisée et fonctions aléatoires stationnaires. Comptes Rendus Académie des Sciences Paris, série A, 274, pp. 257-261.

(8) Blanc-Lapierre A. (1975) : Fonctions certaines admettant des répartitions asymptotiques et fonctions aléatoires stationnaires. "Perspectives in probability and statistics". Papers in honour of M.S. Bartlett on the occasion of his 60th birthday. London, Academic Press Inc.

(1) Loève M. (1963) : Probability theory (3e ed), D. Van Nostrand, New York.

(6) Pham P.H. (1972) : Mesures asymptotiques, thèse, Université de Paris.

(5) Verbitskaya I.N. (1964) : Theory of Probability and its applications, Académie des Sciences Moscou, 9, p. 325 et (1966) : 11, p. 632.

Laboratoire des Signaux et Systèmes
Centre National de la Recherche Scientifique-Ecole Supérieure d'Electricité
Plateau du Moulon
91190 GIF SUR YVETTE (France)

RECENT ADVANCES IN THE METRIC THEORY
OF CONTINUED FRACTIONS

Marius Iosifescu

Bucharest

ABSTRACT

This is a survey of recent results in the metric theory of continued fractions concerning Gauss-Kuzmin-Lévy theorem and extreme value theory.

INTRODUCTION

Each irrational number y in the unit interval has a unique infinite continued fraction expansion of the form

$$y = \cfrac{1}{a_1(y) + \cfrac{1}{a_2(y) + \cdots}} ,$$

where the $a_n(y)$ are natural numbers determined as follows. Put $Ty = 1/y \pmod 1$. Then $a_1(y) = 1/y - Ty$, and $a_{n+1}(y) = a_n(Ty) = a_1(T^n y)$, $n \geqslant 1$. It is obvious that endowing the unit interval with the σ-algebra of Lebesgue measurable sets, the a_n are random variables defined almost everywhere with respect to any probability measure assigning probability 0 to the set of rational numbers (in particular with respect to Lebesgue measure λ).

The metric theory of continued fractions is concerned with the study of the random sequence $(a_n)_{n \geqslant 1}$. The first problem of this theory was raised in 1812 by Gauss who in a letter to Laplace stated that

$$(1) \qquad \lim_{n\to\infty} \lambda(r_n^{-1} < x) = \frac{\log(1+x)}{\log 2}$$

for each x in the unit interval, and asked for an estimate of the error

$$E_n(x,\lambda) = \lambda(r_n^{-1} < x) - \frac{\log(1+x)}{\log 2} .$$

Here $r_n(y) = 1/(T^n y)$, $n \geq 1$, i.e.

$$r_n = a_n + \cfrac{1}{a_{n+1} + \cfrac{1}{a_{n+2} + \cdots}}$$

Kuzmin (1928) first proved (1) giving an error estimate $E_n(x,\lambda) = O(q^{\sqrt{n}})$ as $n \to \infty$ with $0 < q < 1$. Lévy (1929) gave a different proof allowing to improve the error estimate to

$$(2) \qquad E_n(x,\lambda) = O(q^n)$$

with $q < 0.7$. (Subsequently, Szüsz (1961), using Kuzmin's approach, proved that (2) holds with $q < 0.4$.)

After Lévy's work important contributions were made by Hinčin (1935, 1936) and Doeblin (1940). Doeblin's approach is especially interesting since it first revealed the part played in the metric theory of continued fractions by dependence with complete connections, a creation of the Romanian probability school. A rigorous treatment motivated by and clarifying Doeblin's ideas is to be found in Iosifescu (1974).

It has now been recognized the fundamental facts of the metric theory of continued fractions are the following.

1) Under Gauss' measure γ, that is defined by

$$\gamma(A) = (\log 2)^{-1} \int_A \frac{dx}{1+x}$$

for any Borel set A in the unit interval, the sequence $(a_n)_{n\geq 1}$ is a strictly stationary one and

$$\gamma(a_n \geq u) = \frac{\log(1 + 1/u)}{\log 2}$$

whatever the natural integer u.

2) The sequence $(a_n)_{n\geq 1}$ is exponentially ψ-mixing under γ, i.e.

$$|\gamma(A_1 \cap A_2) - \gamma(A_1)\gamma(A_2)| \leq a\, b^n\, \gamma(A_1)\gamma(A_2),$$

for any events A_1 and A_2 defined in terms of the random variables $a_1,\ldots,a_k$ and $a_{k+n},a_{k+n+1},\ldots$ respectively, whatever the natural integer k, with suitable constants $a>0$ and $0<b<1$.

Actually, the exponential ψ-mixing of the sequence $(a_n)_{n\geq 1}$ is a consequence of the validity of the error estimate (2) for all measures $\lambda(\cdot \mid a_1 = i_1,\ldots,a_k = i_k)$, $k \geq 1$, rather than λ.

The present paper is intended to be a survey of recent results in two directions in which research was particularly active : improvements of Gauss-Kuzmin-Lévy theorem and extreme value theory for the sequence $(a_n)_{n\geq 1}$. The only elements of novelty consists of a few immediate but important consequences of already known results.

IMPROVING GAUSS-KUZMIN-LÉVY THEOREM

Let μ denote any non-atomic probability measure on the unit interval and define

$$F_n(x,\mu) = F_n(x) = \mu(r_{n+1}^{-1} < x)$$

for $n \geq 0$, $0 \leq x \leq 1$. Clearly, $F_0(x) = \mu([0,x])$. Since $0 < r_{n+2}^{-1} < x$ iff

$$(x + a_{n+1})^{-1} < r_{n+1}^{-1} < a_{n+1}^{-1}$$

we can write Kuzmin's equation

$$F_{n+1}(x) = \sum_{k \geqslant 1} \left(F_n\left(\frac{1}{k}\right) - F_n\left(\frac{1}{x+k}\right) \right).$$

Assuming F_0' exists and is bounded (e.g. μ has a continuous density function) it is easily seen by induction that F_n' exists and is bounded for any $n \geqslant 1$. (In fact, Kuzmin's equation can be differentiated term by term.) Putting

$$f_n(x) = (x + 1)F_n'(x), \ n \geqslant 0 ,$$

we get

$$f_{n+1} = U f_n ,$$

where

$$(3) \qquad (Uf)(x) = \sum_{k \geqslant 1} \frac{x+1}{(x+k)(x+k+1)} f\left(\frac{1}{x+k}\right)$$

for any measurable bounded function f on the unit interval $[0,1]$. Clearly, U is a (Markov) transition operator since we can write

$$(Uf)(x) = \int_0^1 Q(x,dv)f(v) ,$$

where the transition probability function Q is defined by

$$Q(x,A) = \sum_{(k:(x+k)^{-1} \in A)} \frac{x+1}{(x+k)(x+k+1)}$$

for any Borel set A in $[0,1]$.

Therefore $f_n = U^n f_0$, $n \geqslant 1$, and the asymptotic behaviour of F_n will follow from that of U^n.

The operator U was studied by Szüsz (1968) who in fact reproved in a special case known results from the theory of random systems with complete connections (see Iosifescu and Theodorescu (1969, Chapter 2)).

The present author (Iosifescu (1974)) noticed that the operator U defined by (3) belongs to $\mathcal{L}_1(L,B)$ (see the Appendix), where B = the Banach space of all complex valued, bounded and measurable functions f on $[0,1]$ under the uniform norm

$$|f| = \sup_{x \in [0,1]} |f(x)| ,$$

L = the Banach space of all complex valued Lipschitz functions f on $[0,1]$ under the norm $\|f\| = m(f) + |f|$ with

$$m(f) = \sup_{x' \neq x''} \left| \frac{f(x') - f(x'')}{x' - x''} \right| .$$

Moreover, U is ergodic and $U_1 f = \int_0^1 f(x)\, \gamma(dx)$. Consequently, $U^n = U_1 + V^n$, hence whatever $n \geq 1$ and $f \in L$

$$\|V^n f\| = O(q^n)\, \|f\| \tag{4}$$

for some $0 < q < 1$, the constant involved in O being independent of f.

Since our L is dense in the set C of all continuous complex valued functions on $[0,1]$ it follows that

$$\lim_{n \to \infty} |V^n f| = 0 \tag{5}$$

whatever $f \in C$. Finally, a well known result in the theory of weak convergence of measures implies that (5) still holds for any $f \in B$ which is continuous γ-almost everywhere, i.e. for any $f \in B$ which is Riemann integrable on $[0,1]$.

Notice that Theorem 2.1.24 in Iosifescu and Theodorescu (1969) leads to the following improvement of (5) for $f \in C$:

$$|V^n f| \leq \inf_{1 \leq \ell < m < n} \left[\frac{\operatorname{osc} f \left(3 \sum_{k=0}^{\ell-1} \binom{m}{k} (1-a)^{m-k} + \frac{2\pi^2}{3a} 4^{-\ell}\right) + 3\varepsilon_\ell}{b} + (1-b)^{\frac{n}{m}-1} \operatorname{osc} f \right],$$

where $\operatorname{osc} f = \sup f - \inf f$, $a = 3-2\sqrt{2}$, $b = a^3/4$, and

$$c_n = \sup |f(y') - f(y'')|, \ n \geqslant 1,$$

the upper bound being taken over all irrational $y', y'' \in [0,1]$ such that $a_i(y') = a_i(y'')$, $1 \leqslant i \leqslant s$, $s \geqslant n$, and $a_i(y') = 2$ for at least n values i . (Clearly, $c_n \leqslant \sup | f(y') - f(y'') |$, the upper bound being taken over all irrational $y', y'' \in [0,1]$ such that $|y' - y''| \leqslant 2^{-n+1}$.)

Further results on the operator V appear in the work of Wirsing (1971, 1974). He proves that the operator V' on C defined by $V'g = -(VG)'$, where $G' = g$, namely

$$(V'g)(x) =$$

$$\sum_{k \geqslant 1} \left[\frac{k}{(x+k+1)^2} \int_{x+1/(k+1)}^{x+1/k} g(y)\,dy + \frac{x+1}{(x+k)^3(x+k+1)} \, g\left(\frac{1}{x+k}\right) \right], \ g \in C ,$$

has an eigenvalue $\sigma = 0.3036630028987326586\ldots$ with positive valued eigenfunction $\Phi \in C$ and there exists a positive bounded linear functional F on C such that

$$V'^n g = \sigma^n \Phi F(g) + O(\tau^n |g|)$$

as $n \to \infty$, with $0 < \tau \leqslant \sigma - 0.031$.

It follows that for any $f \in L$ with continuous derivative f'

$$(6) \quad (V^n f)(x) = (-\sigma)^n (x+1) \Psi'(x) F(f') + O(\tau^n |f'|),$$

where the function Ψ is defined by $((x+1)\Psi'(x))' = \Phi(x)$, $\Psi(0) = \Psi(1) = 0$. (Clearly, Φ and Ψ as well as F are determined up to a constant factor.)

The functions Φ and Ψ are in fact holomorphic and can be extended holomorphically into the whole complex plane with a cut along the negative axis from -1 to ∞ . This cut constitutes the natural boundary of these functions. Finally, Ψ verifies the functional equation

$$\Psi(z) - \Psi(z+1) = \frac{1}{\sigma} \Psi\left(\frac{1}{z+1}\right).$$

It is easily seen equation (6) implies that the spectral radius $r_L(V) = \sigma$. Next, the special case $f(x) = x+1$ leads (after suitable norming of F) to an estimate of the form

$$(7) \qquad (-\sigma)^n \Psi(x) + O(x(1-x)\tau^n)$$

for the error $E_n(x,\lambda)$. Thus the optimal q in (2) and (4) equals σ.

It is easily seen estimate (7) is also true for the error $E_n(x,\mu)$ corresponding to a probability measure μ having a continuously differentiable density function h. (One should take $f(x) = (x+1)h(x)$.)

An estimate of $E_n(x,\mu)$ in the case where μ has just a continuous density function can be derived from the improvement of (5) given above. More generally, an estimate of $E_n(x,\mu)$ in the case where μ is absolutely continuous with respect to λ can be derived from Lemma 1 of Philipp (197o). (The convergence of $E_n(x,\mu)$ to 0 as $n \to \infty$ in this case goes back to Lévy (1929).)

To conclude with, we note the problem of determining the entire spectrum of the operator V remains open.

SOME EXTREME VALUE THEORY

Let $M_n^{(s)}$ denote the s-th largest of $a_1,\dots,a_n$ with $1 \leq s \leq n$. Clearly, $M_n^{(1)} = M_n$ is the maximum of $a_1,\dots,a_n$. In his famous 194o paper Doeblin sketched a proof of the following Poisson law. For any $\theta > 0$ the probability (under λ) that $a_k > n/(\theta \log 2)$ for exactly p values k, $1 \leq k \leq n$, approaches $e^{-\theta}\theta^p/p!$ as $n \to \infty$ for any $p = 0,1,2,\dots$. Doeblin's proof contains a gap but as shown by the present author (Iosifescu (1977)) by making use of a method suggested by Galambos (1972) the statement is true under any probability measure μ absolutely continuous with respect to λ. Let us notice that this Poisson law leads immediately to the asymptotic distribution of $M_n^{(s)}$ as $n \to \infty$ for fixed s. Indeed, if we denote by L_n the number of values k, $1 \leq k \leq n$, for which $a_k > n/(\theta \log 2)$, then

$$\mu(M_n^{(s)} \leq \frac{n}{\theta \log 2}) = \mu(L_n < s) =$$

$$= \sum_{j=o}^{s-1} \mu(L_n = j) \longrightarrow e^{-\theta} \sum_{j=o}^{s-1} \theta^j/j!$$

as $n \to \infty$ for any fixed $s = 1,2,\ldots$. In other words

(8) $$\lim_{n\to\infty} \mu(\frac{M_n^{(s)} \log 2}{n} \leq x) = e^{-\frac{1}{x}} \sum_{j=o}^{s-1} \frac{1}{j!\, x^j}$$

for any $x > 0$ and $s = 1,2,\ldots$. In particular

$$\lim_{n\to\infty} \mu(\frac{M_n \log 2}{n} \leq x) = e^{-\frac{1}{x}} .$$

This last result has been proved by Galambos (1972, 1973).

In fact, the present author derived Doeblin's Poisson law as a special case of the following more general

P r o p o s i t i o n 1. Let $(\xi_n)_{n\geq 1}$ be a ψ-mixing strictly stationary sequence of real valued random variables. Set $P(\xi_n > x) = t(x)$, $-\infty < x < \infty$, and assume for some real valued function g on the natural integers there exists the positive finite limit $\lim_{n\to\infty} nt(g(n)) = \theta$, say. Then the probability that $\xi_k > g(n)$ for exactly p values k , $1 \leq k \leq n$, approaches $e^{-\theta}\theta^p/p!$ as $n \to \infty$ for any $p = 0,1,2,\ldots$.

Scrutiny of the proof shows that in the continued fraction expansions case $\xi_n = a_n$, $n \geq 1$, where $P=\gamma$, $t(x)=\log(1+1/([x]+1))/\log 2$ and

(9) $$nt(g(n)) = \theta(1 + O(\tfrac{\theta}{n}))$$

as $n \to \infty$, the constant involved in O being independent of θ, the error term has the form $O(\exp(-(\log n)^{\delta}))$ for any $\delta < 1$. More precisely, whatever $n \geq 1$ we have

(1o) $$\left|\gamma(L_n = p) - e^{-\theta}\, \theta^p/p!\right| \leq K \exp(-(\log n)^{\delta}),$$

for $\theta = O(n^a)$, $0 \leqslant a < 1$, where K depends only, perhaps, on δ and p . The details of computation for the case $p = 0$ can be found in Philipp (1976, p.382), where the proviso $\theta = O(n^a)$, $0 \leqslant a < 1$, does not appear. The slip originates from the use of the equation

$$nt(g(n)) = \theta(1 + O(\tfrac{1}{n}))$$

as $n \to \infty$ which unlike (9) is true for a fixed θ (as needed in Galambos (1972, p.150)).

Apart from the possibility of estimating the error term Proposition 1 is a corollary of Theorem 5.1 of Leadbetter (1974). His paper deals with extreme value theory for strictly stationary sequences satisfying dependence restrictions even weaker than strong mixing (i.e. $|P(A_1 \cap A_2) - P(A_1)P(A_2)| \longrightarrow 0$ as $n \to \infty$ uniformly with respect to k and events A_1 and A_2 defined in terms of $\xi_1, \ldots, \xi_k$ and $\xi_{k+n}, \xi_{k+n+1}, \ldots$ respectively).

The error estimate (10) for $p = 0$ is fundamental in proving (Philipp (1976)) that λ-almost everywhere

$$\liminf_{n \to \infty} \frac{M_n \log \log n}{n} = 1/\log 2. \qquad (11)$$

Apart from the value of the constant this result was conjectured by P.Erdös. (Weaker results are proved in Galambos (1974)). In fact, extending work by Barndorff-Nielsen (1961) for independent identically distributed random variables, Philipp proved a more general result, namely that if $g(n)$ is nonincreasing and $ng(n)$ is nondecreasing then $M_n \leqslant ng(n)/\log 2$ finitely often or infinitely often λ-almost everywhere according as

$$\sum \exp(-1/g(n)) \frac{\log \log n}{n}$$

converges or diverges.

A careful examination of Philipp's proof shows that equation (11) still holds when M_n is replaced by $M_n^{(s)}$ for any fixed $s = 2,3,\ldots$.

There is a natural dual to Proposition 1 that leads to results concerning minimum values. Namely, we have

P r o p o s i t i o n 1'. Let $(\eta_n)_{n\geq 1}$ be a strictly stationary ψ-mixing sequence of positive random variables. Set $P(\eta_n \geq x) = t(x)$, $0 \leq x < \infty$, and assume for some real valued function g on the natural integers there exists the positive finite limit $\lim_{n\to\infty} n(1 - t(g(n))) = \theta$, say. Then the probability that $\eta_k < g(n)$ for exactly p values k, $1 \leq k \leq n$, approaches $e^{-\theta}\theta^p/p!$ as $n \to \infty$ for any $p = 0,1,2,\ldots$.

The above statement obtains by applying Proposition 1 to the random variables $\xi_n = 1/\eta_n$, $n \geq 1$. (Clearly, ψ-mixing of $(\xi_n)_{n\geq 1}$ is ensured by ψ-mixing of $(\eta_n)_{n\geq 1}$.)

Elementary computations show that Proposition 1' applies to a strictly stationary ψ-mixing sequence $(\eta_n)_{n\geq 1}$ such that $P(\eta_n \geq x) = \log(1 + 1/x)/\log 2$ for <u>any</u> $x \geq 1$ with

$$g(n) = 1 + \frac{2\,\theta \log 2}{n}.$$

For such a sequence we can write analogously to (8)

$$\lim_{n\to\infty} P\left(\frac{n(m_n^{(s)} - 1)}{2 \log 2} \geq x\right) = e^{-x} \sum_{p=0}^{s-1} \frac{x^p}{p!} \tag{12}$$

for any $x > 0$ and $s = 1,2,\ldots$, where $m_n^{(s)}$ denotes the s-th smallest of $\eta_1,\ldots,\eta_n$, with $1 \leq s \leq n$. Clearly, we cannot assert that (12) is true for the continued fraction expansion case $\eta_n = a_n$, $n \geq 1$, since the equation $\gamma(a_n \geq x) = \log(1 + 1/x)/\log 2$ holds just for natural values of x. Actually, for this case $\lim_{n\to\infty} m_n^{(s)} = 1$ λ-almost everywhere whatever $s = 1,2,\ldots$. This follows from a well known results going back to Hinčin (1935) according to which the asymptotic relative frequency of 1 among the partial quotients $a_1, a_2, \ldots$ is λ-almost everywhere equal to $\log(4/3)/\log 2 > 0$.

We conjecture that <u>equation (12) holds true for</u> $\eta_n = r_n$, $n \geq 1$, <u>under any P absolutely continuous with respect to</u> λ. Notice that $\gamma(r_n \geq x) = \log(1 + 1/x)/\log 2$ for <u>any</u> $x \geq 1$ but, as easily seen, the sequence $(r_n)_{n\geq 1}$ is <u>not</u> ψ-mixing under γ. Useful suggestions might be found in Lévy (1929, 1936). Notice also that since $0 < r_n - a_n < 1$ the same is true of the difference between the s-th largest of $r_1,\ldots,r_n$ and $M_n^{(s)}$. It trivially follows that equation (8) still holds when $M_n^{(s)}$ denotes the s-th largest of $r_1,\ldots,r_n$.

APPENDIX

For the reader's convenience this appendix gives the fundamentals about a class of operators the study of which was motivated by problems arising in the theory of dependence with complete connections. For details and proofs the books by Iosifescu and Theodorescu (1969) and Norman (1972) should be consulted.

Consider two complex Banach spaces $(B, |\cdot|)$ and $(L, \|\cdot\|)$ with $L \subset B$. Assume that

(i) if $x_n \in L$, $x \in B$, $\lim_{n\to\infty} |x_n - x| = 0$, and $\|x_n\| \leqslant c$ for all n, then $x \in L$ and $\|x\| \leqslant c$.

Let us denote by $\mathcal{L}_k(L,B)$ the set of all linear operators U from L into L which are bounded with respect to both $\|\cdot\|$ and $|\cdot|_L$, where the latter is the restriction of $|\cdot|$ to L, verifying the conditions

(ii) $\sup_{n \geqslant 0} |U^n|_L$;

(iii) there exist two positive constants $a < 1$ and A such that

$$\|U^k x\| \leqslant a\|x\| + A|x|, \quad x \in L;$$

(iv) if L' is a bounded subset of $(L, \|\cdot\|)$, then $U^k L'$ has compact closure in $(B, |\cdot|)$.

The representation of U^n for $U \in \mathcal{L}_k(L,B)$ was obtained by Ionescu Tulcea and Marinescu (1950) in the theorem below.

T h e o r e m . Let $U \in \mathcal{L}_k(L,B)$. The set E of eigenvalues of U of modulus 1 has only a finite number of elements. For each $\rho \in E$, $D(\rho) = (f \in L : Uf = \rho f)$ – the eigenspace corresponding to the eigenvalue ρ – is finite dimensional. There exist bounded linear operators U_ρ, $\rho \in E$, and V on L such that

$$U^n = \sum_{\rho \in E} \rho^n U_\rho + V^n, \quad n \geqslant 1,$$

$U_\rho U_{\rho'} = 0$ if $\rho \neq \rho'$, $U_\rho^2 = U_\rho$, $U_\rho V = VU_\rho = 0$, $U_\rho L = D(\rho)$, $r_L(V) < 1$ (here $r_L(V) = \lim_{n\to\infty} \|V^n\|^{1/n}$ denotes the spectral radius of V).

An important special case is that where E consists of the only eigenvalue $\rho = 1$ and the eigenspace $D(1)$ is one-dimensional. Such an operator U is said to be ergodic. Thus, for an ergodic operator U, we have $U^n = U_1 + V^n$, $n \geq 1$, where U_1 is a bounded linear functional on L.

REFERENCES

Barndorff-Nielsen O.(1961): On the rate of growth of the partial maxima of a sequence of independent identically distributed random variables. Math.Scand. 9(1961), 383-394.

Doeblin W.(194o) : Remarques sur la théorie métrique des fractions continues. Compositio Math. 7(194o), 353-371.

Galambos J.(1972) : The distribution of the largest coefficient in continued fractions expansions. Quart.J.Math.Oxford (2) 23(1972), 147-151.

Galambos J.(1973): The largest coefficient in continued fractions and related problems. In : Diophantine approximation and its applications (Proc.Conf. Washington,D.C., 1972). Academic Press, New York 1973, 1o1-1o9.

Galambos J.(1974) : An iterated logarithm type theorem for the largest coefficient in continued fractions. Acta Arith. 25(1974), 359-364.

Ionescu Tulcea C.T., Marinescu G.(195o) : Théorie ergodique pour des classes d'opérations non complètement continues. Ann.of Math. 52(195o), 14o-147.

Iosifescu M.(1974) : On the application of random systems with complete connections to the theory of f-expansions. In : Progress in statistics (European Meeting of Statisticians, Budapest, 1972), Colloq.Math.Soc.János Bolyai, vol.9. North Holland, Amsterdam 1974, 335-363.

Iosifescu M.(1977) : A Poisson law for ψ-mixing sequences establishing the truth of a Doeblin's statement. Rev.Roumaine Math. Pures Appl. 22(1977), 1441-1447.

Iosifescu M., Theodorescu R.(1969): Random processes and learning. Springer, Berlin-Heidelberg-New York 1969.

Khintchine A.(1935): Metrische Kettenbruchprobleme. Compositio Math. 1(1935), 361-382.

Khintchine A.(1936) : Zur metrische Kettenbruchtheorie. Compositio Math. 3(1936), 276-285.

Kusmin R.O.(1928) : Sur un problème de Gauss. In: Atti.Congr. Internaz. Mat., Bologna 1928, Vol.VI. Zanichelli, Bologna 1932, 83-89.

Leadbetter M.R.(1974): On extreme values in stationary sequences. Z.Wahrscheinlichkeitstheorie und Verw.Gebiete 28 (1974), 289-303.

Lévy P.(1929): Sur les lois de probabilités dont dépendent les quotients complets et incomplets d'une fraction continue. Bull.Soc. Math.France 57(1929), 178-194.

Lévy P.(1936): Sur le développement en fraction continue d'un nombre choisi au hasard. Compositio Math. 3(1936), 286-303.

Lévy P.(1954): Théorie de l'addition des variables aléatoires , 2e édition. Gauthier-Villars, Paris 1954.

Norman M.F.(1972) : Markov processes and learning models. Academic Press, New York-London 1972.

Philipp W.(1970) : Some metrical theorems in number theory II. Duke Math.J. 37(1970), 447-458. Errata ibid. 37(1970), 788.

Philipp W.(1976) : A conjecture of Erdös on continued fractions. Acta Arith. 28(1976), 379-386.

Szüsz P.(1961) : Über einen Kusminschen Satz. Acta Math. Acad. Sci. Hungar. 12(1961), 447-453.

Szüsz P.(1968) : On Kuzmin's theorem II. Duke Math.J. 35(1968), 535-540.

Wirsing E.(1971) : Über den Satz von Gauss-Kusmin. In: Zahlentheorie (Tagung, Math. Forschungsinst., Oberwolfach 1970). Ber. Math. Forschungsinst., Oberwolfach, Heft 5, Bibliographisches Inst., Mannheim 1971, 229-231.

Wirsing E.(1974) : On the theorem of Gauss-Kusmin-Lévy and a Frobenius-type theorem for function spaces. Acta Arith. 24(1974), 507-528.

National Institute of Metrology
Centre of Mathematical Statistics

Str.Stirbei Vodă 174
77104 Bucharest
Romania

A DISCUSSION OF SOME BASIC CONCEPTS IN STATISTICAL INFERENCE

Rashid Ahmad

Glasgow

ABSTRACT

Despite its' distinctive past history, the problem of statistical inference in a broad sense, besides being controversial among statisticians, is continuously stimulating vigorous and rigorous research. There are some interesting recent advances but many problems either remain open and unresolved or need improvements. In order to give a reasonably comprehensive discussion of the basic concepts, ideas and principles, in this paper we concentrate on the various underlying assumptions, statistical decision error structure, various measures of information, sufficiency and ancillarity, and some important and most commonly used principles of statistical inference. It is suggested that sphericity and exchangeability, besides being weaker than normality, are more plausible and tangible, at least from a practical viewpoint. Thus they are perhaps more natural and appealing as a basis for statistical inference. For a large class of standard decisions it is proposed that instead of the classical testing approach, one should minimize a linear combination, with nonnegative weights, of the first and second type errors. After discussing various measures of information, it is pointed out that the family of likelihood functions of various types essentially contain relevant information for statistical problems. Next, along with sufficiency and ancillarity concepts, various inferential principles are discussed. Without assuming discreteness or finiteness of underlying universes, with due regard to the measure-theoretic difficulties, implications, interplay and consequences of various principles such as conditionality, invariance, sufficiency, likelihoods (all types) and their weak versions are clarified.

1. INTRODUCTION

Recently a long and interesting paper entitled 'Statistical Information and Likelihood' was published by Basu (1973) in the proceedings of a Conference on 'Foundational Questions in Statistical Inference' (Aarhus Memoirs No. 1, 1973). Later this paper appeared in Sankhyā Ser. A (1975, 37, 1-71). Besides the Barnard-Basu correspondence and discussion, this paper also contains interesting discussions from Barndorff-Nielsen, Cox, Dempster, Edwards, Kalbfleish, Lauritzen, Martin-Löf and Rasch. Basu's essay is divided into three parts: the first consists of Principles; the second contains Methods; and the third gives Paradoxes. Other recent related work in this area is Birnbaum (1962), Hájek (1967), Godambe-Sprott (eds., 1971) and, of course, Fisher-Neyman-Pearson theories as well as Bayesian and Subjective inferences.

The motivation for this paper is essentially three-fold. Firstly, to present a discussion and critique, in the sense of a brief and relatively easy guided tour and re-evaluation of some basic concepts in statistical inference. The concepts involved are information, likelihood, sufficiency, ancillarity, various inferential principles, assumptions underlying the classsical inference, and decision error structure, not necessarily in that order. For brevity and precision, the choice of topics and the scope of the discussion had to be limited from a very large amount of statistical knowledge in the literature. Secondly, an effort is made to clarify, unify and interrelate the various concepts in the light of the most relevant and recent advances in connection with inferential problems. The reader is assumed to be statistically educated, hence not every entity in the standard statistical terminology is defined, ab ovo. Thirdly, wherever possible restrictions are relaxed, the underlying structure pointed out, and the central issues and implications emphasized. For an easy understanding of the fundamental ideas and principles, the complicated mathematical technicalities and involved proofs are reduced to the minimum as much as possible.

2. NORMALITY, SPHERICITY AND EXCHANGEABILITY

A k-component random vector X with distribution F is said to be a member of the spherical (S) or radial class $S_k(\mu, \Sigma)$ iff it has a characteristic function of the form $\phi(t) = Q(t'\Sigma t) \exp(it'\mu)$ with μ a k-component location parameter and Σ a positive definite k x k matrix, for some function Q. Without loss of generality, one may assume, that $\mu = 0$, then by the Bochner-Khintchine theorem Q is positive semidefinite and is continuous at zero, and hence is continuous everywhere. The spherical class includes many multivariate distributions such as normals, Cauchy, exponentials, double exponentials, some symmetric stable and Student's t-distributions. There are various equivalent definitions of sphericity: (a) X is spherical if all directions are equally likely and the distribution of magnitudes is independent of direction; (b) that the distribution of X is invariant under all k x k orthogonal matrices.

There are two related families to the above class, namely exchangeable (EX) and spherical exchangeable (SEX). The r.vs. $X_1, \dots , X_k$ are called EX if the k! permutations of the X's have the same distribution. An infinite sequence $\{X_j\}$ is EX if each k-element subsequence is EX. A sequence of r.vs. $\{X_j\}$ is called SEX if $\exists$ a function g in R^+ s.t. for each finite set $(i_1, \dots , i_k)$ of natural numbers, the joint c.f. satisfies $\phi(t_1, \dots , t_k) = E \exp (i \sum_{j=1}^{k} t_j X_{i_j}) = g(\sum_{j=1}^{k} t_j^2)$. The families S, SEX, EX are all well-known and have extensively been studied by Haag (1924), DeFinetti (1931), Schönberg (1938), Khintchine (1952), Dynkin (1953), Hewitt-Savage (1955), and many others. Some recent work on these classes was done by Kelker (1968, 1970), Ahmad (1972, 1974, 1975), Eaton-Kariya (1977, Ann. Statist., 5, 206-215), and the references contained therein. It turns out that the EX processes are mixtures of iid r.vs., and SEX processes are zero mean iid Gaussian processes' mixture.

In the classical theory the χ^2, t, F, Hotelling's T^2, the generalized U-statistic, and the Wishart distribution are all derived by taking a random sample from the normal class. In the normal theory 'zero-correlations', 'zero-covariances' and 'independence' are equivalent, and this fact is assumed to play a key role in deriving the classical F-distribution, where F-statistic is defined to be the ratio of two independent chi-square variables divided by their respective degrees of freedom. But this need not be the case, that is, the sample could be from a spherical distribution and the two chi-square variables could be dependent. There are, some other extensions for this class such as Cochran type theorems. It seems that the essential role is played by the fact that in the spherical class the correlation matrices of all members are identical.

There are two very useful transformation groups in statistics, both from theoretical and practical viewpoints. These groups are: (a) the family of finite-dimensional orthogonal matrices, and (b) the class of all finite-dimensional nonsingular matrices. Now, assume that the columns of a k x n random matrix Z are independently spherically distributed with location-scale parameter (μ, Σ), and having a density wrt a σ-finite measure. Denote by θ, the location parameter for Z. Then it can be shown that the density f(z) is invariant under the group of transformations, $(Z, \theta, \Sigma) \to (AZB, A\theta B, A\Sigma A')$, with k x k nonsingular matrices $\{A\}$ and n x n orthogonal matrices $\{B\}$. Furthermore, it turns out that (as pointed above) that the classical F-distribution, the generalized U-statistic, and a Wishart type distribution can be derived by taking a k-variate spherical sample from $S_k(0, \Sigma)$. Moreover, the joint density f(z) can be decomposed into two independent distributions; (a) essentially a Wishart-type distribution, i.e. the distribution of $C = \sum_{j=1}^{n} X_j X_j'$, and (b) the distribution of $\sum_{j=1}^{n} W_j$ where $W_j = X_j(X_j'X_j)^{-\frac{1}{2}}$ is distributed uniformly over the unit hypercube {x: x'x = 1}, see Ahmad (1972, 1975). Notice the application of the above structural results in the theories of hypotheses testing and serial correlation. For further applications, implications and a distribution-free testing structure we refer the reader to Hewitt-Savage (1955), Ahmad (1974, 1975), and Kariya-Eaton (1977).

Thus, from the above considerations and practical viewpoint, it appears natural and quite appealing to replace the normality assumption in the classical inference by sphericity or exchangeable sphericity. The concepts of exchangeability and partial exchangeability besides being applicable in Bayesian and Subjective approaches, are also useful in solving some problems of functional analysis and operator theory via the shift-operators. Since exchangeability in a sense is weaker than independence, perhaps it can be more fruitfully employed in various situations where one introduces white-noise or some kind of randomization in the parameter space, such as in signal detection and electronic control problems. Moreover, many problems possess these kinds of symmetries. Finally, while considering order statistics and rank vectors in various nonparametric problems, one notices that essentially there is 'the same relevant information' in the families of exchangeable and continuous [see Lehmann (1959, p.56 Problem 7) and the many consequences thereof] r.vs.

3. DECISION ERROR STRUCTURE

In the classical Neyman-Pearson theory of hypotheses testing and its various generalizations, and distribution-free testing problems, by using test statistics, one divides the basic space into two regions: the critical region C where one rejects the null hypothesis; and the acceptance region which is complementary to C, say A. In the standard terminology and approach, the statistician wants to make the two error probabilities α and β, as small as possible. Since these error probabilities are increasing and decreasing functions of the region C, respectively, to minimize them simultaneously is not clear. In practice usually one keeps α fixed, and then one minimizes β, thus maximizing the power of the resulting test. A more realistic, useful and general approach would be to minimize some linear function of α and β with positive weights.

Suppose we are testing P_o against P_1, and let $\alpha_N = P_o(T_N \in C)$, $\beta_N = P_1(T_N \in A)$, where T_N is a suitable test statistic based on a random sample of size N. Let $(\mathfrak{X},\mathcal{B})$ be the basic space and assume P_1 possesses a generalized density wrt P_o on R, denote this density by g. Now, consider the problem of minimizing $\varepsilon_N = p_o\alpha_N + p_1\beta_N$, where p_o and p_1 are two strictly positive real numbers, for example prior probabilities. Then it can easily be shown that

$$\inf_{C \in \mathcal{B}} \varepsilon_N = \int_{\mathfrak{X}} \min (p_o, p_1 g) dP_o$$

and that this lower bound is achieved exactly on the set $\bar{C}$ which satifies: $\{p_o < p_1 g\} \subset \bar{C} \subset \{p_o \leq p_1 g\}$. This double inclusion determines $\bar{C}$ uniquely upto P_o-equivalence when $P_o(\{g = p_o | p_1\}) = 0$. On the other hand, $P_o(\{g = a\})$ cannot be different from zero for more than a countable set of values a in R^+. For further details in this respect see Neveu, J. (1975, pp.44-51) [Discrete-Parameter Martingales, North-Holland].

By using a majorization technique via the Choquet capacities, Ahmad (1975) extended Rényi (1966) results and gave a symmetric formulation to the Neyman-Pearson fundamental lemma when the underlying spaces were Polish, i.e. complete separable metric spaces. Here, the testing is based on the <u>standard decision</u>, by which we always decide in favour of the hypothesis having the larger conditional probability given the sample. If these posterior probabilities are equal one decides at random with prior probabilities p_o and p_1. It turns out that the standard decision employes the generalized likelihood ratio test, and that no other decision has smaller error than the standard decision error ε_N. The test is essentially a generalized standard minimax Bayes test. For more details and the interplay between various measures of information and statistical implications we refer the reader to the above reference and later sections. In conclusion, suffice it to say that there does seem to be a great appeal in minimizing ε_N, choosing standard decisions, and letting the null and alternative hypotheses play an essentially equal and symmetric role. This might help reduce the criticism of the classical testing approach.

4. ON MEASURES OF INFORMATION

Consider a vector random variable X in Euclidean space E^k possessing a probability density $f(x, \theta)$ which depends on the m-dimensional parameter vector θ. Assume that thereexists a vector statistic $T = (T_1, \ldots , T_k)$ which is an unbiased estimator of θ and which has the variance-covariance nonsingular matrix $\Sigma = E((T - \theta)(T - \theta)')$. Under quite general regularity conditions, Fisher's information matrix

$$I(\theta) = E\left|\left|\frac{\partial^2 \log f}{\partial\theta_i \, \partial\theta_j}\right|\right| = E \left|\left|\frac{\partial \log f}{\partial\theta_i} \; \frac{\partial \log f}{\partial\theta_j}\right|\right|$$

exists and the Rao-Cramér (1945-46) inequality holds, which has the form $I(\theta) - \Sigma^{-1} \geq 0$. It may be remarked that this inequality apparently was given by

Aitken and Silverstone (1942) [Proc. Roy. Soc. Edin., A61, 186]. This inequality is of basic importance in the theory of parametric estimation. Notice that Fisher's information matrix appeared earlier than Shannon's amount of information, which plays such a key role in the statistical theory of transmission of information and the mathematical theory of communication. These two concepts are related and some connections between them have been pointed out in the paper of Kullback-Leibler (1951), see Kullback (1959). In contrast to the Fisher information, in the other case one considers a random variable with known distribution and wishes to give a measure of the uncertainty of the outcome of a future trial; the various definitions for the entropy provide such measures.

For n = 1,2, ... , define the sets

$$\Delta_n = \{P = (p_1, \dots, p_n) \ : \ p_i \geq 0 \ , \ w(P) = \sum_{i=1}^{n} p_i = 1\},$$

$$\Delta_n^* = \{P = (p_1, \dots, p_n) \ : \ p_i \geq 0 \ , \ w(P) \leq 1\},$$

$$\Delta_n' = \{P = (p_1, \dots, p_n) \ : \ p_i > 0 \ , \ w(P) = 1\},$$

$$\Delta_n^{**} = \{P = (p_1, \dots, p_n) \ : \ p_i > 0 \ , \ w(P) \leq 1\}.$$

Shannon (1948) defined the entropy of the discrete probability distribution ($P \in \Delta_n$) by the measure of its uncertainty in the form: $-\Sigma p_i \log p_i$. For the continuous case he used the so-called H-function of Boltzman: $-\int f(x) \log f(x)dx$, where f(x) is the density. Shannon and other researchers expressed the view that the H-function cannot be considered as a natural extension of the expression given for the discrete case, since it has no direct information-theoretical meaning. This observation consequently lead to the so-called I-divergence, see Rényi (1961), Aczél et al. (1975) and Aczél-Daróczy (1975). Notice that the H-function need not be nonnegative or invariant under 1 - 1 transformations of X. In a series of papers Ingarden and Urbanik [Ingarden-Urbanik (1962), Urbanik (1974), and references contained therein] have axiomatized information theory without probability by using a Boolean algebra. These authors make the concept of information as the primary concept, and then use this to construct the probability measure uniquely except for some degenerate cases. This probability-free approach is useful in statistical physics and thermodynamics. But we shall restrict ourselves to the foundations of the classical information theory as initiated and developed by Shannon, Wiener, and Aczél-Daróczy. Thus all the functionals of the measure of the information yielded by one event A as considered here depend only upon the probability P(A). Of course, there is a subjective aspect of information, which is wholly out of the scope of the classical theory, since the same event does not give the same amount of information to all the observers.

Let the measure of the information given by a single event A with P(A) = p be denoted by I(A) = H(p). Then, from the three natural conditions; (a) nonnegativity:

$H(p) \geq 0$ $p \in (0, 1]$, (b) additivity for independent events: $H(pq) = H(p) + H(q)$, (c) normalization: $H(1/2) = 1$, one gets $H(p) = -\log_2 p$. This result is based on the uniqueness of the nonnegative (or equivalently nondecreasing) solution: $g(x) = cx$, $c > 0$, of the Cauchy functional equation: $g(x + y) = g(x) + g(y)$ on $(0, \infty]$. For convenience set $0 \log 0 = 0$, and let $q_i = 0$ or 1 imply $p_i = 0$ or 1. The Shannon entropy $H_n(P)$, P in Δ_n, was extended by Rényi (1961) as $H_n : \Delta_n^{**} \to R^+$, where $H_n(P) = -\Sigma p_i \log p_i | \Sigma p_i$. Clearly, $H_{mn}(P*Q) \leq H_m(P) + H_n(Q)$, P in Δ_m^{**} and Q in Δ_n^{**}, with equality if P and Q are independent, here P*Q denotes the joint distribution. Furthermore, $H_n(P) \leq \log n$, the Hartley entropy. Aczél-Pfanzagl (1968, Metrika, 11, 91-105) discussed the following functional inequality

$$\sum_{i=1}^{n} p_i g(q_i) \leq \sum_{i=1}^{n} p_i g(p_i) \ , \ 0 < p_i, \ q_i < 1 \ , \ P, \ Q \ \epsilon \Delta_n',$$

and showed that for $n > 2$, the only solutions differentiable in (0, 1) are of the form: $g(x) = a \log_2 x + b$, $a > 0$ and b arbitrary constant. A consequence of this and the previous discussion is that all permutation symmetric, expansible (null events discarded), subadditive and additive entropies are linear combinations of the Shannon and Hartley entropies with nonnegative coefficients.

Consider a probability space $(\mathcal{X}, \mathcal{B}, P)$, called an experiment, and similar to Aczél-Daróczy to every event A in $\mathcal{B}$ associate a real number $I(A) = g(P(A))$, g measurable, the information contained in A. For a given event B in $\mathcal{B}$, to every event A in $\mathcal{B} \cap B$ one can define the conditional information of A wrt B, by $I(A|B) = P(B)\, g\, (P(A)/P(B))$ if $P(B) > 0$ and $= 0$ if $P(B) = 0$. Similarly define $I(A, B) = I(A) + I(B|A^*)$. By substituting $P(A) = p$, $P(B) = q$ from the symmetry axiom for $I(A, B)$ one gets the fundamental equation of information

$$g(p) + (1 - p)\, g\, (q/(1 - p)) = g\, (q) + (1 - q)\, g\, (p/(1 - q)).$$

An information function is any solution of this equation satisfying $g(\frac{1}{2}) = 1$ and $g(0) = g(1)$. Notice the implicit connection of $\{I(.)\}$ with the class of likelihood functions $\{L(.)\}$ in statistics. This aspect we shall explore in more detail in section 6. For some examples and applications see Ahmad (1975b).

5. SUFFICIENCY AND ANCILLARITY

The concepts of sufficiency and ancillarity were introduced by Fisher (1921, 1925). These ideas combined with the concepts of completeness, bounded completeness, symmetric completeness, invariance, similarity and exponential families play a key role in the theory of statistical inference, for example see Lehmann (1959), Birnbaum (1962), Hájek (1967), Basu (1973) Dawid (1977) and others.

Let $(\mathcal{X}, \mathcal{B}_{\mathcal{X}}, P_\mu)$ be the basic space with parameter space Ω, where μ is some σ-finite dominating measure. Denote by $\mathcal{P} = \{P_\theta : \theta \epsilon \Omega\}$, the parameter indexed family. Similar to LeCam (1964, 1974) and Basu (1973) we define the experiment to be

$E = (\mathcal{X}, \Omega, [\mathcal{B}_{\mathcal{X}}, \mathcal{B}_{\Omega}], P_{\mu})$. Two experiments E, E* are said to be similar or isomorphic, written $E \equiv E^*$, if there exists a 1 - 1 onto map $g: \mathcal{X} \rightarrow \mathcal{X}^*$ s.t. $P_{z|\theta} = P^*_{gz^*|\theta}$, where z and z^* are respective generic data points. We call $d = (E, z : z \in \mathcal{X})$ as the data point, and set $\mathbb{D} = \{d\}$. Similarly, for a statistic T we define the marginal and conditional experiments as $E_T = (\mathcal{X}_T, \Omega, P_T)$ and $E_t^T = (\mathcal{X}_t^T \equiv \mathcal{X}_t, \Omega, P_t^T)$, respectively. An experiment E is said to be non-informative about θ in Ω if for all z in $\mathcal{X}$, $P_{z|\theta}$ is constant. Clearly a statistic T(z) is sufficient for θ if E_t^T is non-informative about θ, i.e. if $P_{z|T}$ is independent of θ. Similarly, T is ancillary wrt θ if E_T is non-informative about θ, i.e. if P_T is independent of θ. In the sequel and above, we do not restrict ourselves to discrete and finite universes as Basu does. However, we do assume the usual measure-theoretic niceties so that the various expressions and arguments make sense in our context. Since the underlying subsigma and sigma-fields are natural classes of realizable events, we discuss these when the occasion demands.

An event A is called similar if $P_\theta(A) = P(A)$ for all θ, and a statistic T is called similar if $E_\theta T = \int t(z)P_\theta(dz)$ exists and is independent of θ. Thus events generated by ancillary statistics are similar. The family of similar events is closed under complimentation and countable disjoint unions. These classes are called λ-fields by Dynkin in 1959. If A and B are similar and not disjoint, then $A \cap B$ and $A \cup B$ in general are not similar. The family of similar or ancillary sets is a σ-field iff for every pair of similar sets their intersection is also similar. Though a λ-field is a broader concept than a σ-field, the system of σ-fields contained in a λ-field may not include a largest σ-field. As pointed out by Hájek (1967) these facts cause ambiguity in applications of the conditionality principle. Notice that the family is complete if the only similar statistics are constant. It may be pointed out that similarity is basic to the Neyman structure testsand nonparametric problems, see Ahmad (1974, 1975). The family of ancillary statistics with examples is treated by Basu (1959). In the case of nuisance and incidental parameters, the concepts of partial ancillarity and sufficiency, and S-ancillarity in exponential classes are given by Andersen (1967), Barndorff-Neilsen and Blaesild (1975) and the references contained therein (in particular, see Sverdrup (1966) and Sandved (1967)). In their investigation Barndorff-Neilsen and Blaesild introduce a cut statistic T if the map $C_T: \mathcal{P} \rightarrow \mathcal{P}_T \times \mathcal{P}^T$ defined by $C_T(P) = (P_T, P^T)$ is onto $\mathcal{P}_T \times \mathcal{P}^T$. Notice the appearance of the conditional and marginal experiments in the cut functional, which is essentially a data reduction map without loosing the essential information associated with the problem under consideration. Clearly this functional involves the relevant likelihood function and the appropriate information function, perhaps indirectly but nevertheless these entities are there. Not surprisingly, the authors show that there exists no proper cut, S-ancillary wrt the correlation coefficient ρ; and compare this with the fact that the c.d.f. of the sample correlation coefficient r is invariant under no Lie transformation group, see Brillinger (1963, AMS, 34, 492-500).

It has been shown by Basu that under not too restrictive conditions any statistic independent of a sufficient statistic is ancillary, and the converse holds if the sufficient statistic is complete. This statement can easily be shown to be valid, under some regularity conditions, in a weaker form such as introducing asymptotic (or partial) sufficiency, ancillarity and independence. Let T, T* and $\mathcal{B}_T$, $\mathcal{B}_{T^*}$, respectively, denote (provided they exist) a sufficient statistic, a minimal sufficient statistic and their respective induced subsigma-fields. Clearly, $\mathcal{B}_{T^*} \subset \mathcal{B}_T \subset \mathcal{B}_{\mathfrak{X}}$. Since a minimal sufficient statistic induces a maximal data reduction without loosing sufficiency, and hence the relevant information, if it exists, we consider this in our discussion. We may remark that Fisher said sufficiency but usually he meant minimal sufficiency, which sometime caused confusion in distinguishing between sufficiency and likelihood principles, see the next section.

Recall that in the dominated case the factorization criterion for sufficiency holds. Furthermore, the minimal sufficient σ-field for the dominated family is induced by the log-likelihood statistic for all θ, namely $\log(dP_\theta|d\mu)-\log(dP_{\theta_0}|d\mu)$ [Dynkin (1951, Uspechi Matem. Nauk, 6, 68-90]. More generally, minimal sufficient σ-fields exist for compact families. Since the families which are separable in the supremum-metric are necessarily dominated, minimal sufficient statistics exist for such classes. Bahadur (1955) investigated the problem whether a minimal sufficient statistic induces a minimal sufficient σ-field. He showed indeed this is the case for dominated families on Euclidean spaces. Later on Bahadur and Lehmann (1955) showed that this is not true in general. Motivated by the fact that many cases in practice such as some problems in nonparametric statistics,are excluded by dominated families, Burkholder (1961), Landers (1972) and Rogge (1972) have investigated sufficiency for undominated families. It turns out that for arbitrary families of perfect probability measures, admitting a countably generated sufficient σ- field, each minimal sufficient statistic induces a minimal sufficient σ-field. For the concepts of perfect measures, countably generated spaces and standard Borel spaces see Parthasarathy (1967, pp. 30-32 and 132-137). In particular, we note that any probability measure defined on the Borel-field of a complete, separable and metric space is perfect. Moreover, if $(\mathfrak{X},\mathcal{B})$ is countably generated then there exists a separable metric space $\mathfrak{X}^*$ s.t. $\mathcal{B}$ and $\mathcal{B}^*$ are sigma-isomorphic. The family of complete separable metric spaces are known as Polish spaces. In fact, if $\mathfrak{X}$ is a locally compact space, then the following statements are equivalent: (a) $\mathfrak{X}$ is Polish, (b) $\mathfrak{X}$ is Lusin or Blackwell space, (c) $\mathfrak{X}$ is Suslin space, (d) $\mathfrak{X}$ has a countable base for open sets. Thus, the essential structure of the underlying spaces is that these are either compact or locally compact. An interesting question for the many problems in applied sciences is the following. How far these space structural aspects can be weakened further, and what are the corresponding information-potential concepts parallel to ancillarity, sufficiency, minimal sufficiency and their partial versions?

It is well-known that there is an intrinsic relationship between families of distributions admitting nontrivial sufficient statistics and the exponential family of distributions. Furthermore, recall interplay among the above concepts, completeness property of a family, admissibility of certain tests, and the correspondance between sample and parameter spaces. Some work in this direction was recently done by Johansen (1977) and Soler (1977). We may also point out that the family of infinitely divisible distributions, and the class of distributions which can reasonably be approximated by infinitely-divisible-type kernels, have some relevance here, see Ahmad-Abouammoh (1977)(and LeCam (1960, 1974) in connection with contiguity). Now, similar to Johansen let $G = G(\mathfrak{X})$ denote the semigroup of finite ordered samples from $\mathfrak{X}$, and likewise define $\tilde{G} = \tilde{G}(\Omega)$. For $x = (x_1, \ldots, x_m)$, $y = (y_1, \ldots, y_n)$ with x,y in G define $x \sim y$ if $\prod_{i=1}^{m} f(x_i, \theta) = \prod_{j=1}^{n} f(y_j, \theta)$, $\theta \in \Omega$. That is, two samples x and y are equivalent if they have the same (strict) likelihood functions. Similarly, we define $x \tilde{\sim} y$ if $\exists$ $a > 0$ s.t. $\prod_{i=1}^{m} f(x_i, \theta) = a \prod_{j=1}^{n} f(y_j, \theta)$, $\theta \in \Omega$, that is $x \tilde{\sim} y$ if x and y have the same likelihood function, write $[x] = a[y]$. Thus, we have the following situation: $\mathfrak{X} \overset{\phi_1}{\to} G \overset{\phi_2}{\to} G/\sim \overset{\phi_3}{\to} G/\tilde{\sim}$, and $\phi_1 = \phi_3 \circ \phi_2$. Clearly, ϕ_1 induces the minimal sufficient σ-field. Next, for $\lambda = (\lambda_1, \ldots, \lambda_m)$, $\theta = (\theta_1, \ldots, \theta_n)$ with λ, θ in $\tilde{G}(\Omega)$ define $\lambda \sim \theta$ iff $\prod_{i=1}^{m} f(x, \lambda_i) = \prod_{j=1}^{n} f(x, \theta_j)$, $x \in \mathfrak{X}$. This gives us a mapping $\theta \overset{g_1}{\to} \tilde{G} \overset{g_2}{\to} \tilde{G}/\sim$, where $g_2(g_1 \ (\theta))$ is the maximal identifiable parameter.

Now, recall the regular exponential family $\{\exp[g(\theta)'\phi(x) + b(x) + c(\theta)]\}$ and consider the bilinear form $g(\theta)'\phi(x)$. This has an easy generalization as follows. The semigroups $G|\sim$ and $\tilde{G}|\sim$ are in duality by the bihomomorphism o defined by $[x] \circ [\theta] = \prod_{i=1}^{m} \prod_{j=1}^{n} f(x_i, \theta_j)$. Johansen has employed this duality to represent a family $\{P_\theta, \theta \in \Omega\}$ as a generalized exponential family, and shows how one can extend any family, to a canonical exponential family. Notice how this development links the generic observation, the underlying parameter, and the likelihood function. This, in fact, gives us a clue to the role the likelihood is going to play in various inference principles. Finally, compare this with following observations, see Urbanik (1975). Let $\Sigma(\mu)$ consist of all linear operators A on R^n s.t. $\mu = A\mu * \mu_A$ holds for a certain probability measure (p.m.) μ_A. To every p.m. on the Euclidean space there corresponds a semigroup of linear operators, such as $\Sigma(\mu)$, which is called a decomposibility semigroup. A p.m. μ on R^n is said to be operator-decomposable when there exists an operator A from $\Sigma(\mu)$ s.t. $\mu = A\mu * \mu_A$ and both $A\mu$ and μ_A are not concentrated at a single point. Thus, a p.m. has no operator-indecomposable factor iff it is Gaussian, and the family of Poisson measures are operator-indecomposable. Finally, each p.m. μ can be represented as a finite or infinite convolution $\mu = \mu_o * \mu_1 * \dots$, where μ_o is Gaussian and the measures $\mu_1, \mu_2, \dots$ are operator-indecomposable.

6. SOME PRINCIPLES OF STATISTICAL INFERENCE

In the sequel, we discuss various inferential principles, their implications and interrelationships. The approach is essentially similar to Birnbaum (1962), Hájek (1967), Basu (1975), and Dawid (1977), except that we do not restrict ourselves to finite and discrete universes. Also, underlying spaces are assumed to be Polish in the light of previous discussion, and in some cases one may have undominated families in the inferential structure.

Let experiments $E_i = (\mathcal{X}_i, \Omega, P_i)(i = 1,2, \dots)$ with mixture selection probabilities $\Pi_1, \Pi_2, \dots$ be given. Then the experiment $\{E_N, \mathcal{Z}_N\}$, where N is random integer and $\mathcal{Z}$ is the generic data point, is called a mixture experiment. An experiment E_1 is sufficient in Blackwell sense for E_2 if there exists a transition function $\Pi: \mathcal{X}_1 \times \mathcal{X}_2 \to [0, 1]$ with $\Sigma_{z_2} \Pi(z_1, z_2) = 1$ for z_1 in $\mathcal{X}_1$, which satisfies $p_2(z_2|\theta) = \Sigma_{z_1} p_1(z_1|\theta)\, \Pi(z_1, z_2)$ for all θ in Ω and z_2 in $\mathcal{X}_2$. This means that one can simulate E_2 from E_1 and a randomization process generates z_2 from z_1. Also, if E_1 and E_2 are similar, then each experiment is sufficient for other. Essentially Blackwell sufficiency and its variants can be viewed in terms of the average performance characteristics of statistical decision functions. Assume Ω is nontrivial, that is there exists at least one θ in Ω s.t. P_θ is positive. If the outcome of an experiment E is $\mathcal{Z}$ in $\mathcal{X}$, the map $\theta \to \tilde{P}_{\mathcal{Z}|\theta}$ is called the likelihood function (LF) generated by the data $d = (E, \mathcal{Z})$, and is denoted by $L(\theta) = L(\theta|d)$. Notice the L(.) function is not necessarily finite, is in general non-additive, and is a point function, unless it is transformed to a measure. Equivalently one could define (under the usual measure-theoretic conditions) a standard LF by $\overline{L}(\theta) = L(\theta) / \int_{\theta^* \in \Omega} L(\theta^*)$, where $\int$ denotes the integration or sum. Two LFs are said to be equivalent if there exists a constant $c \in R^+$ s.t. $L_1 = cL_2$ for all θ and we write $L_1 \equiv L_2$. If D is a measurable set in $\mathcal{X}$, we could have defined $L(\theta|\mathcal{Z} \in D) = cP_\theta(\mathcal{Z} \in D)$ for both discrete and continuous cases. Let $\mathcal{L} = \{L(\theta|d)\}$ and $\mathbb{D} = \{d: d = (E, \mathcal{Z}; \mathcal{Z} \in \mathcal{X})\}$. Now, we define the information function, Inf(.), as a map, $\mathrm{Inf}(d): \mathbb{D} \to \mathcal{L}$. Other relevant concepts of information and LF can be realized from $\{q(\mathrm{Inf}(d));\ q: \mathcal{L} \to \mathcal{L}_Q,\ q \in Q\}$, for some appropriate functional class Q.

Next, respectively, denote by $\mathcal{I}$, S, $\mathcal{L}$, C, and * as the principles of Invariance or Similarity, Sufficiency, Likelihood, Conditionality, and their weak versions such as $\mathcal{I}^*$, S^*, etc. Now, we state these principles explicitly.

$\mathcal{I}$: If $E_1 \equiv E_2$, then $\mathrm{Inf}(E_1, \mathcal{Z}_1) = \mathrm{Inf}(E_2, \mathcal{Z}_2)$.

$\mathcal{I}^*$: If $P(\mathcal{Z}\varepsilon D|\theta, E) = P(\mathcal{Z}^*\varepsilon D^*|\theta, E)$, then $\text{Inf}(E,\mathcal{Z}) = \text{Inf}(E,\mathcal{Z}^*)$.

S : A statistic T which is sufficient for E implies that $\text{Inf}(E,\mathcal{Z}) = \text{Inf}(E_T,\mathcal{Z})$.

S^*: That $\text{Inf}(E,\mathcal{Z}) = \text{Inf}(E,\mathcal{Z}^*)$ if there exists a sufficient statistics T s.t. $T(\mathcal{Z}) = T(\mathcal{Z}^*)$.

$\mathcal{L}$: If $L(\theta|z_1, E_1) \equiv L(\theta|z_2, E_2)$ for all $\theta\varepsilon\Omega$, $\mathcal{Z}_i \ \varepsilon \mathcal{X}_i$, then $\text{Inf}(d_1) = \text{Inf}(d_2)$.

$\mathcal{L}^*$: That $\text{Inf}(E,\mathcal{Z}) = \text{Inf}(E,\mathcal{Z}^*)$ if $L(\theta|z, E) \equiv L(\theta|z^*, E)$.

C : If T is an ancillary statistic wrt E, then for $\mathcal{Z}$ in $\mathcal{X}$ and $t = T(\mathcal{Z})$, $\text{Inf}(E,\mathcal{Z}) = \text{Inf}(E_t^T,\mathcal{Z})$.

C_m^*: If E is a mixture of $E_1, \ldots, E_m$, then for any $i \ \varepsilon \ \{1, 2, \ldots, m < \infty\}$ and $\mathcal{Z}_i \ \varepsilon \mathcal{X}_i$, $\text{Inf}(E(i,\mathcal{Z}_i)) = \text{Inf}(E_i, \mathcal{Z}_i)$.

<u>Lemma 1</u>. $\mathcal{I} \Longrightarrow \mathcal{I}^*$, $S \Longrightarrow S^* \Longleftrightarrow \mathcal{L}^* \Longrightarrow \mathcal{I}^*$.

<u>Proof</u>. It is straightforward and follows immediately from the definitions of these principles.

<u>Lemma 2</u>. $C \Longleftrightarrow \mathcal{L} \Longrightarrow \mathcal{I}$, and $\mathcal{L} \Longrightarrow S$.

<u>Proof</u>. By the definition of similar or isomorphic experiments $\mathcal{L}$ implies $\mathcal{I}$. Next, $\mathcal{L}$ implies S since $L(\theta|E,\mathcal{Z}) \equiv L(\theta|E_T,\mathcal{Z})$ if T is a sufficient statistic. Similarly, $\mathcal{L}$ implies C since $L(\theta|E,\mathcal{Z}) \equiv L(\theta|E_t^T,\mathcal{Z})$ if T is an ancillary statistic. The fact that C implies $\mathcal{L}$ follows similar to Hájek (1967, p. 155) arguments.

<u>Lemma 3</u>. (a) $\mathcal{I}^*$ and C_2^* together imply $\mathcal{L}$, (b) S^* and C_2^* together imply $\mathcal{L}$.

<u>Proof</u>. Assume $L(\theta|E_1,\mathcal{Z}_1) = aL(\theta|E_2,\mathcal{Z}_2)$ for a in R^+ and for all θ in Ω. For part (a) we want to show that $\mathcal{I}^*$ and $C_2^* \Longrightarrow \text{Inf}(E_1,\mathcal{Z}_1) = \text{Inf}(E_2,\mathcal{Z}_2)$. To this end consider the mixture experiment E of E_1 and E_2 with mixture selection probabilities $a/(1 + a)$ and $1/(1 + a)$. Then from the above assumption, E, and $\mathcal{I}^*$, one gets

$$(*) \quad \text{Inf}(E, (1,\mathcal{Z}_1)) = \text{Inf}(E, (2,\mathcal{Z}_2)).$$

By applying C_2^* to (*) one completes the proof of (a). Since S^* implies $\mathcal{I}^*$, so S^* and C_2^* imply $\mathcal{I}^*$ and C_2^*, and which in turn implies $\mathcal{L}$ by (a), and this establishes (b).

Note that the above development is essentially similar to that of Basu (1975), Hájek (1967) and Birnbaum (1962), except that we do not restrict ourselves to finite or discrete universes, provided, of course, that various entities involved are defined and meaningful such as the LF function etc. There are some other variants of these principles, but these are essentially mixtures of two or more of these principles. Notice that the likelihood and conditionality principles, if properly understood, indeed look very appealing. However, there are some unpleasant implications of these, which we discuss below. It is common knowledge that if a minimal sufficient statistic exists one should employ it in the considered inference problem, since it has maximal reduction in the data. Fisher consistently said sufficient statistic but in fact meant minimal sufficient statistic. Consequently, Fisher and Barnard thought that S and $\mathcal{L}$ principles are indistinguishable. But, in fact, S^* and $\mathcal{L}$ are indistinguishable.

As Basu and Hájek point out the likelihood principle is the very essence of Bayesianism. All proper Bayesian procedures, that is where the prior distribution is obtained independently of the experiment, are concordant with it. However, any attempt to characterize the prior structure with it violates this principle. The principles of S(C) warn us against any post (pre)-randomization wrt the data analysis stage. The $\mathcal{L}$ principle implies discard a major part of the current theories regarding the analysis of survey data, at the data analysis stage if the data is produced by survey sampling methods. The Maximum Likelihood Estimation confirms to the $\mathcal{L}$-principle provided one does not measure the precision of the estimate. The likelihood interval estimation is alright provided L(.) is unimodal. The likelihood tests of goodness-of-fit, the likelihood ratio methods with many variants, the Neyman-Pearson type arguments, and the nuisance parameters elimination methods all essentially conform to the C-principle. The famous exception in this case is the Wald's theory of sequential tests. Since disjunction of two events is an event, and disjunction of two simple hypotheses is not in general a simple hypothesis; Barnard justly questions the relevance in general of additivity. Some authors take 'additivity' and 'support' as postulates, but this occasionally leads to paradoxes, see Basu (1975). The function L(.) besides being non-additive is s.t. $\Sigma L(\theta)$ is not finite, where the sum is over all $\theta \varepsilon H_i$, H_i's mutually exclusive for $i = 1, 2, \ldots$. Making L(.) an additive measure creates further difficulties and paradoxes, recall L is a point function. Perhaps, one could employ the ideas of sub (super)-additive measures or functionals via the Choquet capacity theory and the relevant transformation group structures. Also, one could enlarge the concept of likelihood function by introducing a topological group structure - in this case the function $L(\theta|E, \mathfrak{Z}, G)$ may not act simply but it will behave relatively 'smoothly and inference-information-potentially' wrt the imposed group structure. This aspect is currently being investigated. It seems that a reasonably rich inference structure would emerge from $\{L(\theta|E, \mathfrak{Z}, G)$; L-independent all relevant prior information; the particular inference problem structure$\}$.

Remark. Some references are mentioned in the text but not given below, as these can easily be found from the cited list. Similarly, some references are very well-known, for example various publications of Neyman, Lévy and Kolmogorov.

REFERENCES

1. Aczél, J. et al. (1975). Entropy and Ergodic Theory. Selecta Statistica Canadiana Vol. 2. Delhi Univ. Press.
2. Aczél, J. - Daróczy, Z. (1975). On Measures of Information and their Characterizations. Math. Sc. & Eng. Vol. 115. Academic Press.
3. Ahmad, R. (1972). Extension of the normal theory to spherical families. Trabajos de Estadistica, 23, 51-60.
4. Ahmad, R. (1974). On the structure of symmetric sample testing: a distribution-free approach. Ann. Inst. Statist. Math., 26, 233-245.

5. Ahmad, R. (1975a). Some characterizations of the exchangeable processes and distribution-free tests. G.P. Patil et al. (eds.), Statistical Distributions in Scientific Work, Vol. 3, 237-248. D. Reidel, Dordrecht-Holland.

6. Ahmad, R. (1975b). The Neyman-Pearson lemma for Choquet capacities in Bayesian problems. Bull. Inter. Statist. Inst., 46(3), 13-18.

7. Ahmad, R. - Abouammoh, A.M. (1977). On the structure and applications of infinite divisibility, stability and symmetry in stochastic inference. Recent Developments in Statistics, J.P. Barra et al. (eds.), 303-317. North-Holland.

8. Bahadur, R. (1954). Sufficiency and statistical decision functions. Ann. Math. Statist., 25, 423-462.

9. Bahadur, R. (1955). Statistics and subfields. Ibid, 26, 490-497.

10. Bahadur, R. - Lehmann, E.L. (1955). Two comments on "Sufficiency and statistical decision functions". Ibid, 26, 139-142.

11. Barnard, G.A. (1967). The use of the likelihood function in statistical practice. 5th Berkeley Symp. Math. Stat. Prob., 1, 27-40.

12. Barndorff-Nielsen, O. - Blaesild, P. (1975). S-ancillarity in exponential families. Sankhyā Ser. A, 37, 354-385.

13. Basu, D. (1959). The family of ancillary statistics. Sankhyā Ser. A, 21, 247-256.

14. Basu, D. (1975). Statistical information and likelihood. Sankhyā Ser. A, 37, 1-71.

15. Birnbaum, A. (1962). On the foundations of statistical inference. JASA, 57, 269-326.

16. Burkholder, D.L. (1961). Sufficiency in the undominated case. Ann. Math. Statist., 32, 1191-1200.

17. Csiszár, I. (1969). On generalized entropy. Stud. Sci. Math. Hung., 4, 401-419.

18. Dawid, A.P. (1977). Conformity of inference patterns. Recent Developments in Statistics, J.R. Barra et al. (eds.), 245-256. North-Holland.

19. Fisher, R.A. (1921). On the mathematical foundations of theoretical statistics. Philos. Trans. Roy. Soc. London Ser. A, 222, 309-368.

20. Fisher, R.A. (1925). Theory of statistical estimation. Proc. Camb. Philos. Soc., 22, 700-725.

21. Fraser, D.A.S. (1968). The Structure of Inference. Wiley.

22. Godambe, V.P. - Sprott, D.A. (1971). Foundations of Statistical Inference (eds.) Holt, Rinehart and Winston, Toronto.

23. Hájek, J. (1967). On basic concepts of statistics. 5th Berkeley Symp. Math. Stat. Prob., 1, 139-162.

24. Hewitt, E. - Savage, L.J. (1955). Symmetric measures on Cartesian products. Trans. Amer. Math. Soc., 80, 470-501.

25. Ingarden, R.S. - Urbanik, K. (1962). Information without probability. Colloq. Math., 9, 131-150.

26. Johansen, S. (1977). Homomorphism and general exponential families. Recent Developments in Statistics, J.R. Barra et al. (eds.), 489-499. North-Holland.

27. Kullback, S. (1959). Information Theory and Statistics. Wiley.

28. Landers, D. (1972). Sufficient and minimal sufficient σ-fields. Z. Wahrsch., 23, 197-207.

29. LeCam, L. (1964). Sufficiency and approximate sufficiency. Ann. Math. Statist., 35, 1419-1455.

30. LeCam, L. (1974). On the information contained in additional observations. Ann. Statist., 4, 630-649.

31. Lehmann, E.L. (1959). Testing Statistical Hypotheses. Wiley.

32. Loève, M. (1963). Probability Theory. 3rd ed. D. van Nostrand, Princeton.

33. Neyman, J. (1954). Sur une famille de tests asymptotiques des hypotheses statistiques composèes. Trabajos de Estadistica, 5, 161-168.

34. Parthasarathy, K. (1967). Probability Measures on Metric Spaces. Academic Press.

35. Rényi, A. (1961). On measures of entropy and information. 4th Berkeley Symp. Math. Stat. Prob., 1, 547-561.

36. Rényi, A. (1966). On the amount of missing information and the Neyman-Pearson lemma. Research Papers in Statistics for Neyman, F.N. David (ed.). Wiley.

37. Rogge, L. (1972). The relations between minimal sufficient statistics and minimal sufficient σ-fields. Z. Wahrsch., 23, 208-215.

38. Savage, L.J. (1976). On rereading R.A. Fisher. Ann. Statist., 3, 441-500.

39. Shannon, C.E. (1948). A mathematical theory of Communication. Bell System Tech. J., 27, 379-423, 623-656.

40. Shannon, C.E. - Weaver, W. (1949). The Mathematical Theory of Communication. Univ. of Illinois Press.

41. Urbanik, K. (1974). On the concept of information. Colloq. Math. Soc. J. Bolyai, 9, 863-868.

SUPPLEMENTARY REFERENCES

42. Andersen, E.B. (1967). On partial sufficiency and partial ancillarity. Skand. Aktuar., 50, 137-152.

43. Dynkin, E.B. (1959). The Foundations of the Theory of Markov Processes. Moscow, Fizmatgiz. (In Russian).

44. Halmos, P.R. (1950). Measure Theory. Van Nostrand, Princeton.

45. Halmos, P.R. - Savage, L.J. (1949). Applications of the Radon-Nikodym theorem to the theory of sufficient statistics. Ann. Math. Statist., 20, 225-241.

46. Hewitt, E. - Stromberg, K. (1967). Real and Abstract Analysis. Springer.

47. Kullback, S. - Leibler, R.A. (1951). On information and sufficiency. Ann. Math. Statist., 22, 79-86.

48. Kuratowski, C. (1933). Topologie I. Warszawa - Lwow.

49. Lehmann, E.L. - Scheffé, H. (1950;1955). Completeness, similar regions and unbiased estimation. Sankhyā; 10, 305-340 ; 15, 219-236.

50. Meyer, P.A. (1966). Probability and Potentials. Blaisdell, New York.

51. Neveu, J. (1975). Discrete-Parameter Martingales. North-Holland.

52. Schwartz, L. (1973). Radon Measures on Arbitrary Topological Spaces and Cylindrical Measures. Tata Institute of Fundamental Research, India.

53. Soler, J.L. (1977). Infinite dimensional exponential type statistical spaces. Recent Developments in Statistics, J.R. Barra et al. (eds.) 269-284 North-Holland.

54. Urbanik, K. (1975). Decomposability properties of probability measures. Sankhyā Ser. A, 37, 530-537.

55. Wilkinson, G.N. (1977). On resolving the controversy in statistical inference (with Discussion). J. Roy. Statist. Soc. B, 39, 119-171.

R. Ahmad,
University of Strathclyde,
Department of Mathematics,
26 Richmond Street,
GLASGOW. G1 1XH
SCOTLAND.

POLYA-TYPE, SCHUR-CONCAVE AND RELATED PROBABILITY DISTRIBUTIONS

Rashid Ahmad, Abdulrahman M. Abouammoh

Glasgow

ABSTRACT

The objective of this paper is to give a unified and an extended structure of the above classes of distributions as well as their many applications. The behaviour of these families is investigated under most commonly occurring functional operations such as closure under convolution, a passage to a limit in the weak sense, reversal and mixing properties. The useful and smooth properties of unimodality, strong unimodality and their variants are found to hold for some subclasses of these applicable probability distributions.

1. INTRODUCTION AND PRELIMINARIES

The above concepts are fundamental and play a basic role in many probabilistic-statistical problems. In particular, these ideas occur in estimation, testing hypotheses and general statistical inference problems. These ideas have predominated in the past many investigations in statistical theory and applications, and solutions for considerably important problems have been sought through these concepts and their various forms, that is some stronger or weaker forms of these concepts. Such approach will be elucidated within the context of the paper. There is a large amount of work in these areas, for example Karlin (1956), Lehmann (1959), Birnbaum et al. (1966), Johnson and Kotz (1972), Barlow and Proschan (1975), Proschan and Sethuraman (1977) and many others.

In the sequel, a generalization of strong unimodality from one dimensional to n-dimensional space is suggested and it is shown that the class of n-dimensional strongly unimodal distributions is closed under convolution, passage to a limit weakly, mixing and reversal. Some other characterization properties are studied through a concavity structure of the underlying probability distributions.

For the classes M_s^n, $-\infty < s < n^{-1}$ of distributions which satisfy a certain convexity property, it turns out that for $s = 0$ the class M_s^n is equivalent to strongly unimodal class in an n-dimensional space. It is realized that the marginal measures (distributions) of an n-dimensional log-concave measures (distributions) is log-concave and hence Ibragimov 1956 result is implied as a special case in one-dimension of n-dimensional strongly unimodal distributions.

In connection with Polya type k (PT_k) distributions, it is observed that PT_2 is equivalent to the monotone likelihood ratio (MLR) property and that these distributions are strongly unimodal when the considered parameter in the definition is location parameter. Furthermore, it is found that PT functions have smooth behaviour as kernel of some integral transformations and are closed under convolution on the positive real line, mixing, reversal and passage to a limit weakly.

The class of increasing failure rate distributions (IFR) is considered and it is found that IFR and IFR average (IFRA) are closed under convolution, passage to a limit weakly but not under reversal and mixing. It is noticed that strongly unimodal PT_2, IFR and IFRA form an increasing sequence of classes of distributions. Thus, IFR is the smallest class containing all exponential distributions.

An n-dimensional density function which is Schur-Concave, that is it satisfies a permutation symmetric property is observed to exhibit essentially exchangeable or symmetric dependence structure. The class of Schur-Concave functions is closed under convolution. In addition we shall show some other structural properties of related exchangeable distributions or variables having some interesting applications. Finally, applications in the context of inferential aspects will be discussed.

2. LOG-CONCAVE PROBABILITY MEASURES AND DISTRIBUTIONS

The well known definition of unimodality as given by Khintchine in 1938 states that a real variable X has a unimodal distribution about a vertex a if its distribution function is convex on $(-\infty,a)$ and concave on (a,∞). An extensive treatment together with many other properties and results concerning these and related type of distributions for the univariate case is given by the authors (1977a). A slightly disturbing property of unimodal distributions was pointed out by Chung in 1953, that is the convolution of two unimodal distributions may not be unimodal. But if one considers the symmetric unimodal distributions, then such class is closed under convolution. Wintner in 1938, Ibragimov in 1956 adapted the strongly unimodal concept in order to preserve the unimodality property under convolution. A distribution F is strongly unimodal if its convolution with any unimodal distribution is again unimodal. Ibragimov gave an interesting characterization of such distributions, namely, a distribution F(x) is strongly unimodal if and only if log f(x) is concave where f(x) is the density of F(x). Such a density is absolutely continuous within the range of its definition and $[\log f(x)]' = f'(x)/f(x)$ is nonincreasing function in x. It can be shown that the normal, exponential, Wishart and uniform distributions are strongly unimodal. Moreover, it follows that if f(x) is symmetric density which is not log-concave then there exists a unimodal distribution with density g, say, such that the convolution $f * g$ is not unimodal. There is another characterization result of strongly unimodal distributions, that is a positive twice differentiable density f is strongly unimodal if and only if

(2.1) $f'^2 > f\,f''$

and this relation (2.1) can be easily derived from the relation $(\log f)'' = [(f''f-f'^2)/f^2] \leq 0.$

Next, we shall show the closure of strongly unimodal class of distributions under mixing. Let f and g be two strongly unimodal densities and $\alpha \in (0,1)$, then by inequality (2.1) one has

$$(2.2) \quad (\alpha f'+(1-\alpha)g')^2 - (\alpha f''+(1-\alpha)g'')(\alpha f+(1-\alpha)g) \geq$$
$$\geq \alpha(1-\alpha)(2f'g'-fg''-gf'')$$
$$\geq fg(f'/f-g'/g)^2 \geq 0.$$

It can be easily realized that such class is closed under convergence to a limit weakly. In fact, Lapin has shown that the class of unimodal distributions is closed under the weak convergence, see Lukacs (1970, p. 97). Furthermore, if we define F(-x) to be the reversal of F(x) then it is not difficult to see that all distributions which are unimodal or strongly unimodal are closed under reversal property. Before we give the multivariate structure we summerize the above results in the following.

Theorem 2.1. The class of one-dimensional strongly unimodal distributions is closed under convolution, mixing, passage to a limit in the weak sense and reversal.

Now, we give the main results of log-concave families of probability measures in R^n. A probability measure P on R^n is log-concave if for all open convex sets $A, B \subset R^n$ and $\alpha \in [0,1]$, we have

$$(2.3) \quad P(\alpha A+(1-\alpha)B) \geq (P(A))^{\alpha} (P(B))^{1-\alpha}.$$

It is clear that (2.3) is also true for all closed convex sets A, B, since any closed convex set is the limit of a decreasing sequence of open convex sets. Another consequence of (2.3) is the inequality

$$P\{(\alpha X+(1-\alpha)Y) \in A \cup B)\} \geq (P(X \in A))^{\alpha} (P(Y \in B))^{1-\alpha}$$

for some vectors $\underline{x}$ and $\underline{y}$ in R^n.

Borell (1975), Prékopa (1973) and Rinott (1976) have dealt with the necessary and sufficient conditions for a density function f(x) in R^n which generates [illegible]bability measure P that satisfies a convexity property of the type

$$(2.4) \quad P(\alpha A+(1-\alpha)B) \geq \begin{cases} [(\alpha P(A))^s+((1-\alpha)P(B))^s]^{1/s} & \text{if } s \in (-\infty,0) \text{ or } = \frac{1}{n} \\ \min(P(A),P(B)) & \text{if } s = -\infty \\ (P(A))^{\alpha} (P(B))^{1-\alpha} & \text{if } s = 0. \end{cases}$$

In case $s = 1/n$, (2.4) holds for Lebesgue measure and is known as the Brunn-Minkowski inequality. A probability measure P in R^n on its corresponding distribution is said to belong to class M_s^n if it satisfies relation (2.3). The class M_s^n is in a sense a generalization of the class of log-concave measures. It is noted that for $s_1 \geq s_2 \geq \ldots \geq -\infty$, one has $M^n_{-\infty}$ is the largest class and $M^n_{s_1} \subseteq M^n_{s_2} \subseteq \ldots \subseteq M^n_{-\infty}$. It is also clear that $M^n_s = \phi$ if $s > 1/n$. Now we shall give the following characterization result which has been proved in a special form by Rinott (1976).

Theorem 2.2. Let P be a probability measure generated by an n-dimensional density f, that is $P(B) = \int_B f(x)dx$, $x \in R^n$, for any Borel set $B \subset R^n$. Then $P \in M^n_s$ if and only if there exist a version h of the density f such that $h^{s/(1-sn)}$ is convex if $s \in [-\infty, 0)$, log h is concave if $s = 0$ and $h^{s/(1-sn)}$ is concave if $s \in (0, 1/n)$.

Proof: The necessary part is easily obtainable, see for example the geometrical argument in Rinott (1976). To prove the sufficiency part, let $s = 0$ and the probability measure P generated by some density f be log-concave. Let $S(r,x)$ to be the sphere in R^n with radius $1/r$ and centre $x \in R^n$. Define $f(r,x) = |S^{-1}(r,x)| \int f(y)dy$, $y \in R^n$, and the integral is taken over $S(r,x)$ and $|S(r,x)|$ is Lebesgue measure. Hence, the log-concavity of P implies that $f(r, \alpha x+(1-\alpha)y) \geq (f(r,x))^{\alpha}(f(r,y))^{1-\alpha}$. Take $h = \liminf_{r\to\infty} f(r,x)$. Thus, h is log-concave and $f = h$ almost everywhere.

Let us now call the density function f defined on R^n to be multivariate strongly unimodal if it is log-concave. In what follows, we shall investigate the behaviour of the class of multivariate strongly unimodal distributions under some functional operations. Denote such a class by U^n where n refers to the underlying dimension. For the proof of the theorem below we need the following lemmas.

Lemma 2.1. (Prékopa (1973, p. 337)). Let $P_1, P_2, \ldots, P_k$ be k probability measures on R^n and let $d\mu(z) = \sup_{\alpha_1 x_1 + \ldots + \alpha_k x_k = z} dP_1(x_1) \ldots dP_k(x_k)$, $z \in R^n$, where $\alpha_i > 0$ are constants and $\alpha_1 + \ldots + \alpha_k = 1$. Then $\mu(z)$ is a probability measure and furthermore

$$\int_{R^n} d\mu(z) \geq \left[\int_{R^n} (dP_1(x_1))^{1/\alpha_1}\right]^{\alpha_1} \ldots \left[\int_{R^n} (dP_k(x_k))^{1/\alpha_k}\right]^{\alpha_k}.$$

Lemma 2.2. (Parthasarathy (1967 p. 40)). Let $\{P_k\}$ be a sequence of probability measures defined on a metric space (Ω, m). Then

(i) $\limsup_{k\to\infty} P_k(A) \leq P(A)$ for every closed set $A \subset \Omega$

(ii) $\lim_{k\to\infty} P_k(B) = P(B)$ for every Borel set B whose boundary has P-measure zero.

Theorem 2.3. The class U^n of distributions is closed under convolution, mixing, reversal and passage to a limit weakly.

Proof: We proceed the proof of various parts in the same order as in the statement of the theorem. Let P_1, P_2 be in U^n, we need to show $\int_{R^n} P_1(x-y)d\,P_2(y)$ is in U^n, that is it is log-concave for $x, y \in R^n$. Since $P_1(x-y)d\,P_2(y)$ is log-concave in R^{2n}. Then by taking $y = (y_1, y_2)$ and $x = (x_1, x_2)$ and applying lemma 2.1. respectively, the result follows.

The proof of the mixture (convex combination) of P_1 and P_2 being in U^n can be obtained by (2.3) and an argument similar to (2.2) which can be applied by taking the partial derivative with respect to the components of the underlying vectors, and hence U^n is closed under mixing. Furthermore, it is easily seen from the definition of U^n that it is closed under reversal. To prove the last part of theorem that U^n is closed under passage to a limit weakly, let $\{P_k\}$ be a sequence of probability measures in U^n for every $k \geq 1$ and P_k converges weakly to a probability measure P. Now, we want to show that P is in U^n, that is, it satisfies (2.3). But (2.3) is satisfied whenever A, B are closed convex non-empty sets whose boundaries have P-measure zero and hence by using lemma 2.2., $P(A) = \lim_{k\to\infty} P_k(A)$ and $P(B) = \lim_{k\to\infty} P_k(B)$. Since any open convex set is the limit of an increasing sequence of closed convex non-empty sets whose boundaries have P-measure zero. Therefore, $P \in U^n$ for any open convex sets A and B. This proves the theorem.

Corollary 2.1. If a distribution $F(x) \in U^n$, $x \in R^n$, then any marginal distribution $F(x_1,\ldots, x_k) \in U^k$ where $k = 1,\ldots, n$.

Remark: the multivariate unimodality has been defined and discussed in many different ways such as generalized unimodal, linear unimodal, monotone unimodal, central convex unimodal and the last two concepts are included in multivariate symmetric unimodality, see Kanter (1977), Ahmad and Abouammoh (1977a) and references mentioned therein for various characterization results.

3. STRUCTURE OF PT AND MLR DISTRIBUTIONS

The smoothness and the nice beahviour of PT and MLR distributions is discussed to examplify the structure and the applications such as in decision theory and other inferential problems in statistics.

The family of distributions $F(x,\theta)$ (or their densities $f(x,\theta)$) of real random variables X depending on a real parameter θ is said to belong to the class PT_k if

$$(3.1) \qquad \Delta_k = \begin{pmatrix} f(x_1,\theta_1) & \cdots & f(x_1,\theta_k) \\ \vdots & & \\ f(x_k,\theta_1) & \cdots & f(x_k,\theta_k) \end{pmatrix}$$

is semi-positive definite matrix for every $k \geq 1$ and all $x_1 < x_2 <\ldots< x_k$ and $\theta_1 < \theta_2 <\ldots< \theta_k$. The family $F(x,\theta)$ belongs strictly to PT_k if Δ_k in (3.1) is positive definite matrix. If $F(x,\theta)$ belongs to PT_k for every $k = 1,2,\ldots$, then $F(x,\theta)$ belongs to PT_∞. The distribution $F(x,y)$ of the two real variables ranging over linearly ordered one-dimensional sets X and Y respectively is said to be totally positive of order k (TP_k) if $\Delta_k(x,y)$ is semi-positive definite, that is, $\Delta_i(x,y) \geq 0$ for all i, $1 \leq i \leq k$, and it is called strictly TP_k if $\Delta_i(x,y) > 0$ for all i, $1 \leq i \leq k$. In fact the PT_k distributions give two familiar and interesting cases for $k = 1,2$.

These are, the family $F(x,\theta)$ is PT_1 that is if and only if $f(x,\theta) \geq 0$ for every x and θ, therefore every distribution identified by a parameter is PT_1 and the family $F(x,\theta)$ is PT_2 if and only if

(3.2) $f(x_1,\theta_1)f(x_2,\theta_2) - f(x_1,\theta_2)f(x_2,\theta_1) \geq 0.$

The later case arises in many inference problems in applied statistics, and any family of distributions whose densities satisfy (3.2) are said to have MLR. One may notice from the definition of PT_k classes of distributions that $PT_\infty \subset \ldots \subset PT_2 \subset PT_1$ and it can be also shown that the exponential family, Lehmann (1959, p. 115), non-central F, noncentral t and noncentral chi-square, Karlin ((1956), and some other families of distributions belong strictly to PT_∞ and hence they have MLR property. In other words, most of the distributions used in statistical inference are PT. However, the most notable example of a distribution which is not PT is Cauchy with density $f(x,\theta) = \{\pi[1+(x-\theta)^2]\}^{-1}$.

It was found that PT functions have very nice property when they are used as kernel of transformation, for example, see Karlin (1957). If $f_1(x,\theta)$ belongs to PT_∞ and is nth differentiable with respect to x for all θ, F_3 is some distribution associated with finite measure, $f_2(\theta)$ is a function of θ which has n sign changes and $f_2(x) = \int f_1(x,\theta)\, f_2(\theta) d\, F_3(\theta)$ is nth differentiable with respect to x inside the integral, then $f_2(x)$ has at most n sign changes. Therefore, if f_1 and f_2 are two continuous differentiable densities of two independent random variables X and Y respectively such that $f_1(x-\theta)$ is strictly PT_∞ and f_2 has m modes, then the density of $Z = X+Y$ that is $f_3(z) = \int f_1(t)\, f_2(z-t)dt$ has at most m modes. Hence, the concavity (or convexity) property of a function is preserved under convolution with any strictly PT_∞ distribution. Furthermore, if f_1^* is the nth convolution of f_1, where f_1 is PT_∞ and f_2 is concave (convex) then $g(.,n) = \int f_1^{*n}(x)\, f_2(x)dx$ is concave (convex). Now we shall give the following lemma which can be proved by the basic composition formula and the direct product of matrices.

Lemma 3.1. The class of PT_k, $k = 1,2,\ldots$ distributions of non-negative random variables is closed under convolutions.

Theorem 3.1. The class of PT_k; $k = 1,2,\ldots$ distributions is closed under, mixing and convergence to a limit in the weak sense.

Proof: It is clear that PT class is closed under reversal since Δ_k defined by (3.1) is semi-definite for any $x_1 < x_2 <\ldots< x_k$ whether positive or negative values of x's. Now, let f_1 and f_2 belong to PT_k, $k \geq 1$ and $f = \alpha f_1+(1-\alpha) f_2$, $0 \leq \alpha \leq 1$. Therefore, by writing Δ_k in the convex combination form, also by taking it as a summation of two matrices the first of f_1 and the second of f_2, and since $\alpha^k \geq 0$, $(1-\alpha)^k \geq 0$ one has PT_k for $k \geq 1$ is closed under mixing. Finally, PT_k, $k \geq 1$, class is closed under convergence to a limit weakly is implied from the fact that the limit of a matrix is defined by the limit of each element of such matrix, which themselves are

of PT form. This completes the proof.
As a consequence from the above we give the following.
Corollary 3.1. The class of distributions with MLR propetty is closed under reversal, mixing and convergence to a limit weakly.

Next, we give two weaker forms of MLR property which are defined by relation (3.2). These are (i) the parameter $\theta \in \Theta \subset R$ is merely a location parameter that is for $x_1 < x_2$ and $\theta_1 < \theta_2$ one has

(3.3) $f(x_1-\theta_1)\, f(x_2-\theta_2) - f(x_1-\theta_2)\, f(x_2-\theta_1) \geq 0,$

and (ii) if $x_1, x_2, x_3, x_4 \in R$ or integers and $\theta_1, \theta_2 \in \Theta$ such that $x_3 < x_1 < x_4$ and $\theta_1 < \theta_2$ we have

(3.4) $f(x_1,\theta_1)\, f(x_2,\theta_2) - f(x_4,\theta_1)\, f(x_3,\theta_2) \geq 0.$

Then one can see that (3.2) implies that $f(x_1,\theta_2)/f(x_1,\theta_1)$ is nondecreasing in x and $f(x_2,\theta_1)/f(x_1,\theta_1)$ is nondecreasing in θ_1 that is $f(x+h,\theta_1)/f(x,\theta_1)$ is nondecreasing in θ_1 for all x and $h > 0$. Also, (3.3) implies $f(x+h,\theta_1)/f(x,\theta_1)$ is nonincreasing in x; $h > 0$ and it implies that f is log-concave function. Further, (3.4) is equivalent to (3.2) if $x_1 = x_3$ and $x_2 = x_4$, and any class of functions satisfying (3.4) is closed under convolution see Ghurye and Wallace (1959). Now we summarize the closure of MLR class under convolutions as below.

Theorem 3.2. Let f_1 and f_2 be two density functions and $f = f_1 * f_2$. Then (3.4) is satisfied by f if it is satisfied by f_1 and f_2; (3.3) is satisfied by f if it is satisfied by f_1 and f_2 and (3.2) is satisfied by f if it is satisfied by f_1 and f_2 for non-negative random variables.

4. IFR TYPE DISTRIBUTIONS

It is assumed sometimes that the distribution of the future life (life distribution) of a device remains the same regardless of the time while it was in use, which is usually known in statistical term by 'new is the same as used', and is characterized by a life distribution of lack of memory such as exponentials. Such a distribution is said to represent no wear phenomenon. Similarly, other classes of distributions may represent the wear out phenomenon, that is new better than used (NBU) and the durability phenomenon that is new worse than used (NWU). Many authors have tackled the problem of finding classes of distributions which reflect these phenomena. To answer this problem Birnbaum et al. (1966) introduced the class of distributions with IFR which was also discussed by Barlow and Proschan (1975), A-Hameed and Proschan (1973), Black and Savits (1976) and many others.

Let the survival probability or reliability be $\bar{F}(x) = P(X > x)) = 1-F(x)$ which is the complement of the life distribution $F(x)$. The conditional reliability for the remaining life given that the device has survived to age t is $\bar{F}(x|t) = \bar{F}(x+t)/\bar{F}(x)$ if $\bar{F}(t) > 0$ and 0 if $\bar{F}(t) = 0$. Similarly the conditional probability failure during a time x for a device of age t is $F(x|t) = (F(x+t) - F(t))/\bar{F}(t) = 1-\bar{F}(x|t)$ if $\bar{F}(t) > 0$ and 0 if $\bar{F}(t) = 0$. It is noticed that no wear characteristic means unfailed device is treated as new that is for all x, t > 0, $\bar{F}(x|t) = \bar{F}(x|0)$, that is the class which satisfies the functional form $\bar{F}(x+t) = \bar{F}(x)\,\bar{F}(t)$ - in other words the class of exponential survival distributions $F(x) = \exp(-\lambda x), \lambda \in [0,\infty]$.

Now we may obtain the conditional failure rate r(t) at time t by

$$r(t) = \lim_{x\to 0} (1/x)\ \{[F(x+t)-F(x)]/\bar{F}(t)\} = f(t)/\bar{F}(t)$$

where $\bar{F}(t) > 0$ and f(t) is the density function of F(t). The failure function (or the cumulative failure rate) is

$$R(x) = \int_0^x r(t)dt = -\log \bar{F}(x)$$

and hence $\bar{F}(x) = \exp\text{-}R(x)$. A distribution function F is IFR if F(x|t) is decreasing in t where t > 0 and x > 0 and this means r(t) is increasing function in t. A distribution function is said to be a decreasing failure rate (DFR) if F(x|t) is increasing in t for all t and x > 0 and this implies that r(t) is decreasing. The inverse of the failure rate function $r^{-1}(t)$ is known by Mills' ratio and has been studied and tabulated by many authors, see Johnson and Kotz (1972) for references. Some wider classes than IFR and DFR are the classes of distributions with IFRA and DFRA. A distribution F belongs to IFRA (DFRA) class if R(x)/x is increasing (decreasing) and hence $[\bar{F}(x)]^{-1/x}$ is increasing (decreasing). Therefore, F is IFRA if and only if $-\log \bar{F}$ is star-shaped, where the non-negative function g on $[0,\infty)$ with g(0) = 0 is star-shaped if $x^{-1}g(x)$ is increasing in $x \in (0,\infty)$ or equivalently $g(\alpha x) \leqslant \alpha\, g(x)$ for $0 \leqslant \alpha \leqslant 1$, $0 \leqslant x \leqslant \infty$. Thus IFRA (DFRA) is characterized by $\bar{F}(\alpha x) \geqslant (\leqslant)\ \bar{F}^{\alpha}(x)$ $0 \leqslant \alpha \leqslant 1$, $x \geqslant 0$.

Next, we give the main classification of life distributions. A distribution F belongs to NBU if $\bar{F}(x) > \bar{F}(x+t)/\bar{F}(t)$ for all x, $t \geqslant 0$, that is $-\log \bar{F}$ is superadditive where the non-negative g defined on $[0,\infty)$ with g(0) = 0 is superadditive if $g(x+y) \geqslant g(x)+g(y)$ for x, $y \geqslant 0$. Similarly F belongs to NWU if $F(x) \leqslant F(x+t)/F(t)$ that is -log F is subadditive where a non-negative function g defined on $[0,\infty)$ with g(0) = 0 is subadditive if $g(x+y) \leqslant g(x)+g(y)$ for x,y $\geqslant 0$. The distribution F is said to belong to the class of new better (worse) than used in expectation NBUE (NWUE) if $\int_0^\infty (F(x+t)/\bar{F}(t))dx \leqslant (\geqslant) \int_0^\infty \bar{F}(x)dx < \infty\ (> -\infty)$ for all t > 0. The distribution F is said to belong to the class of decreasing (increasing) mean residual life DMRL (IMRL) if $\int_0^\infty (\bar{F}(x+t)/\bar{F}(t))dx$ is decreasing (increasing in t), that is, the residual life of an unfailed device of age t has mean which is decreasing (increasing) in t. Finally, if we denote the class of functions which satisfy relation (3.3) by T_2, that is PT_2 with extra condition f(t+x)/f(t) is decreasing in t one can establish the result below.

Theorem 4.1. For the above classes one has: (i) $T_2 \subset$ IFR $\subset$ IFRA $\subset$ NBU $\subset$ NBUE, (ii) IFR $\subset$ DMRL $\subset$ NBUE, (iii) relation (i) and (ii) are the only existing among these classes.

The IFR class is closed under convolution, see Barlow and Proschan (1975, p. 100) and recently, Black and Savits (1976) proved that IFRA class is closed under convolution so one may ask whether other classes of theorem 4.1. are closed under convolution. However, it is realized that IFR is not closed under mixing, and a simple example for this is to note that mixture of two IFR exponential may not be IFR. Moreover, none of these classes is closed under reversal. It is also noticed that DFR class if not closed under convolution in general and this is proved by the counter example; let $F_1 = F_2 = G_\alpha$ where g_α is Γ-density with shape parameter $\alpha \varepsilon (\frac{1}{2},1)$, i.e. $g_\alpha(t) = \Gamma^{-1}(\alpha)\lambda^\alpha t^{\alpha-1}e^{-\lambda t}$ $t > 0$, then F_1 and F_2 are DFR distributions but $F_1 * F_2 = G_{2\alpha}$ which is Γ-distribution with shape parameter more than one and is not DFR distribution. However all classes of distributions included in this section are closed under convergence to a limit weakly.

5. EXCHANGEABILITY AND MAJORIZATION IN DISTRIBUTIONS

Here, we shall explore many results concerning the concepts of exchangeability, rearrangement, majorization and Schur concavity (or convexity) along with many other related ideas arising inevitably in probabilistic and statistical models. We shall discuss the basic properties of these concepts and establish some results concerning preservation property under some functional transforms in a similar manner to those in the earlier sections.

The random variables $X_1,\ldots, X_k$ are called exchangeable if k! permutations $(X_{t_i},\ldots, X_{i_k})$ have the same k-dimensional distribution. The sequence of random variables $X_1,X_2,\ldots$ is said to be spherical exchangeable if there exists a function g on the positive real line such that for each finite set $(i_1,\ldots, i_k)$ of natural numbers the joint characteristic function ϕ of $X_{i_1},\ldots, X_{i_k}$ satisfies $\phi(t_1,\ldots, t_k) = E\exp(i\Sigma t_j X_j) = g(\sum_{j=1}^{k} t_j^2)$. Clearly, each spherical exchangeable process is exchangeable. However, the exchangeable random variables $X_1,\ldots, X_k$ with probabilities of the form $P(X_1-\theta_1,\ldots, X_k-\theta_k) = P(X \varepsilon A+\theta)$, θ is parameter vector, often exhibit a monotonicity property in values of θ partially ordered according to majorizations. Notice that, we denote by X,Y,... and $X_1,X_2,\ldots, Y_1,Y_2,\ldots$ as the random vectors and random variables or components respectively. In fact, this concept has been studied by the authors (1977b) in the context of constructing A-infinitely divisible classes where A refers to the symmetric dependence of the variables.

It was established by Hardy et al. (1952, p. 49) that an n-dimensional vector X is said to be majorized by the vector Y if by rearrangement of the components to obtain $x_1 \geqslant x_2 \geqslant\ldots\geqslant x_n$, $y_1 \geqslant y_2 \geqslant\ldots\geqslant y_n$ one has

$$(5.1) \qquad \sum_{j=1}^{k} x_j \leqslant \sum_{j=1}^{k} y_j,\ k = 1,2,\ldots, n-1 \text{ and } \sum_{j=1}^{n} x_j = \sum_{j=1}^{n} y_j,$$

and we denote this by $X \overset{*}{<} Y$ if relation (5.1) is satisfied. A function f for which $X \overset{*}{<} Y$ implies $f(x) \geq (\leq) f(y)$ is called <u>Schur-concave (convex)</u> and such functions are permutation symmetric, that is, invariant under permutations of the components of the underlying vectors. Therefore, f(x) is Schur-concave implies that the random variables $X_1,\ldots, X_n$ are exchangeable. Thus a differentiable function f(x) of exchangeable random variables is Schur-concave (convex) if and only if

$$(5.2) \qquad \left(\frac{\partial f(x)}{\partial x_i} - \frac{\partial f(x)}{\partial x_j}\right)(x_i-x_j) \leq (\geq)\ 0 \text{ for all } i \neq j,$$

see Schur (1923). The case $X \overset{*}{<} Y$ can be expressed by $X = DY$ for some doubly stochastic matrix D. It can be realized that for any vector x,

$(\sum_{i=1}^{n} x_i/n)(1,1,\ldots, 1) \overset{*}{<} (x_1,x_2,\ldots, x_n)$ and therefore, whenever $\sum_{i=1}^{n} x_i$ is fixed, Schur-concave function attains a maximum (modal) point when the components are equal. Now we give the following lemma due to Marshall and Olkin (1974).

<u>Lemma 5.1.</u> Let f(x) be a Schur-concave function and consider a Lebesgue-measurable set $A \subset R^n$ such that

(5.3) $\quad y \in A$ and $x \overset{*}{<} y$ implies $x \in A$.

Then $P(X \in A+\theta) = \int_{A+\theta} f(x)dx$ is Schur-concave function of θ, where θ is some parameter vector.

In fact, condition (5.3) can be satisfied for every convex set A of exchangeable random variables since $x \overset{*}{<} y$ implies $x = Dy$ for some doubly stochastic matrix D and the set of doubly stochastic matrices is a convex hull of the permutation matrices, whereas (5.3) implies neither convexity nor measurability of A. Moreover, if two sets satisfy (5.3), then so does their union. Actually, Mudholkar (1966) established lemma 5.1 where he generalizes a result of Anderson (1955), see Kanter (1977), but with additional requirement on the set A that is A and $\{y: f(y) \geq c\}$ are convex for each constant c. Also $\{y: f(y) \geq c\}$ is convex, i.e. unimodal in Anderson's sense and f(y) is exchangeable implies condition (5.3). For some related results on exchangeability see Hewitt-Savage (1955) and Ahmad (1974, 1975). Now we give the following interesting result.

<u>Theorem 5.2.</u> The class of Schur-concave (convex) functions is closed under reversal, passage to a limit weakly, mixing and convolution.

<u>Proof</u>: We shall give the proof for Schur-concave functions, whereas a similar argument can be carried out for the Schur-convex case. Clearly, the reversal property is valid that is if f(x) is Schur-concave, then so is f(-x). Now let $\{f_k\}$ be a sequence of Schur-concave functions. Then we can have for any set A satisfying lemma 5.1, that $|f_k| \leq h$ for each k where h is integrable function on $A+\theta$. Next, let f_k converge weakly to a function f then by Lebesgue Dominated Convergence Theorem and lemma (5.1) one gets f as Schur-concave. The closure under mixing is shown if one realizes that the mixture is Schur-concave function of θ.

Finally to show that the class of Schur-concave functions is closed under convolutions, let f_1 and f_2 be two Schur-concave functions, then $f_2(-x)$ is also Schur-concave and we need to prove that

$$(5.4) \quad f(.,\theta) = \int_{R^n} f_1(x)\, f_2(x-\theta)dx$$

for some parameter θ, is Schur-concave in θ. But by using lemma 5.1, $\int_{A+\theta} f_2(-x)dx = \int_{R^n} I_A(-x)\, f_2(\theta-x)dx$ is Schur-concave in θ. Now approximate $f_1(x)$ by an increasing sequence of simple functions $h_k = \Sigma\, \alpha_i I_{A_i}$ where $\Sigma\, \alpha_i = 1$ and the sets A_i satisfy lemma 5.1. Hence by using Lebesgue Monotone Convergence Theorem the required result follows.

One may also see that if $f(x)$ is an exchangeable Schur-concave density function and h is non-negative, exchangeable and Schur-function, then $Eh(X-\theta)$ and $P\{h(X-\theta) \geqslant c\}$ are Schur-function in θ. The following lemma is due to Proschan and Sethuraman (1977), where they show the closure property of Schur-functions under a certain integral transformation.

Lemma 5.2. Let $f_1(x)$ be Schur-concave (convex) function and $f_2(x,\theta)$ is TP_2 and satisfy the semigroup property i.e. for $\theta_1,\theta_2 > 0$: $f_2(x,\theta_1+\theta_2) = \int f_2(x,\theta_1)\, f_2(x,\theta_2-y)dy$, where $0 < \theta, x < \infty$. If

$$(5.5) \quad f(.,\theta) = \int f_1(x) \prod_{i=1}^{n} f_2(x_i,\theta_i)dx, \quad 0 \leqslant x_i < \infty \text{ for every i exists, then it is}$$

Schur-concave (convex) function.

In fact, relations (5.4) and (5.5) look different but coincide when $f_2(x,\theta)$ is of the form $\prod_{i=1}^{n} f(x_i-\theta_i)$ otherwise none of them would imply the other.

If we consider the case when the underlying random variables are independent and identically distributed and they have common marginal density function f_1 say. Thus the joint density function of X is $f(x) = \prod_{i=1}^{n} f_1(x_i)$, and in this case f is Schur-concave (convex) if and only if $\log f_1$ is concave (convex). Therefore, for such random variables Schur-concavity and unimodality are equivalent. There are two other related concepts which have been brought up recently, (i) the concept of positive dependence by mixture (PDM), that is, the class of functions which can be represented by a mixture of densities of exchangeable and independent random variables, Shaked (1977), and (ii) the concept of decreasing in transposition (DT) class of functions, that is, those functions which decrease by rearrangement of the components of the random vector, see Hollander et al. (1977). It is expected from the structure of these two classes to be closed under most of the functional operations studied earlier and some other kernel-type transformations.

6. CONCLUDING REMARKS AND OPEN PROBLEMS

The main purpose of statistical methods is to give the users a better inference for the considered problems and this motive lies behind most of the work of statistical and probablistic theories and their applications. It is nice if one is able to have statistics which are sufficient, asymptotically normal, with some invariant property, complete etc., or tests of hypothesis through test functions which are uniformly most powerful unbiased tests etc. Thus, the different schools of thought in statistics such as the classical, Bayesian, Subjectivists, and nonparametricians are mainly different in stressing either robustness or the efficiency of the statistical models.

For example one may try to approximate the actual model by another one which is effectively very close to the actual model, say by using the contiguity approach in order to attain a very high degree of robustness. These approximations are of exponential or infinitely divisible structures. It is shown by the authors (1977a) that sub-classes of infinitely divisible class such as symmetric stable, stable, symmetric L-functions and some non-symmetric L-functions are unimodal. Further, strong unimodality and universality which are essentially similar, are shown to play and important role in plausibility inference, see Barndorff-Nielsen (1976), which is complimentary to the likelihood inference. Also, PT and MLR functions are well behaved functions and very much utilized in decision problems and testing hypotheses, for example see Lehmann (1959 p. 68), where a test function is constructed in terms of some other function g, say, such that the density function is MLR in g. One may see the importance of IFR and other related classes being closed under some functional operation such as convolution and mixing etc. which mostly arise from the practical considerations. The exchangeability structure and other related classes such as Schur-functions, PDM and DT cover, surprisingly, large families of distributions. Such families are shown to arise mainly and in particular in distribution-free tests, since exchangeable hypotheses are all distribution-free. These families may appear in constructing asymptotically optimal distribution-free tests. The role of exchangeability concepts in rank order statistics and other stochastic inference will be investigated by the authors in a forthcoming paper, whereas other related applications can be found in Ahmad and Abouammoh (1977b), Shaked (1977), Proschan and Sethuraman (1977) and Hollander et al. (1977).

Finally, we conclude by the following unsolved problems. Is it true that the convolution of the multivariate strongly unimodal distribution with any multivariate unimodal implies (the same) multivariate unimodality? What are the conditions other than (3.4) and (3.5) under which PT class is closed under convolution? Which are the classes in theorem 4.1, except those already shown, that are closed under the investigated functional operations? Which practicable subclasses of exchangeable family are closed under such functional operations?

REFERENCES

1. A-Hameed, M.S. and Proschan, F. (1973): Nonstationary shock models. Stoch. Proc. Appl., 1, 383-404.

2. Ahmad, R. (1974): On the structure of symmetric sample testing: a distribution-free approach. Ann. Inst. Statist. Math., 26, 233-245.

3. Ahmad, R. (1975): Some characterizations of the exchangeable processes and distribution-free tests in, Statistical Distributions in Scientific Work (eds. G.P. Patil et al.), Vol. 3, D. Reidel Pub. Co., Dordrecht, 237-248.

4. Ahmad, R. and Abouammoh, A.M. (1977a): On the structure and applications of infinite divisibility, stability and symmetry in stochastic inference, in Recent Developments in Statistics (eds. J.R. Barra et al.), North-Holland, Amsterdam 1977, 303-317.

5. Ahmad, R. and Abouammoh, A.M. (1977b): On infinite A-divisibility. To appear.

6. Barndorff-Nielsen, O. (1976): Plausibility inference. J. Roy. Statist. Soc. Ser. B, 38, 103-131.

7. Barlow, R.E. and Proschan, F. (1975): Mathematical Theory of Reliability and Life Testing. Holt, Rinehart and Winston.

8. Birnbaum, Z.W., Esary, J.D. and Marshall, A.W. (1966): Stochastic characterization for components and systems. Ann. Math. Statist., 37, 316-325.

9. Block, H.W. and Savits, T.H. (1976): The IFRA closure problem. Ann. prob., 4, 1030-1032.

10. Borell, C. (1975): Convex set in d-space. Period. Math. Hungar., 6, 111-136.

11. Ghurye, S.G. and Wallace, D.L. (1959): A convolution class of Monotone Likelihood ratio families. Ann. Math. Statist., 30, 1158-1164.

12. Hardy, G.H., Littlewood, G.E. and Polya, A. (1952): Inequalities. 2nd ed., Cambridge Univ. press. Cambridge.

13. Hewitt, E. and Savage, L.J. (1955): Symmetric measures on Cartesian products. Trans. Amer. Math. Soc., 80, 470-501.

14. Hollander, M., Proschan, F. and Sethuraman, J. (1977): Functional decreasing in transposition and their applications in ranking problems. Ann. Statist., 5, 722-734.

15. Johnson, N.L. and Kotz, S. (1972): Continuous Multivariate Distributions. -2. Houghton Mifflin Co.

16. Kanter, M. (1977): Unimodality and dominance for symmetric random vectors. Trans. Amer. Math. Soc., 227, 65-85.

17. Karlin, S. (1956): Decision theory for Polya type distributions, case of two actions, I. Proceeding of the third Berkeley symposium on Prob. and Statist., Vol. 1, Univ. of California press, 115-129. (1957): Polyá type distributions, Ann. Math. Statist., 23, 231-308.

18. Lehmann, E.L. (1959): Testing Statistical Hypotheses. Wiley, New York.

19. Lukacs, E. (1970). Characteristic Functions. 2nd ed. Griffin, London.

20. Mudholkar, G.S. (1966): The integral of an invariant unimodal function over an invariant convex set - an inequality and applications. Proc. Amer. Math. Soc., 17, 1327-1333.

21. Marshall, A.W. and Olkin, I. (1974): Majorization in Multivariate distributions. Ann. Statist., 2, 1189-1200.

22. Parthasarathy, K.R. (1967): Probability Measures on Metric Spaces. Academic Press. New York.

23. Prékopa, A. (1973): On logarithmic concave measures and functions. Acta Sci. Math. (Szeged), 34, 335-343.

24. Proschan, F. and Sethuraman, J. (1977): Schur functions in statistics I. The preservation theorem. Ann. Statist., 5, 256-262.

25. Rinott, Y. (1976): On convexity of measures. Ann. Prob., 4, 1020-1026.

26. Schur, I. (1923): Ubereine Klasse nom Mittelbildungenmit Anmendungen auf die Determinantentheorie. Sitzber. Berl. Math. Ges, 22, 9-20.

27. Shaked, M. (1977): A concept of positive dependence for exchangeable random variables. Ann. Statist., 5, 505-515.

R. Ahmad and A.M. Abouammoh
University of Strathclyde
Department of Mathematics
26 Richmond Street
Glasgow G1 1XH
Scotland.

RESTRICTED PERMUTATION SYMMETRY AND HYPOTHESES - GENERATING GROUPS IN STATISTICS

Rashid Ahmad and Magnus M. Peterson

Glasgow

ABSTRACT

In a recent paper Rüschendorf (1976) solved an open problem orginally posed by Bell and Haller (1969) in connection with hypotheses - generating groups for hypotheses of multivariate symmetry. This solution applies only when the generic data point has complete permutational symmetry. In many practical applications a more realistic assumption would be a restricted permutational symmetry. In this paper the solution of Rüschendorf is extended to cover some of these cases.

1. INTRODUCTION AND PRELIMINARIES

The family of symmetric probability measures or distributions, that is measures which are invariant under finite permutations of the underlying variables among themselves, are well-known and have many applications to real life problems. For example consider n matched pairs $\{(X_i, Y_i), i = 1,2, \dots , n\}$ with a bivariate distribution F(x, y). If we consider X as 'control' and Y as the treatment response, then $F(x, y) = F(y, x)$ is equivalent to the assumption that there is no treatment effect. The extension to the k-response situation is straightforward.

In a paper Bell and Haller (1969), while considering parametric and nonparametric tests for bivariate symmetry, posed the following problem. Which group G_o of transformations on R^{2n} generates the null hypothesis class $\Omega(H)$ or an appropriately dense subfamily thereof? In this connection, H. Rubin conjectured that in general there is difficulty in finding a transformation which is invertible and which can transform the distribution:

$$H(x, y) = \tfrac{1}{2}\{F(x)F(y) + G(x)G(y)\}$$

into

$$H(x, y) = \tfrac{1}{2}\{F(x)G(y) + F(y)G(x)\}.$$

Recently, Rüschendorf (1976) has solved the original open problem of Bell and Haller to find a group of monotone invertible bimeasurable transformations of R^p which generates a suitable subfamily of the class of totally symmetric distributions on R^p. Here, the distribution F is defined as totally symmetric if almost surely

$$F(x_1, \dots, x_p) = F(\pi(x)) := F(x_{\pi(1)}, \dots, x_{\pi(p)})$$

for all π in the permutation symmetric group S_p. For the structural properties, applications and distribution-free testing structure see Hewitt-Savage (1955), Ahmad (1974a, 1975) and the references contained therein.

Such a generating group is useful in the construction of multivariate symmetry rank and permutation tests. However, in many practical problems an assumption of complete symmetry for the distribution of the generic data point does not lead to a reasonable null hypothesis class. Rather an assumption of partial symmetry would be the natural one. For example consider a four-dimensional distribution F(x, y, w, z) where (X, Y), say, represent the 'mental response' and (W, Z) denote the 'physical response' of a patient under some experimental stress situation. In this case a possible null hypothesis of interest would be:

$$\Omega(H) = \{F : F(x, y, w, z) = F(y, x, z, w)\}.$$

In a more general setting, if n treatments are applied to k samples each of n individuals from k populations known to be essentially different, the generic data point has the form:

$$Z = (X_{11}, X_{12}, \dots, X_{1n}; X_{21}, \dots, X_{2n}; \dots; X_{k1}, \dots, X_{kn}),$$

and a natural null hypothesis is that the distribution F of $\underline{X}$ is partially symmetric in the sense that $F(z) = F(\pi(z))$ for all π in the direct product group $S_n \times S_n \times \dots S_n$ acting in the obvious way. This is a proper subgroup of S_{kn}.

Again if the values of one of the variables $X_1, \dots, X_n$ is measured on a randomly chosen member of a population from time to time with a view to detecting whether the population's characteristics are changing with time a natural null hypothesis for the generic data point

$$Z = (X_{11}, \dots, X_{1k_1}; X_{21}, \dots, X_{2k_2}; \dots; X_{n1}, \dots, X_{nk_n})$$

is that its distribution is symmetric for all permutations in the group

$$S_{k_1} \times S_{k_2} \times \dots \times S_{k_n}.$$

In this paper the methods of Rüschendorf will be extended to yield a hypothesis-generating group G_1 for the class of distributions Ω_Σ^p where $\Sigma = S_{k_1} \times S_{k_2} \times \dots \times S_{k_n} \subsetneq S_p$, $(k_1 + k_2 + \dots + k_n = p)$ and Ω_Σ^p is the dense subclass of the class of all Σ- symmetric distributions with the additional properties that they are continuous and have strictly increasing marginal and conditional distribution functions. In addition we shall construct a hypothesis-generating group G_2 for the multi-sample problem where the natural null hypothesis class is the set of distributions whose symmetry-group is the wreath-product Σ wr S_N, Σ being a group of the form considered previously and the number of samples being N.

2. SOME BASIC RESULTS

Let $(\mathfrak{X},\mathcal{B})$ be a measurable space, and in the usual way define the generated adapted σ-fields by:

(a) $\mathfrak{F}(X_1, \ldots, X_n) = \mathfrak{F}_n$ such that $\mathfrak{F}_n \subset \mathfrak{F}_{n+1}$, and $\bigcup_n \mathfrak{F}_n = \mathfrak{F}$;

(b) $\mathfrak{F}(X_{n+1}, X_{n+2}, \ldots) = \mathfrak{F}_n^*$ such that $\mathfrak{F}_n^* \supset \mathfrak{F}_{n+1}^*$, and $\bigcap_n \mathfrak{F}_n^* = \mathfrak{F}^*$.

As previously defined a probability measure is said to be (totally) symmetric if $P(A) = P(\pi A)$ for all π in S_m, A in $\mathcal{B}_m$. Let $\mathcal{B}_S$ be the sigma-field generated by the symmetric random variables. Then it can easily be seen, see Meyer (1966), that $\mathcal{B}_S \supset \mathfrak{F}^*$. Let P_S be a symmetric probability measure on $(\mathfrak{X},\mathcal{B})$, and let A be in $\mathcal{B}_S$. Then according to the Hewitt-Savage main result there exists a set B in $\mathfrak{F}^*$ such that A = B almost surely. A consequence of this is that if $\{X_j\}$ are independent identically distributed random variables, then the probability law induced by P_S on $\mathcal{B}_S$ is degenerate. A converse of the above statement is the well-known de Finetti theorem, which states that if P_S is a symmetric probability measure, then the random variables $X_1, X_2, \ldots, X_n, \ldots$ are conditionally independent with respect to $\mathcal{B}_S$. That is,

$$P_S\{X_j \in A_j \; ; \; j = 1,2, \ldots, n|\mathcal{B}_S\} = \prod_{j=1}^{n} P_S\{X_j \in A_j|\mathcal{B}_S\}.$$

Notice that the above development can essentially be carried over to the restricted permutational symmetry with some slight modification in terminology. This will be explored elsewhere.

For a generic data point $\mathfrak{Z}$ and the group S^* of symmetries Ω_{S^*} we define the orbit of $\mathfrak{Z}$ as the set $S^*(\mathfrak{Z}) = \{\pi\mathfrak{Z} : \pi \in S^*\}$. It turns out that the orbit is a comple sufficient statistic for the problem of inference. The study of distribution-free or nonparametric tests is essentially the study of tests having Neyman structure, see Lehmann (1959, p. 130), Bell and Haller (1969) and Ahmad (1974a,b, 1975). The construction of these tests consists of selecting a fixed proportion of the points of almost every orbit, and this is accomplished by functions which distinguish among the points of the orbit and their induced statistics which rank the points on the orbit. Since symmetric probability distributions can be constructed as mixtures of independent identically distributed random variables , it is true that under certain continuity conditions, independent identically distributed processes and symmetric processes generate the same Neyman structure tests. Usually in constructing tests practical importance one employs maximal invariants based upon ranks or order statistics. Some further results in this connection will be given in the sequel, but now we return to partial symmetry and related hypothesis-generating (HG) groups.

3. PARTIAL SYMMETRY AND HYPOTHESES-GENERATING GROUPS

3.1 PARTIAL SYMMETRY STRUCTURE

The generic data point whose distribution is considered has the form

$$Z = (X_{11}, X_{12}, \ldots, X_{1k_1}; X_{21}, X_{22}, \ldots, X_{2k_2}; \ldots; X_{n1}, X_{n2}, \ldots, X_{nk_n})$$

where $\sum_{i=1}^{n} k_i = p$ so that $Z \in R^p$. For convenience Z will sometimes be written as $Z = (Z_1, Z_2, \ldots, Z_n)$ where $Z_i = (X_{i1}, X_{i2}, \ldots, X_{ik_i}) \in R^{k_i}$. The null-hypothesis class considered is $\tilde{\Omega}^p_\Sigma$, the class of continuous distributions on R^p which are Σ-symmetric in the following sense. Let π be a permutation in the direct product of symmetric groups

$$\Sigma = S_{k_1} \times S_{k_2} \times \ldots \times S_{k_n}$$

acting in R^p as follows

$$\pi Z = (X_{1\pi_1(1)}, X_{2\pi_1(2)}, \ldots, X_{1\pi_1(k_1)}; \ldots; X_{n\pi_n(1)}, \ldots, X_{n\pi_n(k_n)})$$

where π_i is a permutation of $\{1, 2, \ldots, k_i\}$ for $i = 1, 2, \ldots, n$. Then F is Σ-symmetric if $F(\pi z) = F(z)$ a.e. for all π in Σ.

3.2 THE EXTENDED Σ-ROSENBLATT TRANSFORMATION

The class Ω^p_Σ is the class of continuous Σ-symmetric distributions on R^p having strictly increasing one-dimensional marginal and conditional distributions. The proposed construction of a hypothesis-generating group G_1 for this dense subclass of $\tilde{\Omega}^p_\Sigma$ depends on a modification of Rosenblatt's Transformation, Rosenblatt (1952), which is a close parallel to the modification used by Rüschendorf (1976). Let Ω^p be the class of all continuous distributions on R^p having strictly increasing marginals and conditionals. For $F \in \Omega^p$ the Σ-Rosenblatt transformation corresponding to F is defined by

$$\tau_F(Z) = (\tau^1_F(Z_1), \tau^2_F(Z_2), \ldots, \tau^n_F(Z_n)), \tag{3.1}$$

$$\tau^i_F(Z_i) := (F_{i1}(X_{i1}), F_{i2.1}(X_{i2}|X_{i1}), \ldots, F_{ik_i.1,2,\ldots,k_i-1}(X_{ik_i}|X_{i1}, X_{i2}, \ldots, X_{ik_{i-1}})). \tag{3.2}$$

<u>Lemma 1</u>. If the distribution of Z is $F \in \Omega^p$ then $\tau_F(Z)$ has its components uniformly and independently distributed on $(0, 1)^p$. Also the mapping $\tau_F : R^p \to (0, 1)^p$ is bijective and bimeasurable.

Proof. The first part is straightforward. As all marginals and conditionals of F are proper and continuous τ_F is onto. If two points Z, Z', of R^p have the same image under τ_F then since F_{i1} (.) is a strictly increasing (s.i.) function of X_{i1} only, we have $X_{i1} = X'_{i1}$ (i = 1, ... , n). But then as $F_{i2.1}(.|X_{i1})$ depends on X_{i1} and X_{i2} only and is a s.i. function of X_{i2} and as $X_{i1} = X'_{i1}$ it follows that $X_{i2} = X'_{i2}$ again for i = 1,2, ... , n. Continuing in this fashion it follows that Z = Z'. Hence τ_F is 1 - 1. As F is measurable so also is τ_F and its inverse.

For the hypothesis $\tilde{\Omega}^p_\Sigma$ the natural rank-set would appear to be the set

$$E_\Sigma = \{Z : X_{ij} < X_{ij+1} \; ; \; 1 \leqslant i \leqslant n, \; 1 \leqslant j < k_i\}.$$

A modification of the Σ-Rosenblatt transformation parallel to that used by Rüschendorf (1976) allows the construction of the group G_1.

Let $\Omega^p(E_\Sigma)$ be the class of continuous distributions on $(E_\Sigma, \mathcal{B}^p \cap E_\Sigma)$ which have strictly increasing one dimensional marginal and conditional distribution functions. Then the modified Σ- Rosenblatt transformation defined for F in $\Omega^p(E_\Sigma)$ is defined by (3.1) and (3.2) for all Z in E_Σ.

Lemma 2. (a) For $F \in \Omega^p(E_\Sigma)$, $\tau_F: E \to (0, 1)^p$ is a bijective bimeasurable transformation whose components are uniformly and independently distributed on $(0, 1)^p$ if Z has the distribution F.

(b) For fixed $F_o \in \Omega^p(E_\Sigma)$, $G'_1 = \{\tau_F^{-1} o \tau_F : F \in \Omega^p(E_\Sigma)\}$ is a HG-group for $\Omega^p(E_\Sigma)$.

Proof. The part (a) is proved precisely as Lemma 1.

(b) The proof of this part would copy Smith's (1969) Lemmas 2.2 and 2.3.

For F, $F_o \in \Omega^p(E_\Sigma)$ define $g_F: R^p \to R^p$ by $g_F(Z) = \pi \, o \tau_F^{-1} \; o \; \tau_{F_o} \; o \; \pi^{-1}(Z)$ when $Z \in \pi E_\Sigma$, $\pi \in \Sigma$. Clearly g_F is defined a.e.

Theorem 1 (a) $G_1 = \{g_F : F \in \Omega^p(E_\Sigma)\}$ is a H-G group for Ω^p_Σ.

(b) A maximal invariant under G_1 is the vector of ranks $(R_{11}, R_{12}, \ldots , R_{1k_1}; \ldots \ldots ; R_{n1}, \ldots , R_{nk_n})$ where R_{ij} is the rank of X_{ij} among $(X_{i1}, \ldots , X_{ik_i})$.

Proof. This is identical to the proof of Theorem 1 in Rüschendorf (1976) and is therefore omitted.

4. MULTI-SAMPLE PROBLEMS

The results of section 3 may be extended to various multi-sample problems of which the following is a simple example.

To test whether the distribution functions of m independent Σ-symmetric random variables are identical the null-hypothesis class to be considered is

$\Omega_o = \{F^{(m)} : F \in \tilde{\Omega}^p_\Sigma\}$ where $F^{(m)}$ is the m-fold product measure. It is equivalent to consider the class $\tilde{\Omega}^{pm}_\Gamma$ of all continuous distributions on R^{pm} which are Γ-symmetric where Γ is the wreath-product group Σ wr S_m defined by setting

$$(\pi_1, \pi_2, \dots, \pi_m, \rho)\ Z = (\pi_1 Z^{\rho(1)}, \pi_2 Z^{\rho(2)}, \dots, \pi_m Z^{\rho(m)})$$

for $\pi_q \in \Sigma$, $(q = 1, 2, \dots, m)$ and $\rho \in S_m$ acting in $\{1, 2, \dots, m\}$. Here the generic data point $Z = (Z^1, Z^2, \dots, Z^m)$ and $Z^i \in R^p$ is defined as Z was in section 3.

Let

$$E_\Gamma = \{Z : X^q_{1,j} < X^{q+1}_{1,j}, (q=1, \dots, m-1)\ ;\ X^q_{i,j} < X^q_{i,j+1}, (j=1, \dots, k_i-1)\}.$$

As before for $F \in \Omega^{pm}(E_\Gamma)$ the class of continuous distributions on E_Γ with strictly increasing marginal and conditional distribution functions, the modified Rosenblatt transformation $\tau_F : E_\Gamma \to (0, 1)^{pm}$ is defined following Rüschendorf (1976) by setting the component

$$\tau^q_{Fij}(Z) = F_{q,i,j|c}(X^q_{ij} | X^q_{it}, (t<j), X^r_{is}, (r<q, s=1, \dots, k_i)).$$

where $F_{q,i,j|c}(.|c)$ denotes the marginal-conditional distribution derived from the F under the conditions C (which are void when $q = i = j = 1$).

Then for fixed $F_o \in \Omega^{pm}(E_\Gamma)$ define the mapping

$$h_F : R^{pm} \to R^{pm}$$

by $h_F(Z) := \pi o \tau_F^{-1} o \tau_{F_o} o \pi^{-1}$ when $Z \in \pi E_\Gamma$.

<u>Theorem 2.</u> (a) $G_2 = \{h_F : F \in \Omega^{pm}(E_\Gamma)\}$ is a H-G group for the hypothesis class $\tilde{\Omega}^{pm}_\Gamma$ of continuous Γ- symmetric distributions having strictly increasing marginal and conditional distribution functions.

(b) A maximal invariant for G_2 is the m(p + 1) - dimensional vector of ranks

$$(R^1_1, R^2_1, \dots, R^m_1\ ;\ R^1_2, \dots, R^m_2\ ;\ \dots;\ R^1_n, \dots, R^m_n\ ;\ S^q_{ij}, [q=1, \dots, m; i=1, \dots, n; j=1, \dots, k_i]),$$

where R^q_i is the rank of min $\{X^q_{ij}\ (j=1, \dots, k_i)\}$ among $\{\min \{X^q_{ij} : (j=1, \dots, k_i)\} : (q=1, \dots, m)\}$ and S^q_{ij} is the rank of X^q_{ij} among $\{X^q_{ij} : (j=1, \dots, k_i)\}$.

<u>Proof.</u> This repeats arguments used already and parallel to those in Rüschendorf (1976).

Note that Rüschendorf's theorem 3 is slightly different from our development for the multi-sample case, as he assumes completely symmetric measures while we assume partially symmetric measures.

5. HYPOTHESES TESTING STRUCTURE

Let Ω be the class of continuous distributions. A statistic $T: \Omega \times R^p$ is said to be distribution-free (DF) wrt Ω if there exists a single distribution Q_T s.t.

$P(T \leq t|F) = Q_T(t)$ for all F in Ω and all real t. Since most practicable DF tests in goodness-of-fit and other problems satisfy an additional condition - this introduces the concept strongly (S) DF tests. A statistic T is SDF if it is DF wrt $\Omega \subseteq \Omega_\Sigma^p$ (or Ω_Γ^{pm}) if for every t in $\acute{R}$.

$$P_G(T(Z) \leq t|F) = P_{G*}(T(Z) \leq t|F^*)$$

for each F, F^*, G, G^* in Ω s.t. $\tau_G\tau_F^{-1} = \tau_{G*}\tau_{F*}^{-1}$. The SDF property essentially divides $\Omega \times \Omega$ into subsets over each of which the power of an underlying test, of F against G, is constant.

Let $U_F^{(p)}$ denote the uniform distribution on $(0, 1)^p$ obtained via the Rosenblatt transform from F. Then if a statistic $T(Z|F)$ equals $g(U_F^{(p)}(Z))$ for some measurable function g on $\{(u_1, \ldots, u_p) : 0 \leq u_j \leq 1\}$, then T is a statistic DF wrt Ω or a dense family of it. Notice how this links up neatly with extend Σ- Rosenblatt (or Γ) transformations. In some problems as in this paper one considers a group $\tilde{S}$ of transformations of the sample space, and its induced group S^* of transformations of Ω. Under $\tilde{S}$ the constant power classes are also equivalence classes of S^*. This motivates another equivalent version of SDF definition. A statistic is SDF wrt $\Omega \subseteq \Omega_\Sigma^p$ (or Ω_Γ^{pm}) and $\tilde{S}$ if for all real t.

$$P_G(T(Z) \leq t|F) = P_{G*}(T(Z) \leq t|F^*)$$

for each F, F^*, G, G^* in Ω s.t. $F^* = F_g$ and $G^* = G_g$ for some $g \in \tilde{S}$. In the above development $\tilde{S}$ has the structure $\{g : g = \tau_G^{-1}\tau_{Fo}$ for some F_o, G in $\Omega^* \subseteq \Omega\}$ and F_g is defined by $\tau_{F_g} = \tau_F g^{-1}$.

For convenience in notation let Ω_1, Ω_2, Ω_3, respectively denote the families of null hypotheses distributions for the restricted symmetry, complete symmetry, and independence case, that is for the last case even the Z_i's and their components are independent. Further assume that the underlying random variables are continuous and the distributions have densities. Clearly $\Omega_1 \supset \Omega_2 \supset \Omega_3$. Let Z^i denote the generic data point, and set its order statistics and rank vectors as $Z_{(.)}^i$ and $R_{(.)}^i$ for $i = 1, 2, 3$. If no confusion arizes we shall drop i. Some results parallel to the development below for the classes Ω_2 and Ω_3 have already been obtained by Hájek-Šidák (1967), Bell-Kurotschka(1971), and Ahmad (1974a, b, 1975), also see Lehmann (1959). In what follows we investigate some structural results for the hypothesis Ω_1.

Let $(\mathcal{X}, \mathcal{B})$ be the basic space wrt Ω_1 and let its symmetry group be denoted by S_1^* with cardinality $C(S_1^*)$. Denote by $\mathcal{X}_{(.)}$ the subspace of $\mathcal{X}$ containing points

$\{Z_{(.)}\}$. Let $\mathcal{B}_{(.)}$ be the sigma-field consisting of the Borel subsets of $\mathcal{X}_{(.)}$. Since Z can be reconstructed from $Z_{(.)}$ and $R_{(.)}$, clearly $(Z_{(.)}, R_{(.)})$ form a sufficient statistic for the family Ω_1. Now, by extending arguments of Hájek-Šidák (1967, pp. 36-44) we have the following.

Theorem 3. (a). If Z has a distribution with density h in Ω_1, then the density of $Z_{(.)}$ is given by $h^*(z_{(.)}) = \sum_{\pi \in S_1^*} h(\pi z)$, $z_{(.)} \in \mathcal{X}_{(.)}$. Moreover,

$$P_h(R_{(.)} = r | Z_{(.)} = z_{(.)}) = h(\pi_r z)\left[h^*(z_{(.)})\right]^{-1},\ r \in S_1^*,\ z_{(.)} \in \mathcal{X}_{(.)}.$$

(b) Under the above hypothesis the random vectors $Z_{(.)}$ and $R_{(.)}$ are stochastically independent.

(c) For any density $f \in \Omega_1$ and any statistic g(Z)

$$E_f(g(Z)|R_{(.)} = r) = E_f(g(\pi_r Z))$$

is valid. In particular, if h is a density such that $f(z) = 0$ necessitates $h(z) = 0$, $z \in \mathcal{X}$, then

$$P_h(R_{(.)} = r) = E\{h(\pi_r z) f^{-1}(\pi_r z)\}(C(S_1^*))^{-1}.$$

Proof. Let B be in $\mathcal{B}_{(.)}$. Then from the fact that the correspondence between z and $z_{(.)}$ is 1 - 1 and linear with the Jacobian Unity on each subset $\{R_{(.)} = r\}$, we have

$$\int_{z_{(.)} \in B} h(z)dz = \sum_{\pi_r \in S_1^*} \int_{\{R_{(.)} = r,\ z_{(.)} \in B\}} h(z)dz$$

$$= \sum_{S_1^*} \int_B h(\pi_r z)dz_{(.)} = \int_B h^*(z_{(.)})dz_{(.)}.$$

This proves the first part of (a). Next, since $h^*(z_{(.)}) = 0$ implies $h(\pi_r z) = 0$ for each π_r in S_1^*, one gets

$$P_h(R_{(.)} = r,\ Z_{(.)} \in B) = \int_{\{R_{(.)} = r,\ z_{(.)} \in B\}} h(z)dz$$

$$= \int_B h(\pi_r z)dz_{(.)}$$

$$= \int_B \frac{h(\pi_r z)}{h^*(z_{(.)})} h^*(z_{(.)})dz_{(.)},$$

which proves the second part of (a)

To establish part (b) we notice that from (a) it follows that $P(R_{(.)} = r) = (C(S_1^*))^{-1}$, $r \in S_1^*$ and the density of $Z_{(.)}$ is given by $C(S_1^*)f(z_{(.)})$ for $z_{(.)}$ in $\mathcal{X}_{(.)}$.

To see part (c) let the components of Z be denoted by $Z^1, \ldots, Z^p$. Since

$z^j = Z_{(\pi_{r_j})}$, $j = 1, 2, \ldots, p$, under $R_{(.)} = r$, the independence of $Z_{(.)}$ and $R_{(.)}$ gives

$$E_f(g(Z)|R_{(.)} = r) = E_f(g(\pi_r Z)|R_{(.)} = r) = E_f(g(\pi_r Z)),$$

and this completes the proof of the theorem.

Let $\phi^*(z_{(.)}, \pi)$ be a map on $\mathcal{X}_{(.)} \times S_1^*$ and write (*) $\phi(z^1, \ldots, z^p) = \phi^*(z_{(.)}, \pi)$, if $z^i \neq z^j$ for all $i \neq j$, and define ϕ arbitrarily on the remaining subset. Clearly for each $\phi(.)$ there exists $\phi^*(.)$ such that (*) holds. Now, a necessary and sufficient condition for ϕ to be $\mathcal{B}$-measurable is that ϕ^* is $\mathcal{B}_{(.)}$-measurable for every fixed π in S_1^*. Suppose that the function ϕ^* is $\mathcal{B}_{(.)}$-measurable for every fixed π, $0 \leq \phi^* \leq 1$, and that for some α in (0, 1)

$$(**) \quad \sum_{\pi \in S_1^*} \phi^*(z_{(.)}, \pi) = \alpha (C(S_1^*))^{-1}, \quad z_{(.)} \in \mathcal{X}_{(.)}.$$

The test with critical function ϕ related to ϕ^* by (*) is called a permutation test. If ϕ^* takes values 1 or 0 – that is the permutation test is nonrandomized – then $\alpha = t(C(S_1^*))^{-1}$ for some integer t. A test with critical function ϕ is called similar if $\int \phi dP = \alpha$ for all P in Ω_1.

Theorem 4. (a) For every h the most powerful test of Ω_1 against h is a member of the class of all permutation tests.

(b) In the class of all distribution-free or permutation tests, the most powerful level α test, for Ω_1 against a simple alternative $h'(z_{(.)}, \pi)$ is of the form

$$\phi^*(z_{(.)}, \pi) = \begin{cases} 1, & \text{if } h'(z_{(.)}, \pi) > c(z_{(.)}), \\ \delta, & \text{if } \quad '' \quad = \quad '' , \\ 0, & \text{if } \quad '' \quad < \quad '' , \end{cases}$$

where $c(z_{(.)})$, $z_{(.)} \in \mathcal{X}_{(.)}$, and $\phi^*(., \pi)$ for π such that (**) is satisfied, the critical level and δ are determined by

$$E(\phi(Z)|F) = \alpha \quad \text{for all F in } \Omega_1.$$

Proof. (a) This is essentially similar to the arguments of Lehmann and Stein (1949) though in a slightly modified and extended form to cover restricted symmetry structure. To prove (b) first we observe that (**) can be written equivalently in the form: $E\phi^*(w, R_{(.)}) = \alpha$, $w \in \mathcal{X}_{(.)}$. To establish (b) it is sufficient to maximize the expression

$$E_h(\phi^*(Z_{(.)}, R_{(.)}|Z_{(.)} = z_{(.)})$$

for each $z_{(.)}$ separately, since (**) is valid for each $z_{(.)}$ in $\mathcal{X}_{(.)}$. From the previous theorem the above conditional expectation (power) equals

$$(h^*)^{-1} \sum_{\pi_r \in S_1^*} \phi^*(z_{(.)}, r) h(\pi_r z).$$

The last expression with side restriction (**) is maximized by using the Neyman-

Pearson fundamental lemma to the problem of testing that the distribution on S_1^* is uniform against the alternative with

$$P(R_{(.)} = r) = (h^*)^{-1}h(\pi_r z).$$

Applying this argument on each Z-orbit, that is for each $z_{(.)}$ in $\mathcal{X}_{(.)}$ gives the required form of the test, and the Neyman-Pearson lemma guarantees that the suitable c(.) and δ can be found. This completes the proof of the theorem.

A similar version exists for non-randomized tests, since then ϕ^* takes values 1 or 0 and there exists an integer t such that $\alpha C(S_1^*) = t$. After these structural results, one can easily construct various permutation or permutation-rank-order tests for Ω_1 on the basis of maximal invariants wrt the appropriate hypotheses-generating groups.

6. WITTING'S ALTERNATIVE CONSTRUCTION FOR DF RANK TESTS

At the end of his paper Rüschendorf (1976) poses the following question. Witting (1970) proposes the construction of DF rank tests by defining ranks as maximal invariants under a group G, with the properties that if the null hypothesis class Ω_o is the class of Γ-symmetric distributions where Γ is a (finite) symmetry group then

(i) Ω_o is invariant under G

(ii) Γ has the cross-sectional properties

$$\text{(a) } \forall\, x, y, \exists\, \pi\varepsilon\Gamma \;:\; \pi G(x) = G(y)$$

$$\text{(b) } \forall\, x, \forall\pi\varepsilon\Gamma, \exists\, y \;:\; \pi G(x) = G(y).$$

All known HG-groups {G} have this property and all groups constructed related in this way to a symmetry class of distributions have turned out to be HG-groups. Are the two constructions equivalent in some sense?

A partial answer to this question can be obtained as follows. Note first that if $F\varepsilon\Omega_o$ then F is constant on Γ-orbits by definition of Γ-symmetry. The two cross-sectional properties (a) and (b) may be rephrased to state that G acts as a transitive permutation group on Γ-orbits. Suppose then that G is a HG-group as defined in this paper. Clearly it satisfies (i) by definition.

Moreover if Ω_o contains sufficiently many Γ-symmetric distributions, then for any two orbits γ_1, γ_2 of Γ there will be a distribution $F\varepsilon\Omega_o$ such that if $z_1\varepsilon\gamma_1$ and $z_2\varepsilon\gamma_2$, $F(z_1) \neq F(z_2)$. Consider then the image of Γ-orbit γ under the action of $g\varepsilon G$. If Z, Z' $\varepsilon\gamma$, and g(Z), $g(Z')$ lie in different Γ-orbits, by the remark above G does not satisfy (i). Hence $g(\gamma)$ is a subset of a Γ-orbit γ' say. Similarly $g^{-1}(\gamma')\subseteq\gamma$. But $g o g^{-1}(\gamma') = \gamma' \Rightarrow g^{-1}(\gamma') = \gamma$, and $g(\gamma) = \gamma'$ so that G does act as a permutation group on Γ-orbits. Finally for any two given Γ-orbits γ_1, γ_2 there will exist distributions F_1, $F_2 \varepsilon \Omega_o$ such that $F_1(z_1) = F_2(z_2)$ when $z_1 \varepsilon \gamma_1$, $z_2 \varepsilon \gamma_2$ but $F_1(z_1) \neq F_2(z_2)$ for $z_2 \notin \gamma_2$. Since G is a HG-group there is an element g of G such that $F_1 o g(.) = F_2(.)$, and so in particular $F_1 o g(z_1) = F_2(z_1)$ $\quad g(z_1) \varepsilon \gamma_2 \quad g(\gamma_1) = \gamma_2$. Hence G acts transitively on Γ-orbits.

Thus any HG-group certainly satisfies Witting's criteria. It is still not clear to us whether the converse holds.

ACKNOWLEDGEMENTS

We should like to acknowledge the help and encouragement given to us in our work by our colleagues in the Statistics section and to the typists who have so skilfully coped with the intricacies of our notation.

REFERENCES

1. Ahmad, R. (1974a). On the structure of symmetric sample testing: a distribution-free approach. Ann. Inst. Statist. Math., 26, 233-245.

2. Ahmad, R. (1974b). Characterizations of multivariate distribution-free tests and multivariate randomized tests. Proc. Prague Symp. on Asymptotic Statistics (ed. J. Hájek), Vol. 1, 371-401. Charles University Press, Prague.

3. Ahmad, R. (1975). Some characterizations of the exchangeable processes and distribution-free tests. In Statistical Distributions in Scientific Work (eds. G.P. Patil et al.), Vol. 3, 237-248. D. Reidel Pub. Co., Dordrecht, Holland and Boston USA.

4. Bell, C.B. - Haller, H.S. (1969). Bivariate symmetry tests: parametric and nonparametric. Ann. Math. Statist., 40, 259-269.

5. Bell, C.B. - Kurotschka, V. (1971). Einige Prinzipien zur Behandlung nichtparametrischer Hypothesen. In Studi di probability, statistica e ricerca operativa in onore di Giusseppe Pompilj Oderisi-Gubbio, 165-186.

6. Hájek, J. - Šidák, Z. (1967). Theory of Rank Tests. Academia, Prague.

7. Hewitt, E. - Savage, L.J. (1955). Symmetric measures on Cartesian products. Trans. Amer. Math. Soc., 80, 470-501.

8. Kurotschka, V. (1977). On a general characterization of all and construction of best distribution-free tests. In Recent Developments in Statistics (eds. J.R. Barra et al.), 507-514. North-Holland, Amsterdam.

9. Lehmann, E.L. - Stein, C. (1949). On the theory of some nonparametric hypotheses. Ann. Math. Statist., 20, 28-45.

10. Lehmann, E.L. (1959). Testing Statistical Hypotheses. Wiley, New York.

11. Meyer, P.A. (1966). Probability and Potentials. Blaisdell Pub. Co., New York.

12. Rosenblatt, M. (1952). Remarks on a multivariate transformation. Ann. Math. Statist., 23, 470-472.

13. Rüschendorf, L. (1976). Hypotheses generating groups for testing multivariate symmetry. Ann. Statist., 4, 791-795.

14. Smith, P.J. (1969). Structure of Nonparametric Tests of Some Multivariate Hypotheses. Ph.D. Thesis, Case Western Reserve University, Cleveland Ohio, USA.

15. Witting, H. (1970). On the theory of nonparametric tests. In Nonparametric Techniques in Statistical Inference (ed. M.L. Puri), 41-49. Cambridge University Press.

Rashid Ahmad and Magnus M. Peterson
University of Strathclyde
Department of Mathematics
26 Richmond Street
Glasgow G1 1XH
SCOTLAND

ONE-SAMPLE TESTS FOR
DEPENDENT OBSERVATIONS

Willem Albers

Enschede

ABSTRACT

If the observations are dependent it is well-known that the usual one-sample tests may be invalidated. For the case where the observations come from a moving average process or from an autoregressive process, some methods are discussed to obtain tests which have at least asymptotically the prescribed level.

Let $X_1,\ldots,X_N$ be random variables (r.v.'s) with a common absolutely continuous distribution function (d.f.) $F(x-\theta)$, where F satisfies $F(x) + F(-x) = 1$ for all x, i.e. the distribution determined by F is symmetric about zero. Then we want to test the hypothesis $\theta = 0$ against the alternative $\theta > 0$. If the X_i are not independent it is well-known that the level of the usual tests for this one-sample problem may differ considerably from the desired value if the same critical value as under independence is used. In this note we shall briefly discuss some methods to overcome this difficulty for certain types of dependence.

First we consider the simple case where F is known to be normal and the X_i come from an mth order moving average (MA) or an mth order autoregressive (AR) process. For this situation it is proposed in Albers (1976 a) to use a test based on the sample mean $\bar{X} = N^{-1} \sum_{i=1}^{N} X_i$, divided by an estimator of its standard deviation $\sigma(\bar{X})$. This estimator is obtained as follows : first we evaluate $\sigma^2(\bar{X})$ explicitly as a function of the variance σ^2 of X_1 and of the m MA or AR parameters. Then we replace these (m+1) parameters by suitable estimators. For σ^2 we obviously will apply $S^2 = (N-1)^{-1} \sum_{i=1}^{N} (X_i-\bar{X})^2$; for the MA or AR parameters estimators can be found in Anderson (1971). The resulting test has asymptotically the same critical value for all σ^2 and all values of the MA or AR parameters. Moreover, it is asymptotically equivalent to the optimal t-test for known values of the MA or AR parameters. For m=0 the test reduces to the ordinary t-test, based on $t = N^{\frac{1}{2}}\bar{X}/S$. As concerns the price for the robustness of validity thus obtained, it is shown in Albers (1976 a) that the test described above asymptotically requires mu_{α}^{2} additional observations to match the power of the t-test under independence. Here α is the level of the test and u_{α} is the upper α-point of the standard normal distribution.

In the above we have dealt with the dependence by modifying the statistic of the test considered. Another method is to modify the observations rather than the test statistic. This approach is used in Albers (1976 b) for the case where X_i come from an mth order AR process. Then the X_i are connected with an unobserved sequence of independent identically distributed r.v.'s Z_i through the equation $Z_i = \sum_{k=0}^{m} a_k X_{i-k}$ where the a_k are the -typically unknown- AR parameters. By using again the estimators from Anderson (1971) for these AR parameters, we arrive at the modified observations $\hat{Z}_i = \sum_{k=0}^{m} \hat{a}_k X_{i-k}$. In Albers (1976 b) it is proposed to use one-sample linear rank tests based on these $\hat{Z}_i$ rather than on the original X_i. It is shown that for the asymptotic properties of the tests, both under the hypothesis and under contiguous alternatives, it makes no difference whether the tests are based on the $\hat{Z}_i$ or on the Z_i. As the Z_i are independent, this means that these asymptotic properties are the same as under independence. In particular, the tests are asymptotically distributionfree.

REFERENCES

Albers, W. (1976 a) : Testing the mean of a normal population under dependence. Submitted to the Ann. of Statist.

Albers, W. (1976 b) : One-sample rank tests under autoregressive dependence. To appear in the Ann. of Statist.

Anderson, T.W. (1971) : The statistical analysis of time series. Wiley, New York.

Technological University Twente
Department of Mathematics
P.O. Box 217
Enschede
The Netherlands.

RELATIONS BETWEEN MUTUAL INFORMATION, STRONG EQUIVALENCE, SIGNAL-TO-NOISE RATIO, AND SIGNAL SAMPLE PATH PROPERTIES[1]

Charles R. Baker

Chapel Hill

ABSTRACT

Let (S_t) and (N_t), $t \in [0,T]$, be measurable stochastic processes, with (N_t) Gaussian. Relations are obtained between the average mutual information of (S_t) and $(S_t + N_t)$, strong equivalence of $(S_t + N_t)$ and (N_t), the signal-to-noise ratio of a quadratic-linear test statistic, and the sample path properties of (S_t). These relations generalize previous results due to T. S. Pitcher. For the case when (S_t) and (N_t) are jointly Gaussian, it is shown that these relations cannot be substantially improved.

INTRODUCTION

Let (S_t) and (N_t), $t \in [0,T]$, be zero-mean stochastic processes, with (N_t) Gaussian. We regard (N_t) as noise, and (S_t) as signal. For (S_t) Gaussian and independent of (N_t), Pitcher [1963] proved that the following properties are equivalent: (1) S + N and N are strongly equivalent; (2) the mutual information I(S, S + N) of (S_t) and $(S_t + N_t)$ is finite; (3) almost all sample paths of (S_t) lie in the range of the square root of the covariance operator of (N_t). Pitcher's proof required the hypothesis that $(S_t + N_t)$ and (N_t) are equivalent.

[1]Research partially supported by ONR Grant N00014-75-C-0491.

In this paper, we first discuss results that strengthen Pitcher's result. We then consider the relations that exist between properties (1), (2), and (3) above when (S_t) is not required to be Gaussian. A fourth property is also considered: The existence of a uniform bound on the signal-to-noise ratio of a quadratic-linear test statistic. Extensions of these results are given to more complicated models.

PRELIMINARIES

Let (S_t) and (N_t), $t \in [0,T]$, be measurable zero-mean stochastic processes on (Ω,β,P), with (N_t) Gaussian and $E \int_0^T S_t^2(\omega)dt < \infty$. The path map X of a stochastic process (X_t) is the function that carries ω in Ω into the sample path of (X_t) at ω. The path maps S, N, and S + N induce Gaussian probability measures μ_S, μ_N, and μ_{S+N} on the Borel σ-field Γ of $L_2[0,T]$. For example, $\mu_S(A) = P\{\omega: S(\omega) \in A\}$. Each of these measures has a covariance operator; for a measure μ_X, we denote this operator by R_X. R_X can be represented by an integral operator having the covariance function of (X_t) as its kernel. The path maps (S,N) and (S, S + N) induce probability measures on the product σ-field $\Gamma \times \Gamma$.

Before defining the properties of interest, we recall that if (S_t) and (N_t) are jointly Gaussian, so that μ_{S+N} is a Gaussian measure, then μ_{S+N} and μ_N are either orthogonal ($\mu_{S+N} \perp \mu_N$) or else mutually absolutely continuous. In the latter case, we say that μ_{S+N} and μ_N are equivalent ($\mu_{S+N} \sim \mu_N$). A well-known necessary and sufficient condition for equivalence is that $R_{S+N} = R_N^{1/2}(I + T)R_N^{1/2}$, where I is the identity and T is a Hilbert-Schmidt operator that does not have -1 as an eigenvalue.

For details on the above, we refer to the book by Skorohod [1974].

We can now define the properties of interest, discussed in the Introduction.

(1) Strong Equivalence. μ_{S+N} and μ_N are said to be strongly equivalent ($\mu_{S+N} \overset{s}{\sim} \mu_N$) if $\mu_{S+N} \sim \mu_N$ and if $R_{S+N} = R_N^{1/2}(I + T)R_N^{1/2}$ for a *trace-class* operator T. This terminology is due to Hajek [1962].

(2) Mutual Information. Let μ_{XY} be a probability measure on $\Gamma \times \Gamma$, μ_X and μ_Y its projections, and $\mu_X \otimes \mu_Y$ the product measure on $\Gamma \times \Gamma$. The mutual information I(X,Y), or $I(\mu_{XY})$, of μ_{XY} is defined to be infinite unless μ_{XY} is absolutely continuous with respect to $\mu_X \otimes \mu_Y$ ($\mu_{XY} \ll \mu_X \otimes \mu_Y$).

When $\mu_{XY} << \mu_X \otimes \mu_Y$, then

$$I(\mu_{XY}) \equiv \int \int [\log \frac{d\mu_{XY}}{d\mu_X \otimes \mu_Y} (x,y)] \, d\mu_{XY}(x,y) \ .$$

See Gel'fand and Yaglom [1959].

(3) Sample Path Properties of (S_t). Almost all paths of (S_t) belong to range $(R_N^{\frac{1}{2}})$ is equivalent to $\mu_S[\text{range}\,(R_N^{\frac{1}{2}})] = 1$. If A is any bounded linear operator in $L_2[0,T]$, and (S_t) is Gaussian, then $\mu_S[\text{range}\,(A)] = 1$ if and only if $R_S = ATA^*$ for T trace-class (Baker [1973b]).

(4) Signal-to-Noise Ratio. Let $<\cdot,\cdot>$ denote the inner product in $L_2[0,T]$. Let W be a continuous linear operator in $L_2[0,T]$ and h an element of $L_2[0,T]$. Define $\Lambda(x) = <x,Wx> + <h,x>$. Let $E_i(\cdot)$ denote expectation with respect to the measure μ_i. The deflection (or signal-to-noise ratio if $\mu_1 = \mu_{S+N}$, $\mu_2 = \mu_N$) $D_{12}(W,h)$ is defined by $D_{12}(W,h) = [E_1\,\Lambda(x) - E_2\Lambda(x)]^2 / [E_2\,\Lambda^2(x) - (E_2\,\Lambda(x))^2]$.

RELATIONS WHEN S AND N ARE JOINTLY GAUSSIAN

When (S_t) and (N_t) are jointly Gaussian processes, the measures $\mu_{S,S+N}$ and μ_{S+N} are Gaussian. In this case, we have the following result.

Theorem 1 (Baker [1978b]): If $I(S,N) < \infty$, then the following are equivalent: (1) $\mu_{S+N} \overset{S}{\sim} \mu_N$; (2) $I(S, S + N) < \infty$; (3) $\mu_S[\text{range}\ (R_N^{\frac{1}{2}})] = 1$.

Theorem 1 improves Pitcher's result in two ways: (a) S and N need not be independent; (b) It is not necessary to assume that $\mu_{S+N} \quad \mu_N$. However, one might ask if this result can be further improved. The following result shows that the hypothesis $I(S,N) < \infty$ cannot be omitted.

Proposition 1: (1) There exists a process (S_t), with (S_t) and (N_t) jointly Gaussian, such that $\mu_{S+N} \perp \mu_N$, but $I(S, S + N) = 0$ and $\mu_S[\text{range}\ (R_N^{\frac{1}{2}})] = 1$.

(2) There exists a process (S_t), with (S_t) and (N_t) jointly Gaussian, such that $\mu_{S+N} \overset{S}{\sim} \mu_N$ and $\mu_S[\text{range}\ (R_N^{\frac{1}{2}})] = 1$, but $I(S, S + N) = \infty$.

(3) There exists a process (S_t), with (S_t) and (N_t) jointly Gaussian, such that $I(S, S + N) = 0$, but $\mu_S[\text{range}\ (R_N^{\frac{1}{2}})] = 0$ and $\mu_{S+N} \perp \mu_N$.

Remark: Any trace-class linear operator in $L_2[0,T]$ can be represented by an integral operator (Schatten [1960]). Given such an operator which is also non-negative and self-adjoint (thus a covariance operator), a measurable and separable Gausssian process can be defined. The covariance function of the process will be defined (a.e. dt dt) by the kernel of the operator. Thus, we can prove Proposition 1 entirely in terms of covariance and cross-covariance operators.

Lemma 1 (Baker [1970]): Let μ_{XY} be any zero-mean measure on $\Gamma \times \Gamma$ such that $\int_{L_2 \times L_2} [\langle x,x\rangle + \langle y,y\rangle] d\mu_{XY}(x,y) < \infty$. μ_{XY} has a cross-covariance operator $R_{X,Y}$ defined by $\langle R_{X,Y} u,v\rangle = \int_{L_2}\int_{L_2} \langle x,v\rangle\langle y,u\rangle d\mu_{XY}(x,y)$. $R_{X,Y} = R_X^{1/2} V R_Y^{1/2}$ where $||V|| \leq 1$. Conversely, any operator of this form determines a unique zero-mean Gaussian measure μ_{XY} having μ_X and μ_Y as projections, where μ_X(resp., μ_Y) is the zero-mean Gaussian measure with covariance operator R_X (resp., R_Y). If μ_{XY} is Gaussian, then $I(\mu_{XY}) < \infty$ if and only if $||V|| < 1$ and V is Hilbert-Schmidt. When this condition is satisfied, then $I(\mu_{XY}) = \frac{-1}{2} \sum_n \log(1 - \tau_n)$, where $\{\tau_n, n \geq 1\}$ is the set of eigenvalues of V^*V.

Proof of Proposition 1: To prove part (1), we define R_S and $R_{S,N} = R_S^{1/2} V R_N^{1/2}$ as follows. $R_S^{1/2} = R_N^{1/2} P$, P Hilbert-Schmidt, $||P|| = 1$; this gives $\mu_S[\text{range } (R_N^{1/2})] = 1$. Choose $V^* = -P$. Then $P + V^* = 0$, and $PP^* + PV + V^*P^* + I$ has zero as an eigenvalue. Since $R_{S+N} = R_N^{1/2}(PP^* + PV + V^*P^* + I)R_N^{1/2}$, this implies that $\mu_{S+N} \perp \mu_N$. Moreover, $R_{S,S+N} = R_S + R_{SN} = R_N^{1/2}[PP^* + PV]R_N^{1/2} = 0$, so that $\mu_{S,S+N} = \mu_S \otimes \mu_{S+N}$, and $I(S,S+N)=0$.

For part (2), choose $R_S^{1/2} = R_N^{1/2} P$ where P is *trace-class*, so that $\mu_S[\text{range } (R_N^{1/2})] = 1$. Define $R_{S,N} = R_S^{1/2} V R_N^{1/2}$ by taking V to be any bounded linear operator which is *not* Hilbert-Schmidt but satisfies $||V|| < 1$. Since $R_{S+N} = R_N^{1/2}[PP^*+PV+V^*P^*+I]R_N^{1/2}$, the fact that $\mu_S[\text{range } (R_N^{1/2})] = 1$ and $||V|| < 1$ implies that $\mu_{S+N} \sim \mu_N$ (Baker, 1973a). Moreover, $PP^* + PV + V^*P^*$ is trace-class, since P is trace-class, so that $\mu_{S+N} \overset{S}{\sim} \mu_N$. To see that $I(S, S + N) = \infty$, we note that $R_{S,S+N} = R_S^{1/2} U R_{S+N}^{1/2} = R_S + R_{S,N} = R_S + R_S^{1/2} V R_N^{1/2}$, or $U R_{S+N}^{1/2} = P^* R_N^{1/2} + V R_N^{1/2}$. Since $\mu_{S+N} \sim \mu_N$, range $(R_{S+N}^{1/2})$ = range $(R_N^{1/2})$, and so $R_{S+N}^{1/2} = R_N^{1/2} B$, where B is a bounded linear operator with bounded inverse. Thus, $UB^* = P^* + V$, or $U = (P^* + V)B^{*-1}$. Since P^* is Hilbert-Schmidt and V is not Hilbert-Schmidt, and B^{*-1} has bounded inverse, it is not possible for $(P^* + V)B^{*-1}$ to be Hilbert-Schmidt. Since U is not Hilbert-Schmidt, $I(S, S + N) = \infty$.

To prove (3), let $R_S^{1/2} = R_N^{1/2}P$, where P is an operator which is compact, not Hilbert-Schmidt and has $||P|| = 1$. Define R_{SN} and $R_{S,S+N}$ as above, and set $V = -P^*$. Then $R_{S+N} = R_N^{1/2}(PP^* + PV + V^*P^* + I)R_N^{1/2} = R_N^{1/2}(I - PP^*)R_N^{1/2}$. Since $||P|| = 1$, PP^* has 1 as an eigenvalue, and so $\mu_{S+N} \perp \mu_N$. Also, $R_{S,S+N} = R_S^{1/2}UR_{S+N}^{1/2} = R_S + R_{SN} = R_S + R_S^{1/2}VR_N^{1/2}$, so that $UR_{S+N}^{1/2} = R_S^{1/2} + VR_N^{1/2} = (P^* + V)R_N^{1/2}$, or $UR_{S+N}^{1/2} = 0$. This shows that $U \equiv 0$ on $\overline{\text{range } (R_{S+N}^{1/2})}$, and thus we may assume that $U \equiv 0$ on all of $L_2[0,T]$. Hence $I(S, S + N) = 0$. $\mu_S[\text{range } (R_N^{1/2})] = 0$ since P is not Hilbert-Schmidt.

RELATIONS FOR NON-GAUSSIAN SIGNALS

In this section (S_t) is not required to be Gaussian. However, we assume that (S_t) and (N_t) are independent. We obtain a weakened version of Theorem 1.

Proposition 2: If $\mu_{S+N} \overset{s}{\sim} \mu_N$, then $I(S, S + N) < \infty$ and $\mu_S[\text{range } (R_N^{1/2})] = 1$. $\mu_S[\text{range } (R_N^{1/2})] = 1$ does not imply $\mu_{S+N} \overset{s}{\sim} \mu_N$. However, $\mu_S[\text{range } (R_N^{1/2})] = 1$ implies $\mu_{S+N} \sim \mu_N$ and $\mu_{S,S+N} \sim \mu_S \otimes \mu_{S+N}$.

Proof: $\mu_{S+N} \overset{s}{\sim} \mu_N$ implies $R_S = R_N^{1/2}TR_N^{1/2}$ for T trace-class. Let (X_t) be a zero-mean Gaussian process with the same covariance function as (S_t). Then $I(S, S + N) \leq I(X, X + N)$, and $I(X, X + N) < \infty$, by Theorem 1. Also, $R_S = R_N^{1/2}TR_N^{1/2}$ for T trace-class implies $\mu_S[\text{range } (R_N^{1/2})] = 1$ (Baker [1973a]). To see that $\mu_S[\text{range } (R_N^{1/2})] = 1$ does not imply $\mu_{S+N} \overset{s}{\sim} \mu_N$, let μ_S be a purely discrete measure giving probability p_n to the vector e_n, where $\{e_n, n \geq 1\}$ are o.n. eigenvectors of R_N corresponding to non-zero eigenvalues $\{\lambda_n, n \geq 1\}$. Set $p_n = \lambda_n(\sum_k \lambda_k)^{-1}$. Then $R_S = (\sum_n \lambda_n)^{-1}R_N$. Hence, it is not true that $\mu_{S+N} \overset{s}{\sim} \mu_N$. However, $\mu_S[\text{range } (R_N^{1/2})] = 1$ implies $\mu_{S+N} \sim \mu_N$ (Baker [1973a]). The fact that $\mu_{S,S+N} \sim \mu_S \otimes \mu_{S+N}$ whenever $\mu_S[\text{range}(R_N^{1/2})] = 1$ has also been previously shown (Baker [1976]).

RESULTS ON SIGNAL-TO-NOISE RATIO

For the signal-to-noise ratio $D_{12}(W,h)$ of a quadratic-linear test statistic, defined above, we set $\mu_1 = \mu_{S+N}$, $\mu_2 = \mu_N$. Then, if (S_t) and (N_t) are jointly Gaussian, $\mu_{S+N} \sim \mu_N$ implies $\sup_{(W,h)} D_{12}(W,h) < \infty$. This follows from the fact (Baker, [1969]) that $\sup_{(W,h)} D_{12}(W,h) < \infty$ if and only if $R_{S+N} = R_N^{1/2}(I + T)R_N^{1/2}$, for T Hilbert-Schmidt. From this, it follows also that $\mu_S[\text{range}(R_N^{1/2})] = 1$ implies $\sup_{(W,h)} D_{12}(W,h) < \infty$.

However, if (S_t) is not required to be Gaussian, we obtain a quite different result. That is, for any (N_t) we can define a process (S_t), independent of (N_t), such that $\sup_{(W,h)} D_{12}(W,h) = \infty$, while $\mu_{S+N} \sim \mu_N$ and $\mu_S[\text{range}(R_N^{1/2})] = 1$. For example, one can define μ_S as the discrete measure defined in the proof of Proposition 2.

If (S_t) is independent of (N_t), then it is clear that $\mu_{S+N} \overset{s}{\sim} \mu_N$ implies $\sup_{(W,h)} D_{12}(W,h) < \infty$.

APPLICATION TO STATIONARY PROCESSES

Proposition 3: Suppose that (S_t) and (N_t) are segments of wide-sense stationary processes, with (S_t) having spectral density f_S and (N_t) having rational spectral density f_N. If $\int_{-\infty}^{\infty} [f_S(\lambda)/f_N(\lambda)]d\lambda < \infty$, then $\mu_S[\text{range } (R_N^{1/2})] = 1$ and the signal-to-noise ratio $D_{12}(W,h)$ is uniformly bounded over all (W,h). Moreover, if (S_t) and (N_t) are independent, then $\mu_{S+N} \overset{s}{\sim} \mu_N$ and $I(S, S + N) < \infty$.

Remark: (S_t) in Proposition 3 is not required to be Gaussian.

Proof: Pinsker [1964] has shown that if (S_t) and (N_t) are independent and wide-sense stationary, (S_t) is Gaussian, (S_t) has spectral density f_S, and (N_t) has rational spectral density f_N, then $I(S, S + N) < \infty$ whenever $\int_{-\infty}^{\infty} [f_S(\lambda)/f_N(\lambda)]d\lambda < \infty$. From Theorem 1, this implies $R_S = R_N^{1/2}TR_N^{1/2}$ for T trace-class, which gives $\mu_S[\text{range } (R_N^{1/2})] = 1$ without requiring (S_t) to be Gaussian. Since $R_S^{1/2} = R_N^{1/2}P$ for P Hilbert-Schmidt, $R_{S+N} = R_N^{1/2}(I + PV + V^*P^* + PP^*)R_N^{1/2}$, where $R_{S,N} = R_S^{1/2}VR_N^{1/2}$ is the cross-covariance operator of $\mu_{S,N}$. Since P is Hilbert-Schmidt, $PV + V^*P^* + PP^*$ is Hilbert-Schmidt, and so $\sup_{(W,h)} D_{12}(W,h) < \infty$, without requiring (S_t) to be either Gaussian or independent of (N_t). Application of Proposition 2 completes the proof.

EXTENSIONS

Some extensions of the preceding results can be given. First, let A be any bounded linear operator in $L_2[0,T]$. Suppose that (S_t) is Gaussian, independent of (N_t). Then $I(S, AS + N) < \infty$ if and only if $\mu_{AS}[\text{range } (R_N^{1/2})] = 1$ (Baker [1976]). Next, let A be any Borel-measurable mapping of $L_2[0,T]$ into $L_2[0,T]$, with (S_t) allowed to be non-Gaussian, independent of (N_t). Let R_{AS} be the fixed covariance operator of $([AS]_t)$. We assume that range (R_N) is infinite-dimensional. Then, $I(S, AS + N)$ is maximized (over all mappings A and processes (S_t) such that μ_{AS} has

R_{AS} as covariance operator) when μ_{AS} is Gaussian (Baker [1978a]). Since this maximum can be attained by a Gaussian (S_t) and continuous linear A, application of the preceding result and Theorem 1 shows that $I(S, AS + N) < \infty$ whenever $R_{AS} = R_N^{1/2} T R_N^{1/2}$ for T trace-class.

Finally, we consider a model suitable for the most complicated communication channels. In the additive independent Gaussian channel, the condition $\mu_S[\text{range}(R_N^{1/2})] = 1$ is the same as the condition $\mu_{x+N} \sim \mu_N$ a.e. $d\mu_S(x)$, where μ_{x+N} is the measure obtained by translating (N_t) by a fixed L_2 function x. We have seen that $\mu_S[\text{range}(R_N^{1/2})] = 1$ (or $\mu_{x+N} \sim \mu_N$ a.e. $d\mu_S(x)$) implies $\mu_{S,S+N} \sim \mu_S \otimes \mu_{S+N}$ (without requiring (S_t) to be Gaussian). We generalize this as follows. Let B be any Borel-measurable mapping of $L_2[0,T] \times L_2[0,T]$ into $L_2[0,T]$. Let (S_t) and (X_t) be any pair of independent measurable stochastic processes such that $E\int_0^T S_t^2(\omega)dt < \infty$ and $E\int_0^T X_t^2(\omega)dt < \infty$. Define a stochastic process (Y_t) by $(Y_t) = ([B(S,X)]_t)$. We consider conditions for $\mu_{SY} \sim \mu_S \otimes \mu_Y$. The form of (Y_t) given here permits the maximum generality in specifying how a channel output (Y_t) is determined as a function of signal (S_t) and noise (X_t); $B(u,v) = u + v$ is perhaps the simplest non-trivial example. Fix an element u in $L_2[0,T]$, and let ν_u be the measure on Γ defined by $\nu_u[A] = P\{\omega: B[u,X(\omega)] \in A\}$. Then we have the following result: $\mu_{SY} \sim \mu_S \otimes \mu_Y$ if $\nu_u \sim \mu_X$ a.e. $d\mu_S(u)$ (Baker, [1976], Theorem 1). This is a key result in determining the capacity of the non-linear additive Gaussian channel.

REFERENCES

Baker, C. R. (1969), On the Deflection of a Quadratic-Linear Test Statistic, *IEEE Trans. on Information Theory*, IT-15, 16-21.

Baker, C. R. (1970), Mutual Information for Gaussian Processes, *SIAM J. Applied Math.*, 19, 451-458.

Baker, C. R. (1973a), On Equivalence of Probability Measures, *Annals of Probability*, 1, 690-698.

Baker, C. R. (1973b), Zero-One Laws for Gaussian Measures on Banach Space, *Trans. American Math. Soc.*, 186, 290-308.

Baker, C. R. (1976), Absolute Continuity and Applications to Information Theory, *Lecture Notes in Mathematics (Probability in Banach Spaces)*, 526, 1-11.

Baker, C. R. (1978a), Capacity of the Gaussian Channel Without Feedback, *Information and Control*, 37.

Baker, C. R. (1978b), Relations Between Mutual Information, Strong Equivalence, and Sample Path Properties for Gaussian Processes, to appear.

Gel'fand, I. M. and Yaglom, A. M. (1959), Calculation of the Amount of Information About a Random Function Contained in Another Such Function, *Amer. Math. Soc. Translations* (2); 12, 199-246.

Hajek, J. (1962), On Linear Statistical Problems in Stochastic Processes, *Czech. Math. J.*, *87*, 404-444.

Pinsker, M. S. (1964), *Information and Information Stability of Random Variables and Processes*, Holden-Day, San Francisco.

Pitcher, T. S. (1963), On the Sample Function of Processes Which Can be Added to a Gaussian Process, *Ann. Math. Stat.*, *34*, 329-333.

Schatten, R. (1960), *Norm Ideals of Completing Continuous Operators*, Springer, Berlin.

Skorohod, A. V. (1974), *Integration in Hilbert Space*, Springer-Verlag, Berlin.

Department of Statistics
University of North Carolina
Chapel Hill, North Carolina 27514
U.S.A.

OPTIMUM EXPERIMENTAL DESIGN FOR A BAYES ESTIMATOR IN LINEAR REGRESSION

Hans Bandemer, Jürgen Pilz

Freiberg (Sachs)

ABSTRACT

We consider the problem of optimum experimental design for a Bayes estimator of the response surface in a linear regression model. We give a characterization of optimal designs which minimize the Bayes risk of this estimator. On the basis of this characterization we obtain a sufficient condition on the information matrix of an optimal design and a condition for the existence of optimal one-point designs. These conditions are then used to determine optimal designs in a simple linear regression model.

INTRODUCTION

Usually, the response surface in a given linear regression model is estimated by the method of least squares and optimal designs are chosen so as to minimize some functional of the information matrix. Here we consider the problem of optimum experimental design for a Bayes estimator of the response surface with respect to a quadratic loss function and independent normally identically distributed errors of unknown variance. This problem has been treated by Lindley (1968) and Brooks (1972), (1974) for the case where the regression parameter and the variance follow a noninformative prior distribution. In Brooks (1976) the case of a conjugate prior distribution and a spheric experimental region is dealt with and the response surface is assumed to be a hyperplane. In this paper we are concerned with the case of a general (continuous) response surface and an arbitrary (compact) experimental region. For this purpose the designing problem is reformulated in terms of approximate designs as they were introduced by Kiefer (1959). The convexity of the design criterion (minimum Bayes risk) then permits the application of the general equivalence theorem from Whittle (1973) to derive an equivalent characterization of the optimality of designs. On the basis of this characterization we obtain a sufficient condition on the optimal information matrix so that the optimal design can be computed as a solution of a corresponding system of nonlinear equations. We use this condition, which coincides with that given in Brooks (1976) if the response surface happens to be a hyperplane, to determine optimal designs for a simple linear regression.

FORMULATION OF THE PROBLEM

Let be $B \subseteq R^k$ and $f(\cdot) = (f_1(\cdot), \ldots, f_r(\cdot))'$ a vector of real-valued, linearly independent and continuous functions. The family $\{Y(x)\}_{x \in B}$ of random variables is assumed to satisfy the linear regression model

(1) $$E\,Y(x) = f(x)'\vartheta \;; \qquad \vartheta \in \Theta \subseteq R^r \,.$$

Here $\vartheta = (\vartheta_1, \ldots, \vartheta_r)'$ is an unknown parameter from a given set Θ. Suppose the statistician to be able to observe realizations of $Y(x)$ at the points of a compact experimental region $V \subseteq B$; any n-tuple $V_n = (x_1, \ldots, x_n) \in V^n$ is called an <u>exact design</u>. If $Y(V_n) = (Y(x_1), \ldots, Y(x_n))'$ and $F(V_n) = (f(x_1), \ldots, f(x_n))'$ denote the vector of observations and the design matrix, respectively, then we have the observation model

(2) $$Y(V_n) = F(V_n)\vartheta + \varepsilon(V_n); \; E\,\varepsilon(V_n) = 0, \; \vartheta \in \Theta, \; V_n \in V^n$$

where $\varepsilon(V_n) = (\varepsilon(x_1), \ldots, \varepsilon(x_n))'$ is an unobservable vector of random errors.

<u>Assumption 1</u>: For any $V_n \in V^n$ let $\varepsilon(V_n)$ have a normal distribution with expectation $E\,\varepsilon(V_n) = 0$ and covariance matrix $\mathrm{Cov}\,\varepsilon(V_n) = \lambda I_n$, where λ is unknown, $\lambda \in K \subseteq (0, \infty)$. (Here I_n denotes the $(n \times n)$-identity matrix).

We are interested in estimating the response

(3) $$\eta(x,\vartheta) := f(x)'\vartheta \;; \quad x \in X \subseteq B,$$

in all points x of a compact set $X \subseteq B$. By D we denote a set of estimators $\hat{\eta}(\cdot, Y(V_n)), V_n \in V^n$, for the response surface $\eta(\cdot, \vartheta)$. The goodness of estimation is valued by a risk function $R(\vartheta, \lambda, x; \hat{\eta}, V_n)$ which is generated by a quadratic loss function according to

<u>Assumption 2:</u> $R(\vartheta,\lambda,x;\hat{\eta},V_n) = E_{Y(V_n)|\vartheta,\lambda}(\eta(x,\vartheta)-\hat{\eta}(x,Y(V_n)))^2$

Thus, the problem of optimum experimental design can be interpreted as a statistical decision problem

(4) $$G = [\Theta \times K \times X,\ D \times V^n,\ R],$$

where the elements $(\vartheta,\lambda,x) \in \Theta \times K \times X$ are the states of nature and the elements $(\hat{\eta},V_n) \in D \times V^n$ are the strategies of the statistician, who is to choose an estimator and an exact design.

Now we assume that we have some prior knowledge about the possible values of the parameters ϑ, λ and x which we can express by a prior distribution. Thus, the parameters ϑ, λ and x become random variables $\underline{\vartheta}$, $\underline{\lambda}$ and $\underline{x}$.

<u>Assumption 3:</u> $(\underline{\vartheta},\underline{\lambda})$ and $\underline{x}$ are assumed to be independent.

Let $P_{\underline{\vartheta},\underline{\lambda}}$ and $P_{\underline{x}}$ denote the prior distributions of $(\underline{\vartheta},\underline{\lambda})$ and $\underline{x}$, respectively. The distribution $P_{\underline{\vartheta},\underline{\lambda}}$ may be motivated, for example, by previous observations and the prior distribution $P_{\underline{x}}$ can be considered a weight function on X evaluating the frequency and/or the accuracy with which the response $\eta(x,\vartheta)$ will have to be estimated for the different points $x \in X$.

We are interested in a bayesian solution $(\hat{\eta}^*,V_n^*)$ of the decision problem (4). If

(5) $$\begin{aligned}\rho(\hat{\eta},V_n) &= E_{\underline{x}}\, E_{\underline{\vartheta},\underline{\lambda}}\, R(\underline{\vartheta},\underline{\lambda},\underline{x};\hat{\eta},V_n)\\ &= \int_X \int_{\Theta\times K} R(\vartheta,\lambda,x;\hat{\eta},V_n)\,d\,P_{\underline{\vartheta},\underline{\lambda}}(\vartheta,\lambda)\,d\,P_{\underline{x}}(x)\end{aligned}$$

denotes the Bayes risk of the strategy $(\hat{\eta},V_n) \in D \times V^n$, then $(\hat{\eta}^*,V_n^*)$ must satisfy

$$\rho(\hat{\eta}^*,V_n^*) = \inf_{(\hat{\eta},V_n)\in D\times V^n} \rho(\hat{\eta},V_n).$$

<u>Assumption 4:</u> Let $P_{\underline{\vartheta},\underline{\lambda}}$ be such that the conditional distribution $P_{\underline{\vartheta}|\lambda}$ is normal with mean $\mu \in R^r$ and covariance matrix λS^{-1}, where S is a positive definite matrix.

<u>Lemma 1:</u> Under the assumptions 1 to 4

(6) $$\hat{\eta}^*(x,Y(V_n)) = f(x)'(F(V_n)'F(V_n)+S)^{-1}(F(V_n)'Y(V_n)+S\mu)$$

is a Bayes estimator for $\eta(x,\vartheta)$. (cf. Pilz (1978a), Corollary 4.1).

Note that we do not require a marginal prior distribution for $\underline{\lambda}$ as it is the case if we employ the usual conjugate prior for $(\underline{\vartheta},\underline{\lambda})$ where $\underline{\lambda}^{-1}$ must have a gamma distribution (see e.g. Raiffa/Schlaifer (1961)).

We remark that the estimator (6) is robust, to a certain extent, against violations of the assumptions 1,2, and 4: the results, formulated in Pilz (1978a), show that $\hat{\eta}^*$ is also bayesian with respect to more general loss functions and/or wider classes of error and prior distributions (see also Rao (1976)). Moreover, the estimator $\hat{\eta}^*$ is also optimum, in some sense, relative to certain forms of nonbayesian prior knowledge, $\hat{\eta}^*$ is minimax, for example, if the regression parameter belongs to an ellipsoid (cf. Bandemer et. al. (1977), Näther/Pilz (1978)).

CHARACTERIZATION OF OPTIMAL DESIGNS FOR THE BAYES ESTIMATOR

We will now deal with the designing problem for the Bayes estimator $\hat{\eta}^*$ given by formula (6).

Lemma 2: Under the assumptions 1 to 4 it holds for all $x \in X$ and $V_n \in V^n$:

$$E_{\underline{\vartheta},\underline{\lambda}} R(\underline{\vartheta},\underline{\lambda},x;\hat{\eta}^*,V_n) = (E\underline{\lambda})\, f(x)'(F(V_n)'F(V_n)+S)^{-1}f(x)$$

(cf. Brooks (1976), Näther/Pilz (1978)).

Remark 1: To prove lemma 2 it is not necessary to require normality in the assumptions 1 and 4. Because of the linearity of the estimator $\hat{\eta}^*$ it suffices that the error distribution and the conditional prior distribution $P_{\underline{\vartheta}|\lambda}$ have first and second order moments as indicated in these assumptions.

An optimal design $V_n \in V^n$ is to be determined so that it minimizes the Bayes risk $\varrho(\hat{\eta}^*,\cdot) = E_{\underline{x}}\, E_{\underline{\vartheta},\underline{\lambda}} R(\underline{\vartheta},\underline{\lambda},\underline{x};\hat{\eta}^*,\cdot)$ of $\hat{\eta}^*$, i.e. V_n^* must satisfy

(7) $$\int_X f(x)'(F(V_n)'F(V_n)+S)^{-1}f(x)\, dP_{\underline{x}}(x) = \inf_{V_n\in V^n} \int_X f(x)'(F(V_n)'F(V_n)+S)^{-1}f(x)\, dP_{\underline{x}}(x)$$

This problem is similar to the problem of finding an I-optimal design for the least-squares estimator

(8) $$\tilde{\eta}(x,Y(V_n)) = f(x)'(F(V_n)'F(V_n))^{-1}F(V_n)'Y(V_n)$$

for $\eta(x,\vartheta)$, which leads to the minimization of

$$\int_X f(x)'\, M(V_n)^{-1}f(x)\, p(x)\, dx$$

where $M(V_n) = n^{-1}F(V_n)'F(V_n)$ is the information matrix and $p(\cdot)$ is a weight function on X (e.g. the density function of a prior distribution $P_{\underline{x}}$ with respect to Lebesgue measure). Now, in the designing problem for the Bayes estimator $\hat{\eta}^*$, the information matrix is replaced by the matrix

(9) $$M_b(V_n) = n^{-1}(F(V_n)'F(V_n)+S)$$

We call $M_b(V_n)$ the <u>bayesian information matrix</u> of the exact design V_n. It can be shown that the inverse of this matrix is proportional to the covariance matrix of the preposterior distribution of $\underline{\vartheta}$.

<u>Lemma 3:</u> Under the assumptions 1,2 and 4 we have

$$n^{-1}(E\,\underline{\lambda})\; M_b(V_n)^{-1} = E_{Y(V_n),\underline{\lambda}}\, \mathrm{Cov}(\underline{\vartheta}\,|\,Y(V_n),\underline{\lambda})$$

(see Pilz (1978c), lemma 2.2).

Because of the invariance of $M_b(V_n)$ with respect to permutations of the components of V_n the optimization problem (7) can be embedded into a generalized problem by introducing approximate designs.

Let Ξ be the set of all probability measures (approximate designs) on the experimental region and define

$$(10) \qquad M_b(\xi) = M(\xi) + n^{-1}S = \int_V f(x)f(x)'d\,\xi(x) + n^{-1}S$$

to be the bayesian information of the approximate design $\xi \in \Xi$.

Then, according to (7), the generalized designing problem takes the form

$$\Delta(\xi) := \int_X f(x)'M_b(\xi)^{-1} f(x)\, dP_{\underline{x}}(x) \longrightarrow \underset{\xi\in\Xi}{\mathrm{Min}}$$

Any solution to this problem we call a <u>bayesian design.</u>

Introducing the matrix

$$(11) \qquad U := \int_X f(x)f(x)'\, d\,P_{\underline{x}}(x),$$

the functional Δ can be written as

$$\Delta(\xi) = \mathrm{tr}\; U\, M_b(\xi)^{-1}, \quad \xi \in \Xi\,.$$

(Here tr A denotes the trace of a matrix A).

<u>Lemma 4:</u> The functional Δ is convex on Ξ.

This follows immediately from the fact that for any two positive semidefinite matrices M_1 and M_2 and for any $\alpha \in (0,1)$ the difference $\{\alpha M_1^{-1} + (1-\alpha)M_2^{-1}\} - \{\alpha M_1 + (1-\alpha)M_2\}^{-1}$ is positive semidefinite.

<u>Lemma 5:</u> There always exists a bayesian design $\xi^* \in \Xi$ the support of which does not contain more than $r(r+1)/2+1$ points.

The proof would follow the ideas of Kiefer (1959) (cf. Pilz (1978b), theorem 2.3).

The convexity of the functional Δ makes it possible to obtain an equivalent characterization of the Bayes-optimality of a design by use of the general equivalence theorem given in Whittle (1973).

<u>Theorem 1:</u> The design $\xi^* \in \Xi$ is bayesian if and only if

$$\mathrm{tr}\; U\, M_b(\xi^*)^{-1}(I_r - n^{-1}S\,M_b(\xi^*)^{-1}) = \sup_{x\in V} f(x)'M_b(\xi^*)^{-1}\,U\,M_b(\xi^*)^{-1}f(x).$$

<u>Proof:</u> It can be shown that for any two designs $\xi, \xi' \in \Xi$ the directional derivative of the functional Δ,

$$\Phi_\Delta(\xi,\xi') := \lim_{\alpha\to 0}\, (d\,\Delta((1-\alpha)\xi + \alpha\xi')/d\alpha)$$

is given by

$$\Phi_\Delta(\xi,\xi') = \text{tr}\, U\, M_b(\xi)^{-1}(I_r - M_b(\xi')\, M_b(\xi)^{-1}).$$

Furthermore, $\Phi_\Delta(\cdot,\cdot)$ is linear, i.e. it holds

$$\Phi_\Delta(\xi,\xi') = \int_V \Phi_\Delta(\xi,\delta_x)\, d\,\xi'(x)$$

with δ_x the one-point measure giving probability 1 to the point $x \in X$ (see Pilz (1978b), lemma 3.3). Hence, by a theorem due to Whittle (1973), ξ^* is bayesian if and only if $\inf_{x\in V} \Phi_\Delta(\xi^*,\delta_x) = 0$. With $M_b(\delta_x) = f(x)f(x)' + n^{-1}S$ this is equivalent to the condition given in the above theorem.

On the basis of theorem 1 we can deduce a sufficient optimality condition, which suggests a method for the construction of bayesian designs. In what follows we suppose the validity of

<u>Assumption 5</u>: Let $U = \int_X f(x)f(x)'\, d\,P_{\underline{x}}(x)$ be of full rank.

This assumption is satisfied, for example, if the region X and the prior distribution $P_{\underline{x}}$ satisfy the following, rather weak condition.

<u>Lemma 6</u>: If it holds supp $P_{\underline{x}} = X$ and if X contains at least r points $x_1,\dots,x_r$ for which the vectors $f(x_1),\dots,f(x_r)$ are linearly independent, then $U = \int_X f(x)f(x)'\, d\,P_{\underline{x}}(x)$ is positive definite.

We now give a sufficient condition on the bayesian information matrix of optimal designs. To do so we distinguish between the case where the response surface $\eta(\cdot,\vartheta) = f(\cdot)'\vartheta$ does not contain an absolute term, and the alternative case where $\eta(\cdot,\vartheta)$ does, i.e. $f_1(\cdot) = 1$. In the latter case define $\tilde{S}$ to be the transformed matrix

(12) $$\tilde{S} = (U^{-1/2})'\, S\, U^{-1/2} .$$

We further define

(13) $$b = r\, n^{-1}\, \text{tr}\, S\, U^{-1} + \sup_{x\in V} f(x)'\, U^{-1}\, f(x).$$

<u>Theorem 2</u>: a) Let be $f_1(\cdot) \not\equiv 1$. If there exists a design $\xi^* \in \Xi$ with the bayesian information matrix $M_b(\xi^*) = b\, r^{-1}\, U$, then ξ^* is a bayesian design.

b) Let be $f_1(\cdot) \equiv 1$ and $\tilde{s}_{11}$ the leading subelement of the matrix $\tilde{S}$. If $\xi^* \in \Xi$ is such that

$$M_b(\xi^*) = (U^{1/2})'\, C\, U^{1/2} ,$$

where $C = \text{diag}\,(c_1,\dots,c_r)$ is a diagonal matrix with elements

$$c_1 = 1 + \tilde{s}_{11}/n \; ; \quad c_2 = \dots = c_r = (b - c_1)(r-1)^{-1} ,$$

then ξ^* is a bayesian design.

The proof follows by verifying that the above given matrices satisfy the optimality condition established in theorem 1 (see also Pilz (1978b), theorem 4.1).

Thus, in case of $f_1(\cdot) \not\equiv 1$ a bayesian design ξ^* can be obtained as a solution of the system

$$M(\xi^*) = b r^{-1} U - n^{-1} S$$

and in case of $f_1(\cdot) \equiv 1$ any design ξ^* satisfying the system

(14) $$M(\xi^*) = (U^{1/2})' C U^{1/2} - n^{-1} S$$

is bayesian. Correspondingly, it is eventually possible to obtain exact optimal designs: If $f_1(\cdot) \not\equiv 1$, then an exact optimal design $V_n^* = (x_1^*, \ldots, x_n^*)$ should satisfy

$$\sum_{i=1}^{n} f(x_i^*) f(x_i^*)' = n b r^{-1} U - S$$

and if $f_1(\cdot) \equiv 1$, then V_n^* should satisfy

(15) $$\sum_{i=1}^{n} f(x_i^*) f(x_i^*)' = n (U^{1/2})' C U^{1/2} - S.$$

For the special case in which $f(x) = f(x_1, \ldots, x_k) = (1, x_1, \ldots, x_k)'$, i.e. $\eta(\cdot, \vartheta)$ is a hyperplane, and $V = X$ is a hypersphere $V = X = \{x \in R^k | x'x \leqq a\}$ with some given $a > 0$, the system (15) coincides with the system of equations derived in Brooks (1976) by minimizing $\operatorname{tr} U(F(V_n)'F(V_n) + S)^{-1} = n^{-1} \operatorname{tr} U M_b(V_n)^{-1}$ with the help of the method of Lagrange multipliers. This can be readily verified by observing that with $f(x) = (1, x_1, \ldots, x_k)'$ the matrix U has the special structure

$$U = \begin{pmatrix} 1 & \mu' \\ \mu & Q \end{pmatrix}; \quad \mu = E\underline{x}, \ Q = E\underline{x}\underline{x}' = \operatorname{Cov} \underline{x} + \mu\mu'$$

so that we obtain

$$U^{1/2} = \begin{pmatrix} 1 & 0' \\ \mu & B \end{pmatrix}; \quad B' = (\operatorname{Cov} x)^{1/2}$$

Remark 2: A necessary condition for the existence of designs ξ^* for which $M_b(\xi^*)$ has the structure indicated in theorem 2 is that the maximal eigenvalue q of $\tilde{S}$ is bounded from above. For the case $f_1(\cdot) \not\equiv 1$ it is necessary to have $rq \leqq nb$ and in case of $f_1(\cdot) \equiv 1$ it must hold $q \leqq \min(n + \tilde{s}_{11}, (nb - n - \tilde{s}_{11})/(r-1))$.

Remark 3: The matrix $U^{-1/2}$ implies a basis change of the components of the function $f(\cdot)$ such that the components of the transformed function $\tilde{f}(\cdot) = U^{-1/2} f(\cdot)$ form an orthonormal system with respect to the prior measure $P_{\underline{x}}$, i.e.

$$\int_X \tilde{f}(x)\tilde{f}(x)' \, d P_{\underline{x}}(x) = I_r .$$

Thus, the matrix $U^{-1/2}$ can be easily obtained by application of the usual orthogonalization process for functions with the scalar product $(f_i, f_j) := \int_X f_i(x) f_j(x) \, d P_{\underline{x}}(x); \ i,j = 1, \ldots, r.$

It can be shown that for a response surface without an absolute term a bayesian design $\xi^* \in \Xi$ satisfying $M_b(\xi^*) = b\, r^{-1} U$ also maximizes the determinant of the bayesian information matrix:

<u>Theorem 3</u>: Let be $f_1(\cdot) \not\equiv 1$ and $\xi^* \in \Xi$ such that $M_b(\xi^*) = b\, r^{-1} U$. Then it holds: $\det M_b(\xi^*) = \max_{\xi \in \Xi} \det M_b(\xi)$.

The proof follows from theorems 4.4 and 4.5a) in Pilz (1978c).

We are now going to formulate a criterion for the existence of a bayesian design the support of which contains only a single point which receives the probability p = 1. Such one-point designs, say δ_x, are of special interest because they lead to exact designs: If δ_x turns out to be optimal, then $V_n^* = (x,\ldots,x)$ is an exact bayesian design. Let be $h(\cdot,\cdot)$ the function defined by

(16) $h(x,v) = f(x)' M_b(\delta_v)^{-1} U M_b(\delta_v)^{-1} f(x); \quad (x,v) \in V \times V$

<u>Theorem 4</u>: The one-point design δ_v; $v \in V$; is bayesian if and only if

$$\sup_{x \in V} h(x,v) = h(v,v)$$

(see Pilz (1977), corollary 4.20).

By use of these results we determine bayesian designs for the response surface

(17) $\eta(x,\vartheta) = \vartheta_1 + \vartheta_2 x; \quad \vartheta = (\vartheta_1,\vartheta_2)' \in \Theta \subseteq R^2, \; x \in R^1$

and we assume $V = X = [-1,1]$ and $P_{\underline{x}}$ to be a uniform distribution on X.

Thus, we obtain with $f(x) = (1,x)'$

$$U = \frac{1}{2} \int_{-1}^{1} \begin{pmatrix} 1 & x \\ x & x^2 \end{pmatrix} dx = \begin{pmatrix} 1 & 0 \\ 0 & 1/3 \end{pmatrix},$$

$$U^{1/2} = \begin{pmatrix} 1 & 0 \\ 0 & 1/\sqrt{3} \end{pmatrix}; \quad U^{-1/2} = \begin{pmatrix} 1 & 0 \\ 0 & \sqrt{3} \end{pmatrix}$$

It is easy to verify that the set of designs, whose support is a subset of $V_0 = \{-1,1\}$, forms an essentially complete class. Hence, any optimal design ξ^* takes the form

$$\xi^* = \{(-1,p_1),(1,p_2)\}; \quad p_1 + p_2 = 1$$

and the information matrix is given by

$$M(\xi^*) = \begin{pmatrix} 1 & p_2 - p_1 \\ p_2 - p_1 & 1 \end{pmatrix}$$

Further, if the elements of the matrix S are denoted by s_{ij} $(i,j = 1,2)$, then we have

$$b = n^{-1}\,\mathrm{tr}\, SU^{-1} + \sup_{x\in[-1,1]} f(x)'U^{-1}f(x)$$
$$= n^{-1}(s_{11} + 3s_{22}) + 4$$

and $$C = \mathrm{diag}\,(1 + s_{11}/n,\ 3(1 + s_{22}/n))$$

(note that $\tilde{s}_{11} = s_{11}$). According to theorem 2, ξ^* is a bayesian design if it holds

$$M(\xi^*) = (U^{1/2})'CU^{1/2} - n^{-1}S$$
$$= \begin{pmatrix} 1 & -n^{-1}s_{12} \\ -n^{-1}s_{12} & 1 \end{pmatrix}$$

Consequently, if $|s_{12}| \leqq n$, then ξ^* defined by

$$\xi^* = \{(-1,p_1),(-1,p_2)\};\ p_1 = (1+s_{12}/n)/2,\ p_2 = (1-s_{12}/n)/2$$

is bayesian. If $|s_{12}|$ happens to be greater than n, we can find bayesian one-point designs. From theorem 4 it follows that δ_{-1} is bayesian if $s_{12} \geqq n$ and δ_{+1} is bayesian if $s_{12} \leqq -n$. In the special case $s_{12} = 0$, which occurs if the regression coefficients $\underline{\vartheta}_1$ and $\underline{\vartheta}_2$ are uncorrelated a-priori, the bayesian design gives the weight $p = \frac{1}{2}$ to each of the points -1 and +1. This design is a G- (D- and A-) optimal design for the least squares estimator (8), i.e. ξ^* minimizes $\sup_{x\in[-1,1]} f(x)'M(\cdot)^{-1}f(x)$ over Ξ. In other words, for the simple linear regression model the minimax design is bayesian if there is no correlation between $\underline{\vartheta}_1$ and $\underline{\vartheta}_2$.

REFERENCES

Bandemer, H. et al. (1977): Theorie und Anwendung der optimalen Versuchsplanung I. Handbuch zur Theorie. Akademie-Verlag Berlin 1977.

Beckenbach, E.F./Bellman, R. (1965): Inequalities. Springer-Verlag Berlin 1965.

Brooks, R.J. (1972): A decision theory approach to optimal regression designs. Biometrika 59 (1972), 563-571.

——— (1974): On the choice of an experiment for prediction in linear regression. Biometrika 61 (1974), 303-311.

Brooks, R.J. (1976): Optimal regression designs for prediction when prior knowledge is available. Metrika 23 (1976), 221-230.

Fedorov, V.V. (1972): The Theory of Optimum Experiments. Translation of the Russian edition (Izd. Nauka, Moskva 1971) by W.J.Studden and E.M. Klimko. Academic Press, New York 1972.

Kiefer, J. (1959): Optimum experimental design. Journ. Roy. Statist. Soc., Ser. B, 21 (1959), 272-319.

Lindley, D.V. (1968): The choice of variables in multiple regression. Journ. Roy Statist. Soc., Ser. B, 30 (1968), 31-66.

Näther, W./Pilz, J. (1978): Estimation and experimental design in a linear regression model using prior information. Zastosowania Matematyki (to appear).

Pilz, J. (1978a): Das bayessche Schätzproblem im linearen Regressionsmodell. In: Beiträge zur optimalen Versuchsplanung V, Freiberger Forschungshefte Ser. D, Deutscher Verlag für Grundstoffindustrie, Leipzig (to appear).

——— (1978b): Konstruktion von optimalen diskreten Versuchsplänen für eine Bayes-Schätzung im linearen Regressionsmodell. In: Beiträge zur optimalen Versuchsplanung V (s. Pilz (1978a)).

——— (1978c): Optimalitätskriterien, Zulässigkeit und Vollständigkeit im Planungsproblem für eine bayessche Schätzung im linearen Regressionsmodell. In: Beiträge zur optimalen Versuchsplanung V (s. Pilz (1978a)).

——— (1977): Bayessche Schätzung und Versuchsplanung im linearen Regressionsmodell. Thesis, Bergakademie Freiberg.

Raiffa, H./Schlaifer, R. (1961): Applied statistical decision theory. Harvard Univ., Boston 1961.

Rao, C.R. (1976): Characterization of prior distributions and solution to a compound decision problem. Ann. Statist. 4 (1976), 823-835.

Whittle, P. (1973): Some general points in the theory of optimal experimental design. Journ. Roy. Statist. Soc., Ser. B, 35 (1973), 123-130.

Bergakademie Freiberg
Sektion Mathematik
DDR-92 Freiberg, PF 47

A COMPARISON OF TWO EXPONENTIAL FAMILIES

Ludwig Baringhaus, Detlef Plachky

Münster

ABSTRACT

This paper explores the asymptotic behaviour of the error probabilities of second kind of uniformly most powerful (UMP) unbiased tests in the two sample case. It is shown that, if the sample sizes increase to infinity, the error probabilities of second kind of the UMP unbiased tests at a fixed level for the comparison of two one-parameter exponential families tend to zero exponentially fast. The result is applied to Fisher's exact test for the comparison of two binomial distributions. Moreover, properties of the associated measure of efficiency are given.

1. INTRODUCTION

Let $\mathcal{P} = \{P_\theta;\ \theta \in \Theta\}$ be a one-parameter exponential family of probability distributions on $(\mathbb{R}^1, \mathcal{B}^1)$ with densities given by

$$(1) \qquad dP_\theta(x) = c(\theta)\exp(\theta x)d\nu(x),$$

where ν denotes a σ-finite measure on $(\mathbb{R}^1, \mathcal{B}^1)$. The parameter space $\Theta \subset \mathbb{R}^1$ is assumed to be open. Let $X^{(n)} = (X_1,\dots,X_n)$ and $Y^{(m)} = (Y_1,\dots,Y_m)$ be independent samples from the distributions $P_\lambda \in \mathcal{P}$ and $P_\mu \in \mathcal{P}$ $(\lambda \in \Theta, \mu \in \Theta)$. Then the joint distribution of $X^{(n)}$ and $Y^{(m)}$ has the density

$$(2)\qquad p_{\lambda,\mu}^{(n,m)}(x_1,\dots,x_n,y_1,\dots,y_m) = c(\lambda)^n c(\mu)^m \exp(\lambda\sum_{i=1}^{n} x_i + \mu\sum_{j=1}^{m} y_j)$$

with respect to $\nu^{(n,m)}$, which is the (n+m)-fold product measure of ν. Putting $\eta := \lambda-\mu$, $\xi := \mu$, $c(\eta,\xi) := c(\lambda)^n c(\mu)^m$ and $U_{n,m}(x_1,\dots,x_n, y_1,\dots,y_m) = \sum_{i=1}^{n} x_i$, $V_{n,m}(x_1,\dots,x_n,y_1,\dots,y_m) = \sum_{i=1}^{n} x_i + \sum_{j=1}^{m} y_j$, this density can be written as

$$(3)\qquad p_{\eta,\xi}^{(n,m)} = c(\eta,\xi)\exp(\eta U_{n,m} + \xi V_{n,m}).$$

Defining $\Theta-\Theta := \{\lambda-\mu;\ \lambda\in\Theta,\ \mu\in\Theta\}$, therefore

$$(4)\qquad \mathcal{P}^{(n,m)} = \{P_{\eta,\xi}^{(n,m)};\ dP_{\eta,\xi}^{(n,m)} = p_{\eta,\xi}^{(n,m)}\, d\nu^{(n,m)},\ \eta\in\Theta-\Theta,\ \xi\in\Theta\}$$

denotes a two-parameter exponential family. For testing the hypothesis $H:\lambda\le\mu$ against the alternative $K:\lambda>\mu$ (equivalently $H:\eta\le 0$ against $K:\eta>0$) there exists a UMP unbiased test at level $\alpha\in(0,1)$. Depending on the sufficient statistic $(U_{n,m}, V_{n,m})$, this test is given by

$$(5)\qquad \varphi_{n,m}(u,v) = \begin{cases} 1, & u > c_{n,m}(v) \\ \gamma_{n,m}(v), & u = c_{n,m}(v) \\ 0, & u < c_{n,m}(v) \end{cases},\quad (u,v)\in\mathbb{R}^2,$$

where the functions $c_{n,m}$ and $\gamma_{n,m}$ may be determined by

$$(6)\qquad E_{0,\cdot}(\varphi_{n,m}(U_{n,m}, V_{n,m})\mid v) = \alpha,\quad v\in\mathbb{R}^1,$$

[see Lehmann (1970)]. Hence $c_{n,m}(v)$ can be choosen as the $(1-\alpha)$th quantile of the conditional distribution $P_{0,\cdot}^{U_{n,m}\mid v}$ of $U_{n,m}$ for given $V_{n,m} = v$. (Note that for $\eta = 0$ the conditional distribution of $U_{n,m}$ for given $V_{n,m} = v$ does not depend on $\xi\in\Theta$.) According to a result of Chernoff (1956) in the one sample case, it is shown that under certain conditions the error probabilities of second kind of the UMP unbiased tests $\varphi_{n,m}(U_{n,m}, V_{n,m})$ tend to zero exponentially fast, provided that the sample sizes n and m increase to infinity in such a way that $\lim_{\substack{n\to\infty\\ m\to\infty}} \frac{n}{n+m} = h\in(0,1)$ exists.

In particular, when the sample sizes n and m are large, one avoids difficulties by the calculation of the critical values $c_{n,m}(v)$, if one replaces the UMP unbiased tests $\varphi_{n,m}(U_{n,m}, V_{n,m})$ by the tests $\psi_{n,m}(U_{n,m}, V_{n,m})$, where $\psi_{n,m}$ is given by

$$(7)\qquad \psi_{n,m}(u,v) = \begin{cases} 1, & u - E_{o,\cdot}(U_{n,m}|v) > u_\alpha \sqrt{Var_{o,\cdot}(U_{n,m}|v)} \\ 0, & u - E_{o,\cdot}(U_{n,m}|v) \le u_\alpha \sqrt{Var_{o,\cdot}(U_{n,m}|v)} \end{cases}, \quad (u,v) \in \mathbb{R}^2.$$

(u_α denotes the $(1-\alpha)$th quantile of the normal distribution $N(0,1)$. Note that $E_{o,\cdot}(U_{n,m}|V_{n,m}) = \frac{n}{n+m} V_{n,m}$ $P^{(n,m)}_{o,\xi}$ - a.e.). This is a well-known proceeding in applications (see, for example, Fisher's exact test, which will be discussed in Section 3). It is motivated by the fact that, if $\eta=0$, the random variables $(U_{n,m} - \frac{n}{n+m} V_{n,m})/\sqrt{Var_{o,\cdot}(U_{n,m}|V_{n,m})}$ tend in distribution to the normal distribution $N(0,1)$. It is shown that the error probabilities of second kind of the tests $\psi_{n,m}(U_{n,m}, V_{n,m})$ have the same asymptotic behaviour as that of the UMP unbiased tests.

2. MAIN RESULTS

We shall prove our theorems by deriving upper and lower bounds for the error probabilities of second kind. For this purpose, we state first the following lemma, due to Krafft and Plachky (1970).

Lemma 1. Let P and Q be probability distributions possessing densities p and q, respectively, with respect to a σ-finite measure μ, and let $\tilde{\varphi}$ be a most powerful test at level $\alpha \in (0,1)$ for testing the simple hypothesis H: $\{P\}$ against the simple alternative K: $\{Q\}$. Then the inequality

$$(8)\qquad 1 - E_Q(\tilde{\varphi}) \ge (1-\alpha)^{w/(w-1)} \exp(-I_w(P:Q))$$

holds for all $w>1$, where $I_w(P:Q) = \frac{1}{w-1} \log \int p^w q^{1-w}\, d\mu$ denotes Rényi's information number of order w.

In what follows we need an extension of the function $c: \Theta \longrightarrow \mathbb{R}^1$ given in (1): Let $c(w) := (\int \exp(wx)\, d\nu(x))^{-1}$ for all $w \in \mathbb{R}^1$ for which $\int \exp(wx)\, d\nu(x)$ is finite. Note that the domain $\tilde{\Theta}$ of c is then an interval (open, half-open, closed, finite, infinite), and that c is a convex function. ($\tilde{\Theta}$ is the natural parameter space of the exponential family.) We can prove now

Lemma 2. Let $\varphi_{n,m}(U_{n,m}, V_{n,m})$ be the UMP unbiased level α test determined by (5). Then it holds for any pair $(\eta,\xi)\in(\Theta-\Theta)\times\Theta, \eta>0$, that

(9) $$\liminf_{\substack{n\to\infty\\ m\to\infty}} [1-E_{\eta,\xi}(\varphi_{n,m}(U_{n,m},V_{n,m}))]^{\frac{1}{n+m}} \geq \frac{c(\lambda)^h c(\mu)^{1-h}}{c(h\lambda+(1-h)\mu)},$$

where $\lambda = \eta+\xi$, $\mu=\xi$ and $\lim_{\substack{n\to\infty\\ m\to\infty}} \frac{n}{n+m} = h\in(0,1)$.

Proof: Let $(\eta,\xi)\in(\Theta-\Theta)\times\Theta, \eta>0$, $(s,t)\in(\mathring{\Theta}-\mathring{\Theta})\times\mathring{\Theta}$, $s\leq 0$, and let $\tilde{\varphi}_{n,m;s,t}$ be a most powerful test at level $\alpha\in(0,1)$ for testing the simple hypothesis $\tilde{H}$: $\{P^{(n,m)}_{s,t}\}$ against the simple alternative $\tilde{K}$: $\{P^{(n,m)}_{\eta,\xi}\}$. Then we obtain from Lemma 1 that

(10) $$1 - E_{\eta,\xi}(\varphi_{n,m}(U_{n,m}, V_{n,m})) \geq 1 - E_{\eta,\xi}(\tilde{\varphi}_{n,m;s,t})$$

$$\geq (1-\alpha)^{w/(w-1)} \cdot \frac{c(s+t)^{nw/(1-w)}\cdot c(t)^{mw/(1-w)}\cdot c(\lambda)^n c(\mu)^m}{c(sw+\eta(1-w)+tw+\xi(1-w))^{n/(1-w)}\cdot c(tw+(1-w)\xi)^{m/(1-w)}}$$

$$, \quad 1 < w < 1+\delta,$$

where $\delta>0$ is choosen so that $t(1+\delta)-\delta\xi$ and $s(1+\delta)-\eta\delta+t(1+\delta)-\xi\delta$ belong to $\tilde{\Theta}$. This implies

(11) $$\liminf_{\substack{n\to\infty\\ m\to\infty}} [1 - E_{\eta,\xi}(\varphi_{n,m}(U_{n,m}, V_{n,m}))]^{\frac{1}{n+m}}$$

$$\geq \frac{c(s+t)^{hw/(1-w)}\cdot c(t)^{(1-h)w/(1-w)}\cdot c(\lambda)^h\cdot c(\mu)^{1-h}}{c(sw+\eta(1-w)+tw+\xi(1-w))^{h/(1-w)}\cdot c(tw+\xi(1-w))^{(1-h)/(1-w)}}, 1<w<1+\delta.$$

For $w \longrightarrow 1$ we obtain

(12) $$\liminf_{\substack{n\to\infty\\ m\to\infty}} [1 - E_{\eta,\xi}(\varphi_{n,m}(U_{n,m}, V_{n,m}))]^{\frac{1}{n+m}}$$

$$\geq \frac{c(\lambda)^h c(\mu)^{1-h}}{c(s+t)^h c(t)^{1-h}} \cdot \exp\{-[(1-h)g(t)(t-\mu)+hg(s+t)(s+t-\lambda)]\},$$

where $g(w) = -\frac{d}{dw}\log c(w)$. Putting $s=0$ and $t = h\lambda+(1-h)\mu$, this gives the inequality (9).

Remark. The function on the right sight of (12) in fact takes its maximum at s=0 and $t = h\lambda + (1-h)\mu$. Moreover, one can show that

$$(13)\quad \sup_{\substack{(s,t)\in(\mathring{\tilde{\Theta}}-\mathring{\tilde{\Theta}})\times\mathring{\tilde{\Theta}}\\ s\le 0}} \liminf_{\substack{n\to\infty\\ m\to\infty}} [1- E_{\eta,\xi}(\tilde{\varphi}_{n,m,s,t})]^{\frac{1}{n+m}} = \frac{c(\lambda)^h c(\mu)^{1-h}}{c(h\lambda+(1-h)\mu)} .$$

Hence the bound given in (9) is the best possible.

The next theorem states that, if the random variables X_i and Y_j are bounded, the equality holds in (9).

Theorem 1. Let $\mathfrak{P}$ be an exponential family with distributions $P\in\mathfrak{P}$ having a common bounded support. Then

$$(14)\quad \lim_{\substack{n\to\infty\\ m\to\infty}} [1- E_{\eta,\xi}(\varphi_{n,m}(U_{n,m}, V_{n,m}))]^{\frac{1}{n+m}} = \frac{c(\lambda)^h c(\mu)^{1-h}}{c(h\lambda+(1-h)\mu)}$$

for all $(\eta,\xi)\in(\Theta-\Theta)\times\Theta$, $\eta>0$.

Proof: We note first that

$$(15)\quad c_{n,m}(V_{n,m}) \le E_{0,\cdot}(U_{n,m}|V_{n,m}) + \sqrt{\tfrac{1}{\alpha}\,\mathrm{Var}_{0,\cdot}(U_{n,m}|V_{n,m})}\quad P_{0,\cdot}^{(n,m)}\text{-a.e.}$$

This follows easily from Chebyshev's inequality. If the random variables X_i and Y_j are identically distributed, it holds that

$$0 = \mathrm{Var}(\sum_{i=1}^{n} X_i + \sum_{j=1}^{m} Y_j \,|\sum_{i=1}^{n} X_i + \sum_{j=1}^{m} Y_j)$$

$$= (n+m)\mathrm{Var}(X_1 |\sum_{i=1}^{n} X_i + \sum_{j=1}^{m} Y_j) + (n+m)(n+m-1)\mathrm{Cov}(X_1,Y_1 |\sum_{i=1}^{n} X_i + \sum_{j=1}^{m} Y_j)$$

P- a.e. This implies $\mathrm{Cov}(X_1,Y_1 |\sum_{i=1}^{n} X_i + \sum_{j=1}^{m} Y_j)\le 0$ P- a.e. Hence it may be concluded that

$$\mathrm{Var}(\sum_{i=1}^{n} X_i |\sum_{i=1}^{n} X_i + \sum_{j=1}^{m} Y_j) \le n\,\mathrm{Var}(X_1 |\sum_{i=1}^{n} X_i + \sum_{j=1}^{m} Y_j)\ \text{P- a.e.}$$

Now since the distribution of X_1 has a compact support, there exists some $K\ge 0$ such that

$$(16)\quad c_{n,m}(V_{n,m}) \le \frac{n}{n+m} V_{n,m} + \sqrt{\frac{n}{\alpha}}\,K \quad P_{0,\cdot}^{(n,m)}\text{- a.e.}$$

Therefore we have

$$1 - E_{\eta,\xi}(\varphi_{n,m}(U_{n,m},V_{n,m})) \le P^{(n,m)}_{\eta,\xi}(U_{n,m} \le c_{n,m}(V_{n,m}))$$

$$\le P^{(n,m)}_{\eta,\xi}(U_{n,m} \le \frac{n}{n+m} V_{n,m} + \sqrt{\frac{n}{\alpha}}\ K)$$

$$\le E_{\eta,\xi}(\exp(t\ [\frac{n}{n+m} V_{n,m} - U_{n,m}]))\cdot\exp(t\sqrt{\frac{n}{\alpha}}\ K)$$

for all $t>0$. Putting $t=\eta$, it follows

$$\text{(17)}\qquad \limsup_{\substack{n\to\infty\\ m\to\infty}} [1- E_{\eta,\xi}(\varphi_{n,m}(U_{n,m},V_{n,m}))]^{\frac{1}{n+m}} \le \frac{c(\lambda)^h c(\mu)^{1-h}}{c(h\lambda + (1-h)\mu)},$$

which together with Lemma 2 yields the desired result.

For the tests determined by (7) we get

<u>Theorem 2.</u> Let Ψ be given as in Theorem 1. Then

$$\text{(18)}\qquad \lim_{\substack{n\to\infty\\ m\to\infty}} [1 - E_{\eta,\xi}(\psi_{n,m}(U_{n,m},V_{n,m}))]^{\frac{1}{n+m}} = \frac{c(\lambda)^h c(\mu)^{1-h}}{c(h\lambda + (1-h)\mu)}$$

for all $(\eta,\xi)\in(\Theta-\Theta)\times\Theta$, $\eta>0$.

<u>Proof:</u> If $\eta = 0$ and $\xi' \in \mathring{\Theta}$ we have

$$\lim_{\substack{n\to\infty\\ m\to\infty}} P^{(n,m)}_{0,\xi'}(U_{n,m} - \frac{n}{n+m} V_{n,m} > u_\alpha \sqrt{\mathrm{Var}_{0,\cdot}(U_{n,m} \mid V_{n,m})}) = \alpha.$$

Hence, if the sample sizes n and m are large enough, $\psi_{n,m}(U_{n,m},V_{n,m})$ denotes a test at level $\alpha'\in(0,1),\alpha'>\alpha$, for testing the simple hypothesis H: $\{P^{(n,m)}_{0,\xi'}\}$ against any simple alternative K: $\{P^{(n,m)}_{\eta,\xi}\}$, $\eta\in\Theta-\Theta,\eta>0,\xi\in\Theta$. Then arguing as in the proof of Lemma 2 we obtain the inequality (9) with $\varphi_{n,m}$ replaced by $\psi_{n,m}$. Since

$$P^{(n,m)}_{\eta,\xi}(U_{n,m} - \frac{n}{n+m} V_{n,m} \le u_\alpha \sqrt{\mathrm{Var}_{0,\cdot}(U_{n,m} \mid V_{n,m})})$$

$$\le E_{\eta,\xi}(\exp(t[\frac{n}{n+m} V_{n,m} - U_{n,m}])\cdot\exp(tu_\alpha \sqrt{n}\ K)$$

for all $t>0$, also we get the inequality (17) with $\varphi_{n,m}$ replaced by $\psi_{n,m}$.

Remark. 1) One can show that Theorem 1 and Theorem 2 are valid for exponential families for which the conditional variance $\mathrm{Var}_{0,\cdot}(U_{n,m}|V_{n,m})$ is a polynomial in $V_{n,m}$ at most of degree 2.In fact, it follows from a theorem of Bolger and Harkness (1965) that under the stated condition the exponential family will be a family of normal, binomial, Poisson, Gamma or negative binomial distributions. It can be shown easily in all of these cases that the inequality (17) holds (also with $\varphi_{n,m}$ replaced by $\psi_{n,m}$).

2) In Theorem 1 and Theorem 2 we started from a fixed level $\alpha\in(0,1)$. Instead of this, we may assume also that $\alpha = \alpha_{n,m}$ depends on n and m in such way that $\alpha_{n,m}^{-1} = \mathcal{O}(n+m)$ and $\alpha_{n,m} \leq 1-\varepsilon$ for some $0<\varepsilon<1$.

3) The methods described above can be applied also to one sample problems. For example, let $X^{(n)} = (X_1,\ldots,X_n)$ be a sample of the trinomial distribution $M(1;p_1,p_2,p_3)$ and let φ_n denote a UMP unbiased level α test based on $X^{(n)}$ for testing the hypothesis H: $p_1\leq p_2$ against the alternative K: $p_1>p_2$. Then one can show using inequalities corresponding to (10) and (15) that

$$\lim_{n\to\infty} [1 - E_{p_1,p_2}(\varphi_n)]^{\frac{1}{n}} = (1 - (\sqrt{p_1} - \sqrt{p_2})^2)$$

for $p_1>p_2$.

4) The limit

$$e(h;\lambda,\mu) = \frac{c(\lambda)^h c(\mu)^{1-h}}{c(h\lambda + (1-h)\mu)} \qquad ,\lambda\in\Theta,\mu\in\Theta,\lambda>\mu,h\in(0,1),$$

can serve as a measure of efficiency for sequences of UMP unbiased tests $\varphi_{n,m}$, where the sample sizes n and m tend to infinity in such a way that $h = \lim\limits_{\substack{n\to\infty\\ m\to\infty}} \frac{n}{n+m}$. For instance, one can determine that $h = \tilde{h}(\lambda,\mu)\in(0,1)$ for which $e(h;\lambda,\mu)$ takes its minimum. It is worthwhile to note that $\lim\limits_{\substack{\lambda\to\mu\\ \lambda>\mu}} \tilde{h}(\lambda,\mu) = \frac{1}{2}$. Further, asking for $\tilde{h}(\lambda,\mu) = \frac{1}{2}$ for all

$\lambda,\mu\in\Theta,\lambda>\mu$, one obtains a characterization of the exponential family $\mathfrak{P}$ as a family of normal distributions. Characterizations of $\mathfrak{P}$ as a family of normal distributions or gamma distributions are obtained also by the requirement that $e(h;\lambda,\mu)$ depends from λ and μ only through $(\lambda-\mu)$ or $\frac{\lambda}{\mu}$, respectively, for all $h\in(0,1)$.

3. AN EXAMPLE

As an application of Theorem 1 and Theorem 2 we consider Fisher's exact test. Let $X^{(n)} = (X_1,\ldots,X_n)$ and $Y^{(m)} = (Y_1,\ldots,Y_m)$ be independent samples from the binomial distributions $b(1,p_1)$ and $b(1,p_2)$, where $p_1\in(0,1)$ and $p_2\in(0,1)$ are unknown. The binomial distributions $b(1,p)$, $p\in(0,1)$, constitute an exponential family with densities of the form (1), where $\theta = \log\frac{p}{1-p}\in\mathbb{R}$, $c(\theta) = \frac{1}{1+e^{\theta}} = (1-p)$ and ν denotes the counting measure on $\{0,1\}$. Then from Theorem 1 and Theorem 2 it follows that

$$(19)\quad \lim_{\substack{n\to\infty\\ m\to\infty}} [1 - E_{p_1,p_2}(\varphi_{n,m}(U_{n,m}, V_{n,m}))]^{\frac{1}{n+m}}$$

$$= \lim_{\substack{n\to\infty\\ m\to\infty}} [1 - E_{p_1,p_2}(\psi_{n,m}(U_{n,m}, V_{n,m}))]^{\frac{1}{n+m}}$$

$$= p_1^{\,h}\, p_2^{\,1-h} + (1-p_1)^h(1-p_2)^{1-h}$$

for $p_1>p_2$, where $\varphi_{n,m}(U_{n,m}\ V_{n,m})$ denotes the UMP unbiased level α test determined by (5) for testing the hypothesis H: $p_1\le p_2$ against the alternative K: $p_1>p_2$. Note that the rejection region of the test $\psi_{n,m}(U_{n,m}, V_{n,m})$ is given by

$$\{U_{n,m} - \frac{n}{n+m}V_{n,m} > u_\alpha\sqrt{\frac{n\,m}{(n+m)^2}\cdot\left(\frac{n+m-V_{n,m}}{n+m-1}\right)\cdot V_{n,m}}\},$$

which means

$$\{\frac{1}{n}\sum_{i=1}^{n}X_i - \frac{1}{m}\sum_{j=1}^{m}Y_j > u_\alpha\,[\,\frac{1}{nm}(\sum_{i=1}^{n}X_i + \sum_{j=1}^{m}Y_j)\,\frac{n+m-\sum_{i=1}^{n}X_i - \sum_{j=1}^{m}Y_j}{n+m-1}\,]^{\frac{1}{2}}\}.$$

REFERENCES

Bolger E., Harkness W. (1965): Characterizations of some distributions by conditional moments. Ann. Math. Statist., 36, 703-705.

Chernoff H. (1956): Large sample theory: Parametric case. Ann. Math. Statist., 27, 1-22.

Krafft O., Plachky D. (1970): Bounds for the power of likelihood ratio tests and their asymptotic properties. Ann. Math. Statist., 41, 1646-1654.

Lehmann E. Testing statistical hypotheses. Fifth printing, John Wiley & Sons, New York, 1970.

Institut für Mathematische Statistik
der Universität Münster
4400 Münster
Roxeler Str. 64
Federal Republic of Germany

MINIMAX INSPECTION STRATEGIES
IF EXPECTED SYSTEM LIFETIME IS KNOWN

Frank Beichelt

Halle - Neustadt

ABSTRACT

A model for proper scheduling of inspections is considered, if system failures can be detected only by checking.

On condition that the probability distribution of the system lifetime is unknown but the mathematical expectation, optimum inspection strategies are constructed with respect to certain cost and availability criterions.

FORMULATION OF THE BASIC PROBLEM

At time $t = 0$ a system starts working. The time till failure of the system (lifetime) is a random variable X with the probability distribution function $F(t) = P(X < t)$, $F(+0) = 0$. It is assumed that system failure is known only by inspecting. Immediately on discovery of a failure the system is replaced by a new one with the same distribution of lifetime. Each inspection entails a fixed cost c_1 and takes negligible time. On the other hand, a downtime t of the system (=time between happening of system failure and its detection plus replacement time) gives rise to costs $c_2 t$. Each replacement requires a fixed time d and fixed costs c_3.

This situation has been first analyzed by Barlow, Hunter and Proschan (1963) in case of completely known lifetime distribution.

Here we consider the case that this distribution is unknown with exception of the mathematical expectation $\mu = E(X)$, $0 < \mu < \infty$. That means F is element of the set $\mathcal{F}_\mu$ defined by

$$\mathcal{F}_\mu = \left\{ F;\ F \in \mathcal{F},\ \mu = \int_0^\infty t dF(t) \right\},$$

where $\mathcal{F}$ is the set of all probability distribution functions H(t) with the property H(+0) = 0. In the paper results are summarized obtained by Beichelt (1975), (1976), and (1977).

COST CRITERION

Let $S = \{t_k\}$ be an inspection strategy (short: strategy) that is an unbounded strictly increasing sequence of numbers (at time t_k occurs the k-th inspection when no failure has been detected before). Then the expected total costs in each cycle (=time between two neighbouring replacements) amount to

$$Q(S,F) = \sum_{k=0}^{\infty} \int_{t_k}^{t_{k+1}} \left[(k+1)c_1 + c_2(t_{k+1} - t + d) \right] dF(t) + c_3 ,$$

where $t_0 = 0$. The expected length of a cycle is given by

$$(1) \qquad L(S,F) = \sum_{k=0}^{\infty} t_{k+1} \left[F(t_{k+1}) - F(t_k) \right] + d.$$

Hence we have the average cost per unit time.

$$K(S,F) = \frac{Q(S,F)}{L(S,F)} .$$

Since it is only known that $F \in \mathcal{F}_\mu$, we consider the criterion

$$K(S) = \sup_{F \in \mathcal{F}_\mu} K(S,F) .$$

Our aim is to find a strategy $S = S^*$ with the property

$$K(S^*) = \inf_{S \in \mathfrak{S}} K(S),$$

where $\mathfrak{S}$ is the set of all inspection strategies. We call S^* "partial minimax strategy" and $K(S^*)$ "partial minimax loss costs".

Let be for $x \geq 0$

$$D(S,F,x) = Q(S,F) - xL(S,F)$$

and

$$D(S,x) = \sup_{F \in \mathfrak{F}_\mu} D(S,F,x).$$

The following lemma informs on the structure of a possible partial minimax strategy.

Lemma 1: Assume for all x, $0 \leq x \leq c_2$, the existence of a unique strategy $S(x)$ and of a number x_0, $0 < x_0 < c_2$, so that

$$D(S(x),x) = \inf_{S \in \mathfrak{S}} D(S,x) \text{ and } D(S(x_0),x_0) = 0.$$

Then $S^* = S(x_0)$ is the unique partial minimax strategy and $x_0 = K(S^*)$ are the corresponding partial minimax loss costs.

Proof: The equation $D(S(x_0),x_0) = 0$ yields

$$(2) \qquad x_0 = \sup_{F \in \mathfrak{F}_\mu} \frac{Q(S(x_0),F)}{L(S(x_0),F)} = K(S(x_0)).$$

By definition of $S(x)$ we have for any $S \in \mathfrak{S}$

$$0 = D(S(x_0),x_0) < D(S,x_0).$$

But $0 < D(S,x_0)$ implies

$$(3) \qquad x_0 \leq \sup_{F \in \mathfrak{F}_\mu} \frac{Q(S,F)}{L(S,F)} = K(S) .$$

From (2) and (3) we have for any $S \in \mathfrak{S}$ $K(S(x_0)) \leq K(S)$. Therefore, $S(x_0)$ is partial minimax strategy

Now let S^* be a partial minimax strategy, $K(S^*) < c_2$, and let x^* be defined by $D(S^*,x^*) = 0$. If $S^* \neq S(x_0)$, then by definition of $S(x)$ there holds

$$D(S(x^*),x^*) < D(S^*, x^*) = 0.$$

Hence it would be $K(S(x^*)) < x^* = K(S^*)$, condratictory to the fact that S^* is partial minimax strategy. Thus, it must be $S^* = S(x_0)$, and the lemma is proved.

We assume now

(4) $$(c_1 + c_3)/\mu < c_2 .$$

Otherwise, the expected loss costs per unit downtime of the system would be smaller or equal than the mean loss per unit time, who arises by "ideal inspection and replacement" (i.e. failure of the system is detected immediately and replacement takes negligible time). But then inspection and replacement are uneconomically from the first.

Let us for any $S = \{t_k\}$ the integer $s = s(S)$ and expressions $G_{ij}(S,x)$ define by $t_s \leq \mu < t_{s+1}$ and

$$G_{ij}(S,x) = \frac{t_j-\mu}{t_j-t_i}\Big[(i+1)c_1 + (c_2-x)\delta_i\Big] + \frac{\mu-t_i}{t_j-t_i}\Big[(j+1)c_1 + (c_2-x)\delta_j\Big] ,$$

respectively; $\delta_k = t_{k+1}-t_k$, $k = 0,1,\ldots$. There holds

(5) $$D(S,x) = \sup_{\substack{(i,j)\\ i \leq s < j}} G_{ij}(S,x) - x(\mu + d) + c_2 d + c_3 .$$

This is essentially due to the fact that we can replace the set $\mathfrak{F}_\mu$ (with respect to the definition of $D(S,x)$) by that subset of $\mathfrak{F}_\mu$, whose elements have exactly two points of increase on $[0,\infty)$, see Hoeffding (1955).

Definition: A strategy $S = \{t_k\}$ is called strictly periodic with the inspection interval δ, if $\delta_k = t_{k+1}-t_k = \delta$, $k = 0,1,\ldots$.

Theorem 1: For every x, $0 \leq x < c_2$, there holds $S(x) = S_{\delta(x)}$, where $\delta(x)$ is given by

(6) $$\delta(x) = \sqrt{\mu c_1/(c_2-x)} .$$

Proof: We write $G_{ij}(S,x)$, $S = \{t_k\}$, in the form

$$G_{ij}(S,x) = (i+1)c_1 + \bar{c}_2\delta_i + (\mu - t_i)c_1 a_{ij} ,$$

where $\bar{c}_2 = c_2 - x$ and

$$a_{ij} = \frac{(j-i)c_1 - \bar{c}_2(\delta_j - \delta_i)}{(t_j-t_i)c_1} , \quad 0 \leq i \leq s < j .$$

Let further be $a_i = \sup_{j > s} a_{ij}$ and $a = \min_{i \leq s} a_i$, $s = s(S)$.

Simple calculations yield

(7) $$G_{ij}(S_\delta, x) = (1 + \mu/\delta)c_1 + (c_2 - x)\delta \ .$$

Supposing there exists a strategy $S = \{t_k\}$ satisfying

(8) $$D(S,x) < D(S_{\delta(x)},x) \ .$$

Then we have because of (7)

$$(i+1)c_1 + \bar{c}_2\,\delta_i + (\mu - t_i)c_1 a < c_1 + \frac{\bar{c}_2}{a} + \mu c_1 a,$$

$0 \leq i \leq s$. From this inequalities we obtain by induction

(9) $$\delta_i < 1/a \ , \ 0 \leq i \leq s \ .$$

Let i_0 be defined by $a_{i_0} = a$. It is surely

(10) $$\frac{(j-i_0)c_1 + \bar{c}_2(\delta_j - \delta_{i_0})}{(t_j - t_{i_0})c_1} \leq a \ , \quad j > s \ .$$

Starting in (10) with $j = s + 1$ we inductively get from (9) that $\delta_j < \delta_{i_0} < 1/a$ for all $j > s$. Therefore, it must be $a = \sup a_{i_0 j} \geq$ $\geq 1/\delta_{i_0}$, condratictory to (9). Hence there exists no strategy S with the property (8). Moreover, the given proof yields the uniqueness of $S(x) = S_{\delta(x)}$. Thus the proof of the theorem is complete.

From this theorem and the lemma we get that the partial minimax strategy is strictly periodic with the inspection interval $\delta(x_0)$, where $\delta(x_0)$ is given by (6) with $x = x_0$. To obtain the partial minimax loss costs $x_0 = K(S^*)$, we compute

$$D(S_{\delta(x)},x) = 2\sqrt{\mu c_1(c_2 - x)} - x(\mu + d) + c_1 + c_2 d + c_3 \ .$$

Of course, it is $D(S_{\delta(0)},0) > 0$ and because of (4) also $D(S_{\delta(c_2)},c_2) < 0$. Hence a unique solution x_0 of $D(S_{\delta(x)},x) = 0$ exists. Solving this equation we get

$$x_0 = \frac{1}{\mu + d}\left[c_1 + c_2 d + c_3 - \frac{2\mu c_1}{\mu + d} + 2\sqrt{\frac{\mu c_1}{\mu + d}\left(c_2\mu - c_1 - c_3 + \frac{\mu c_1}{\mu + d}\right)} \right] .$$

AVAILABILITY

According to (1) the (long-run) availability A(S,F) of the system by applying the strategy $S = \{t_k\}$ is given by

$$A(S,F) = \frac{\mu}{\sum_{k=0}^{\infty} t_{k+1}(F(t_{k+1}) - F(t_k)) + d} .$$

Naturally, an availability of the system arbitrarily close to its maximum value $\mu/(\mu + d)$ we may secure by choosing strategies $S = \{t_k\}$ with $\sup_k \delta_k$ being sufficiently small. But in this case the expected inspection costs would tend to infinity. Hence the problem consists in achieving maximum availability under restrictions on the expected inspection costs by applying a proper inspection strategy.

Let be C(S,F) the expected inspection costs during a cycle. Then we have

$$C(S,F) = c_1 \sum_{k=0}^{\infty} \left[(k+1)\, F(t_{k+1}) - F(t_k)\right] .$$

For a given constant K, $c_1 < K < \infty$, and known F the problem of maximizing A(S,F) on condition that $C(S,F) \leq K$ has been solved by Beichelt (1977). It is, evidently, equivalent to the following one:

$$Z(S,F) = \sum_{k=0}^{\infty} t_{k+1}(F(t_{k+1}) - F(t_k)) \longrightarrow \min.$$

$$C(S,F) \leq K .$$

Since we have got only the information "$F \in \mathcal{F}_\mu$" on the lifetime distribution, we define $Z(S) = \sup_{F \in \mathcal{F}_\mu} Z(S,F)$ and $C(S) = \sup_{F \in \mathcal{F}_\mu} C(S,F)$ and state the following problem:

$$Z(S) \xrightarrow[S \in \mathcal{K}]{} \min., \tag{11}$$

where $\mathcal{K} = \{S;\ S \in \mathcal{S},\ C(S) = K\}$. We solve this problem by means of the LAGRANGE multiplier approach.

Lemma 2: Let be $M(S,y) = Z(S) + y(C(S) - K)$. For every y, $0 \leq y < c_2$, there exist a strategy $S^{(y)}$ satisfying

$$M(S^{(y)},y) = \min_{S \in \mathcal{S}} M(S,y).$$

If there exists $y_o \in \mathcal{Y} = \{y;\ S^{(y)} \in \mathcal{K}\}$ so that

$$M(S^{(y_o)},y_o) = \min_{y \in \mathcal{Y}} M(S^{(y)},y),$$

then $S^{(y_o)}$ is solution of optimization problem (11).

Proof: If S^* is solution of (11), then we have for every $y \geq 0$

$$Z(S^*) = \min_{S \in \mathcal{K}} Z(S) = \min_{S \in \mathcal{K}} M(S,y) \geq \min_{y \in \mathcal{Y}} \min_{S \in \mathcal{S}} M(S,y)$$
$$= \min_{y \in \mathcal{Y}} M(S^{(y)},y) = M(S^{(y_0)},y_0) = Z(S^{(y_0)}) \; .$$

By definition of S^* and because of $S^{(y_0)} \in \mathcal{K}$, it must be $S^* = S^{(y_0)}$, and the lemma is proved.

Analogous to (5) we have (without loss of generality be $c_1=1$)

$$Z(S) = \sup_{\substack{(i,j) \\ i \leq s < j}} \left[\frac{t_j-\mu}{t_j-t_i}\delta_i + \frac{\mu-t_i}{t_j-t_i}\delta_j \right] + \mu \quad \text{and}$$

$$C(S) = \sup_{\substack{(i,j) \\ i \leq s < j}} \left[\frac{t_j-\mu}{t_j-t_i}(i+1) + \frac{\mu-t_i}{t_j-t_i}(j+1) \right] .$$

Especially, we have for a strictly periodic strategy S_δ .

$$Z(S_\delta) = \delta + \mu \quad \text{and} \quad C(S_\delta) = \mu/\delta + 1.$$

It is of importance that $M(S,y)$ has the same functional structure as $D(S,x)$. Hence theorem 1 also secures the existence of a unique strictly periodic strategy $S^{(y)} = S_{\delta(y)}$ for all y, $0 \leq y < c_2$. Moreover, there exists evidently one strictly periodic strategy $S_{\delta(y_0)} \in \mathcal{K}$ (that means the set $\mathcal{Y}$ is not empty but contains only the element y_0). Thus the assumptions of lemma 2 are fulfiled. We summarize the results in the following theorem.

Theorem 2: There exists a strictly periodic inspection strategy S^* with the inspection interval $\delta^* = \mu/(K-1)$, which is solution of optimization problem (11). The corresponding "partial minimax availability" $A(S^*)$ is given by

$$A(S^*) = \frac{\mu}{\mu K/(K-1) + d} \; .$$

It is interesting that in case $d = 0$ the availability $A(S^*)$ does not depend on the expected system lifetime μ.

Within the same model, Platz (1976) has studied the behaviour of system availability in case of completely known lifetime distribution and without restrictions on inspection costs.

REFERENCES

Beichelt F. (1975) : Minimax inspection strategies for replaceable systems with partial information on lifetime distribution. Math. Operationsforsch. und Statistik 6(1975), No. 3, 479-492.

(1976) : Prophylaktische Erneuerung von Systemen Akademie-Verlag, Berlin 1976

(1977) : Maximum availability under cost restrictions Zeitschr. f. Angew. Mathematik und Mechanik 57(1977), No. 6, 347-349.

Hoeffding W.(1955) : The extrema of the expected value of a function of independent random variables. Ann. Math.Stat. 26(1955), No. 2, 268-275.

Platz, O. (1976) : Availability of a renewable, checked system. IEEE Transact. Reliability R-25(1976), No. 1, 56-58.

Proschan F. (1963) : Optimum checking procedures. Journ. Soc. Ind. Appl. Math. 4(1963), No.6, 1078-1095.
Hunter L. C.
Barlow R. E.

VEB Braunkohlenkombinat "Gustav Sobottka"
Department of Rationalization
GDR - 4256 Röblingen am See

BESTIMMUNG DES STICHPROBENUMFANGES FÜR TESTS IN DER MULTIPLEN LINEAREN REGRESSIONSANALYSE MODELL II

Jürgen Bock

Rostock

ABSTRACT

In multiple linear regression model II the regression coefficients are compared with zero by the t-test and simultaneously compared with zero by the F-test. A series for the computation of the risk of second kind β for this tests is derived, whose terms can be calculated bv recurrence formulas.

In a normal probability plot of β as a function of $\sqrt{f_2}$ (f_2-denominator degrees of freedom) for fixed risk of first kind α, number f_1 of regressors and coefficient of determination (square of the multiple are partial - multiple correlation coefficient) we get nearly straigth lines in the domain $0,01 \leq \beta \leq 0,5$, hence easy nomograms to determine the sample size.

This proves specially the goodness of the approximation formula

$$f_2 = \frac{(u_{1-\frac{\alpha}{2}} + u_{1-\beta})^2}{z^2}$$

in the case $f_1 = 1$, where z denotes Fishers z-transformation of the partial correlation coefficient.

ZUSAMMENFASSUNG

Der Test von einzelnen oder mehreren Regressionskoeffizienten gegen Null erfolgt in der multiplen linearen Regressionsanalyse Modell II mit dem t-Test bzw. F-Test. In der vorliegenden Arbeit wird eine Reihe zur Bestimmung des Risikos 2. Art ß dieser Tests abgeleitet, deren Glieder rekursiv errechnet werden können. Stellt man ß für vorgegebene Irrtumswahrscheinlichkeit α, vorgegebene Anzahl von Regressoren und vorgegebenen multiplen bzw. partiell-multiplen Korrelationskoeffizienten bzw. vorgegebenes theoretisches Bestimmtheitsmaß als Funktion von $\sqrt{f_2}$ (f_2-Nennerfreiheitsgrade) im Normalverteilungspapier dar, so ergeben sich in dem für die Versuchsplanung relevanten Bereich $0{,}01 \leq ß \leq 0{,}50$ näherungsweise Geraden und damit einfache Nomogramme zur Bestimmung des Stichprobenumfanges.

Speziell erweist sich die Approximationsformel $f_2 = (u_{1-\frac{\alpha}{2}} + u_{1-ß})^2 / z^2$, wobei z der nach der Fisherschen z-Transformation transformierte Wert des entsprechenden partiellen Korrelationskoeffizienten ist, als hinreichend genau, wenn ein einzelner partieller Regressionskoeffizient getestet wird.

1. EINLEITUNG

In der Versuchsplanung zur Regressionsanalyse Modell I wurden in den letzten Jahrzehnten bedeutsame Fortschritte erreicht. Das betrifft vor allem Fragen der optimalen Wahl der Versuchspunkte, aber auch die Probleme bei der Bestimmung des Stichprobenumfanges. Letztere werden detailliert in den Büchern von Cohen J.(1969), Odeh R.E. and M. Fox (1975) und Rasch D., Herrendörfer G., Bock J. und K. Busch (1978) behandelt, wobei auch auf die praktische Bedeutung dieser Ergebnisse verwiesen wird.

Obwohl die Regressionsanalyse Modell II viele praktische Anwendungen findet, wurde der zugehörigen Versuchsplanung keine allzu große Aufmerksamkeit geschenkt. Im Falle des Modells II geht es um den Zusammenhang zwischen den Komponenten eines Zufallsvektors $\underline{\mathfrak{z}}' = (\underline{x}_0, \underline{x}_1, \ldots, \underline{x}_k)$, $\mathfrak{z} \in R^{k+1}$. Dabei wird eine, das sei ohne Beschränkung der Allgemeinheit $\underline{y} = \underline{x}_0$, als Zielgröße (Regressand) herausgegriffen. Die Gestalt des Zusammenhanges wird durch die Regressionsfunktion $y(x_1,\ldots,x_k) = E(\underline{y}|x_1,\ldots,x_k)$ beschrieben, seine Stärke durch verschiedene Korrelationsmaße. Wir beschränken uns hier auf den Fall, daß $\underline{\mathfrak{z}}$ mit dem Erwartungswertvektor $\mu' = (\mu_0,\ldots,\mu_k)$ und der nichtsingulären Kovarianzmatrix $\Sigma = (\sigma_{ij})$, $\sigma_{ii} = \sigma_i^2$, $(i,j = 0,1,\ldots,k)$, normalverteilt ist. Dann ist die Regressionsfunktion linear:

$$y(x_1,\ldots,x_k) = ß_0 + ß_1 x_1 + \ldots + ß_k x_k \tag{1}$$

Da die Einflußgrößen (Regressoren) $\underline{x}_1,\ldots,\underline{x}_k$ Zufallsvariable sind, ist hier die Frage nach der optimalen Allokation der Versuchspunkte nicht sinnvoll. Die Versuchsplanung beschränkt sich auf die Wahl der Schätzfunktionen und Tests sowie die Bestimmung der zugehörigen Stichprobenumfänge. Diese Arbeit bezieht sich auf die verschiedenen Varianten des F-Tests.

2. TEST AUF UNABHÄNGIGKEIT VON DEN REGRESSOREN

Für n unabhängige Beobachtungen $\underline{\mathfrak{z}}'_l = (\underline{y}_l, \underline{x}_{1l}, \dots, \underline{x}_{kl})$, $(l=1,\dots,n)$ des Zufallsvektors $\underline{\mathfrak{z}}$ gilt unter den obigen Voraussetzungen das Modell

$$\underline{\mathfrak{y}} = \underline{X}\,\vec{\beta} + \underline{\mathfrak{e}} \tag{2}$$

mit dem Vektor $\underline{\mathfrak{y}}' = (\underline{y}_1, \dots, \underline{y}_n)$ der Beobachtungen des Regressanden, der zufälligen Matrix

$$\underline{X} = \begin{pmatrix} 1 & \underline{x}_{11} & \cdots & \underline{x}_{k1} \\ \\ 1 & \underline{x}_{1n} & \cdots & \underline{x}_{kn} \end{pmatrix} \tag{3}$$

dem Vektor $\vec{\beta} = (\beta_0, \dots, \beta_k)$ der Regressionskoeffizienten und dem mit dem Erwartungsvektor $\mathcal{O}$ und der Kovarianzmatrix $\sigma^2 E_n$ normalverteilten Vektor der Zufallsabweichungen $\underline{\mathfrak{e}} = (\underline{\varepsilon}_1, \dots, \underline{\varepsilon}_n)$. Dabei bezeichnet E_n die Einheitsmatrix der Ordnung n, $\sigma^2 = \frac{\Lambda}{\Lambda_{oo}}$ die Restvarianz, Λ die Determinante von Σ und Λ_{oo} den Kofaktor von σ_o^2 in Λ. Wir beschränken uns auf den Bereich $\det X'X \neq 0$.

Der Vektor der Regressionskoeffizienten wird durch

$$\underline{\mathfrak{b}} = (\underline{X}'\underline{X})^{-1}X'\underline{\mathfrak{y}} \quad , \quad \underline{\mathfrak{b}}' = (\underline{b}_o, \dots, \underline{b}_k) \tag{4}$$

geschätzt.

Da wir Normalverteilung vorausgesetzt haben, ist $\underline{y}$ genau dann von den Regressoren unabhängig, wenn der multiple Korrelationskoeffizient

$$\rho = \rho_{yx_1 \dots x_k} = \sqrt{1 - \frac{\Lambda}{\sigma_o^2 \Lambda_{oo}}} \tag{5}$$

bzw. sein Quadrat, das sogenannte theoretische multiple Bestimmtheitsmaß, verschwindet. Als Schätzung für ϱ^2 verwendet man das (geschätzte) multiple Bestimmtheitsmaß

$$(6) \qquad \underline{B} = \frac{\underline{SQ}_{RA}}{\underline{SQ}_y}$$

mit

$$\underline{SQ}_y = \underline{y}'(E_n - \frac{1}{n} e_n e'_n)\underline{y} \qquad e'_n = (1,\dots,1)$$

und

$$\underline{SQ}_{RA} = \underline{y}'(\underline{X}(\underline{X}'\underline{X})^{-1} \underline{X}' - \frac{1}{n} e_n e'_n)\underline{y}$$

Der Test der Nullhypothese $H_o : \varrho^2 = 0$ gegen die Alternativhypothese $H_A : \varrho^2 > 0$ erfolgt anhand der Teststatistik

$$(7) \qquad \underline{F} = \frac{\underline{B}}{1-\underline{B}} \quad \frac{n-k-1}{k}$$

Dieser Test wird auch zum Prüfen der Nullhypothese $H_o : \beta_1 = \dots = \beta_k = 0$ gegen die Alternativhypothese, daß mindestens einer der Regressionskoeffizienten von Null verschieden ist, verwandt.

Unter H_o ist $\underline{F}$ mit $f_1 = k$ und $f_2 = n-k-1$ Freiheitsgraden zentral F-verteilt, so daß als kritischer Wert zur Irrtumswahrscheinlichkeit α das $(1-\alpha)$-Quantil $F(f_1,f_2,1-\alpha)$ dieser Verteilung zu verwenden ist.

3. BESTIMMUNG DES STICHPROBENUMFANGES

Zur Bestimmung des Stichprobenumfanges benötigt man die nichtzentrale Verteilung der Teststatistik.

In der Arbeit von Bock J.(1977) wurde gezeigt, daß im Falle $k = 1$ die Teststatistik

$$(8) \quad \underline{t} = \frac{\underline{b}_1 - \beta'_1}{\underline{s}_R} \sqrt{\underline{SQ}_x}, \quad \underline{SQ}_x = \sum_{l=1}^{n} \underline{x}_l^2 - \frac{(\sum_{l=1}^{n} \underline{x}_l)^2}{n} \quad (\underline{x} = \underline{x}_1)$$

wobei $\underline{s}_R^2$ die geschätzte Restvarianz bezeichnet, eine Verteilung mit der Dichtefunktion

$$(9) \qquad f(t,n-2) = C_{n\Delta} \int_0^1 \frac{w^{n-2}}{\sqrt{1-w^2}\,(1 - \frac{\Delta}{\sqrt{\Delta^2+1}} \; \frac{t}{\sqrt{n-2+t^2}} \, w)^{n-1}} \, dw$$

mit

$$C_{n\Delta} = \frac{(n-2)^{\frac{n-2}{2}}\ (n-2)!}{\sqrt{\pi}\, 2^{n-3}\, \Gamma\left(\frac{n-1}{2}\right)\Gamma\left(\frac{n-2}{2}\right)(\Delta^2+1)^{\frac{n-1}{2}}\ (n-2+t)^{2\ \frac{n-1}{2}}}$$

besitzt. β' ist eine vorgegebene Konstante, im Falle $\beta' = 0$ gilt $\underline{t}^2 = \underline{F}$ und $\Delta = \rho_1 \mid \sqrt{1-\rho_1^2}$, ansonsten

$$\Delta = \frac{(\beta_1 - \beta'_1)}{\sigma}\sigma_x$$

mit der Varianz σ_x^2 von $\underline{x}$. ρ_1 bezeichnet den einfachen Korrelationskoeffizienten. Daraus folgt, daß

$$\underline{r} = \underline{t}/\sqrt{n-2+\underline{t}^2}$$

die Dichte

$$f(r) = \frac{n-2}{\pi}(1-r^2)^{\frac{n-4}{2}}(1-\rho^{*2})^{\frac{n-2}{2}}\int_0^1 \frac{v^{n-2}}{\sqrt{1-v^2}(1-\rho^* vr)^{n-1}}dv \tag{10}$$

mit $\rho^* = \Delta/\sqrt{\Delta^2+1}$ besitzt. Diese hat dieselbe Gestalt wie die Dichte des geschätzten einfachen Korrelationskoeffizienten.

Im Falle $\beta_1 = 0$ ist $\rho^* = \rho_1$. Die Fishersche z-Transformation

$$z(\underline{r}) = \frac{1}{2}\ln\frac{1+\underline{r}}{1-\underline{r}} \tag{11}$$

führt zu einer approximativ normalverteilten Zufallsvariablen mit dem Erwartungswert $z(\rho^*) + \frac{\rho^*}{2(n-1)}$ und der Varianz $1/(n-3)$. Geht man wie in der genannten Arbeit vor, und vernachlässigt man Glieder höherer Ordnung, so erhält man für den Stichprobenumfang die Approximationsformel

$$f_2 = n-2 \approx \frac{(u_{1-\frac{\alpha}{2}} + u_{1-\beta})^2}{z^2(\rho^*)} \tag{12}$$

Im Spezialfall $\beta'_1 = 0$ liefert (12) den Stichprobenumfang für den in Abschnitt 2 beschriebenen Test (für $k = 1$).

Prinzipiell kann die nichtzentrale Verteilung von $\underline{F}$ für alle ganzzahligen $k \geq 1$ aus der Verteilung des geschätzten multiplen Korrelationskoeffizienten $\underline{R} = \sqrt{\underline{B}}$ hergeleitet werden. Diese wurde bereits 1928 von R.A. Fisher angegeben. Zu ihrer numerischen Behandlung wurden eine Anzahl von Approximationen entwickelt, über die Yoong-Sin Lee (1970) einen Überblick gibt.

Wir wollen den Stichprobenumfang in Abhängigkeit vom Risiko 1. Art α, vom Risiko 2. Art β und von ρ^2 bestimmen. Daher benötigen wir nur den Wert der Verteilungsfunktion der nichtzentralen Verteilung von $\underline{F}$ an der Stelle $F(f_1,f_2,1-\alpha)$. Dafür werden wir eine Reihenentwicklung ableiten, indem wir zunächst das Risiko 2. Art β_X für den Test in der bedingten Verteilung für festes X bestimmen und dann den Erwartungswert bilden.

Die bedingte Verteilung von $\underline{F}$ für feste Matrix X ist - wie aus der Theorie für das Modell I der Regressionsanalyse hervorgeht - eine nichtzentrale F-Verteilung mit f_1 und f_2 Freiheitsgraden und dem Nichtzentralitätsparameter

(13)
$$\lambda_X = \frac{1}{\sigma^2}\beta' X'(X(X'X)^{-1}X' - \frac{1}{n} e_n e_n')X\beta = \frac{1}{\sigma^2}\beta' X'(E_n - \frac{1}{n} e_n e_n') X \beta \quad .$$

Aus einer Reihenentwicklung der Verteilungsfunktion der nichtzentralen F-Verteilung mit Hilfe der unvollständigen Betafunktion ergibt sich

(14)
$$\beta_X = P(\underline{F} < F(f_1,f_2,1-\alpha) \mid X) = P(\underline{B} < B_{1-\alpha} \mid X) = \sum_{j=0}^{\infty} \frac{e^{-\frac{\lambda_X}{2}} \lambda_X^j}{2^j \, j!} I_{B_{1-\alpha}}(\frac{f_1}{2} + j, \frac{f_2}{2})$$

mit

$$B_{1-\alpha} = \frac{f_1 F(f_1,f_2,1-\alpha)}{f_2 + f_1 F(f_1,f_2,1-\alpha)} \quad .$$

$I_x(p,q)$ bezeichnet die unvollständige Betafunktion

$$I_x(p,q) = \frac{1}{B(p,q)} \int_0^x t^{p-1}(1-t)^{q-1} dt$$

mit der Betafunktion

$$B(p,q) = \int_0^1 t^{p-1}(1-t)^{p-1} dt$$

Die Komponenten $\beta_o + \sum_{i=1}^{k} \beta_i \underline{x}_{il}$ $(l=1,\ldots,n)$ des Vektors $\underline{X}\beta$ sind unabhängig voneinander mit dem Erwartungswert μ_o und der Varianz

$$\delta^{*2} = \sum_{i=1}^{k} \sum_{j=1}^{k} \beta_i \beta_j \sigma_{ij} = \sigma_o^2 - \sigma^2 = \sigma^2 \frac{\rho^2}{1-\rho^2}$$

(siehe z.B. Cramer H. (1951) normalverteilt. Daher ist $\frac{\sigma^2}{\delta^{*2}} \lambda_X$ mit n-1 Freiheitsgraden zentral χ^2-verteilt.

Nutzt man das aus, so ergibt sich

$$(15) \qquad w_j = E\left(\frac{e^{\frac{-\lambda_X}{2}} \lambda_X^j}{2^j j!}\right) = (1-\rho^2)^{\frac{n-1}{2}} \frac{\Gamma\left(\frac{n-1}{2}+j\right)}{\Gamma\left(\frac{n-1}{2}\right)} \frac{\rho^{2j}}{j!}$$

und damit durch gliedweise Integration aus (14) die Reihe

$$(16) \qquad \beta = E(\underline{\beta}_X) = \sum_{j=0}^{\infty} w_j \, I_{B_{1-\alpha}}\left(\frac{f_1}{2}+j, \frac{f_2}{2}\right) .$$

für das Risiko 2. Art, in der die Koeffizienten w_j Einzelwahrscheinlichkeiten einer negativen Binomialverteilung darstellen. Für sie gelten die Rekursionsgleichungen

$$(17) \qquad \begin{aligned} w_o &= (1-\rho^2)^{\frac{n-1}{2}} \\ w_j &= \frac{\left(\frac{n-1}{2}+j-1\right)}{j} \rho^2 \, w_{j-1} \qquad (j = 1,2,\ldots). \end{aligned}$$

Aus partieller Integration folgt die rekursive Beziehung

$$(18) \qquad I_x(p+j,q) = I_x(p+j-1,q) - A_j$$

mit

$$(19) \qquad A_j = \frac{x^{p+j-1}(1-x)^q}{(p+j-1)\, B(p+j-1,q)}$$

und daraus

$$(20) \qquad I_x(p+j) = I_x(p,q) - (A_1 + \ldots + A_j)$$

Verwendet man die Gleichung

$$(21) \qquad B(p+j-1,q) = \frac{(p+j-2)}{(p+q+j-2)} B(p+j-2,q)$$

so können die A_j rekursiv aus

$$(22) \qquad A_1 = \frac{x^p(1-x)^q}{p\, B(p,q)}, \quad A_j = x\frac{p+q+j-2}{p+j-1} A_{j-1} \quad (j = 2,3,\ldots)$$

berechnet werden. Auch B(p,q) kann rekursiv gewonnen werden. Wie einne Umformung des entsprechenden Integrals zeigt, ist

$$I_{B_{1-\alpha}}\left(\frac{f_1}{2}, \frac{f_2}{2}\right)$$

gleich dem Wert der Verteilungsfunktion der zentralen F-Verteilung mit f_1 und f_2 Freiheitsgraden an der Stelle $F(f_1, f_2, 1-\alpha)$, also gleich $1-\alpha$. Damit folgt aus (16) und (20) für $p = \frac{f_1}{2}$, $q = \frac{f_2}{2}$ und $x = B_{1-\alpha}$

$$\beta = 1 - \alpha - \sum_{j=1}^{\infty} w_j (A_1 + \ldots + A_j) \; . \tag{23}$$

Da die Glieder dieser Reihe rekursiv berechnet werden können, ist sie leicht auf dem Computer zu handhaben. Mit einem von K.D.Feige erarbeiteten Programm wurde eine Tabelle $\alpha = 0{,}05$, $\rho^2 = 0{,}001$, $0{,}01$, $0{,}05(0{,}05)0{,}5$, $f_1 = 1(1)10$ und $f_2 = 2(2)50(5)100(10)200(20)300(50)500$, 1000 berechnet.

Die Darstellung der Ergebnisse erfolgte in Normalverteilungspapier, da aus (12) $\beta \approx \Phi(u_{1-\frac{\alpha}{2}} - |z(\rho_1)| \sqrt{f_2})$ folgt ($\Phi(u)$ bezeichnet die Verteilungsfunktion der standardisierten Normalverteilung) und sich zumindest im Falle $k = 1$ näherungsweise eine Gerade ergeben muß, wenn die Abszisse nach $\sqrt{f_2}$ geteilt wird. In Abbildung 1 sind einige der Kurven dargestellt.

Es zeigt sich, daß (12) hinreichend genau den Stichprobenumfang approximiert. Auch für $k = 2, \ldots, 10$ ergeben sich in dem für die Versuchsplanung interessanten Bereich $0{,}01 \leq \beta \leq 0{,}5$ in guter Näherung Geraden, so daß der Stichprobenumfang leicht abgelesen werden kann. Wegen der Monotonie der Gütefunktion bzw. des Risikos erfüllt der Test zu dem aus der Abb.1 abgelesenen Stichprobenumfang die folgende Genauigkeitsforderung: Ist das theoretische Bestimmtheitsmaß größer als der vorgegebene Wert ρ^2, so sind die Risiken 1. und 2. Art kleiner gleich den vorgegebenen Werten α bzw. β.

Aufgrund der Beziehung $\sum_{i=1}^{k} \sum_{j=1}^{k} \beta_i \beta_j \sigma_{ij} = \sigma^2 \frac{\rho^2}{1-\rho^2}$

kann die Genauigkeitsforderung auch für die Regressionskoeffizienten formuliert werden.

4. TESTS FÜR PARTIELLE HYPOTHESEN

Greift man $m < k$ der Regressionskoeffizienten $\beta_1, \ldots, \beta_k$ heraus,

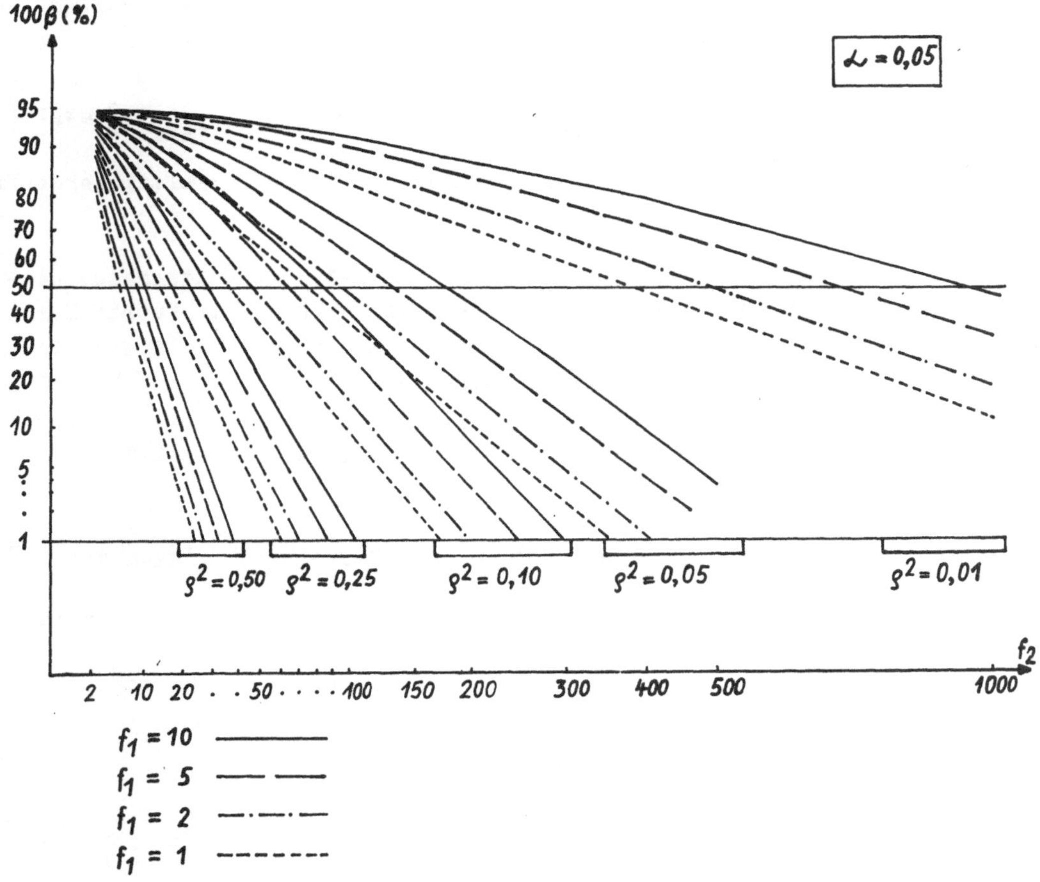

ABB. 1

Risiko 2. Art β für den F-Test, mit den Freiheitsgraden f_1, f_2 zum Vergleich von Regressionskoeffizienten mit Null.

und soll geprüft werden, ob diese sich von Null unterscheiden, so lautet die Nullhypothese nach entsprechender Umnumerierung der Regressoren und Regressionskoeffizienten $H_o = \beta_{k-m+1} = \dots = \beta_k = 0$. Die Alternativhypothese ist die, daß mindestens einer der betrachteten Koeffizienten von Null verschieden ist. Es wird ein F-Test mit der Teststatistik

$$(24) \qquad \underline{F} = \frac{n-k-1}{m} \quad \frac{\underline{y}'(\underline{X}(\underline{X}'\underline{X})^{-1}X' - \underline{X}_o(\underline{X}'_o\underline{X}_o)^{-1}\underline{X}'_o)\underline{y}}{\underline{y}'(E_n - \underline{X}(\underline{X}'\underline{X})\ \underline{X})\underline{y}}$$

durchgeführt. Dabei geht man von einer Unterteilung der Matrix X in der Form $(X_o|X_1)$ aus, in der X_o die aus den ersten k-m+1 Spalten von X gebildete Teilmatrix bezeichnet. (Wir beschränken uns auf den Bereich det $X'X \neq 0$ und det $X_o'X_o \neq 0$.) Mit diesem Test werden auch die Hypothesen $H_o: \varrho_{y|x_1...x_{k-m}} = 0$ und $H_A: \varrho^2_{y|x_1...x_{k-m}} > 0$ geprüft. $\varrho^2_{y|x_1...x_{k-m}}$ bezeichnet den partiell multiplen Korrelationskoeffizienten.

Unter H_o ist $\underline{F}$ mit $f_1 = m$ und $f_2 = n-k-1$ Freiheitsgraden zentral verteilt. Die bedingte Verteilung von $\underline{F}$ für feste Matrix X ist unter H_A eine nichtzentrale F-Verteilung mit f_1 und f_2 Freiheitsgraden und dem Nichtzentralitätsparameter

$$\lambda_{X,X_o} = \frac{1}{\sigma^2} \vec{\beta}'X' (X(X'X)^{-1}X' - X_o(X_o'X_o)^{-1}X'_o) X \vec{\beta} \tag{25}$$

$$= \beta^{*\prime}X'_1(E_n - X_o(X'_oX_o)^{-1} X'_o)X_1\beta^*, \; \beta^* = (\beta_{k-m+1},...,\beta_k)$$

Die Komponenten des Zufallsvektors $\underline{X}_1\beta^*$ sind unter der Bedingung, daß X_o fest ist, mit dem Erwartungswert $\mu^* = \beta_{k-m+1} x_{k-m+1} + ... + \beta_k x_k$ und der Varianz

$$\tilde{\sigma}^2 = \sum_{i=k-m-1}^{k} \sum_{j=k-m-1}^{k} \mathrm{cov}(\underline{x}_i, \underline{x}_j \mid x_1,...,x_{k-m}) \, \beta_i\beta_j$$

unabhängig voneinander normalverteilt. Da die Matrix $E_n - X_o(X'_oX_o)^{-1}X_o'$ idempotent ist, ist die bedingte Verteilung von $\underline{\lambda}_{X,X_o}$ für festes X_o eine χ^2-Verteilung mit dem Nichtzentralitätsparameter

$$\lambda = \mu^{*2} e_n' (E_n - X_o(X_o'X_o)^{-1} X_o') e_n$$

Das Gleichungssystem $X_o'X_o c = X_o' e_n$ hat die eindeutige Lösung $c = (1,0,...,0)$. Daher folgt aus $n = e'_n X_o c = e_n' X_o(X_o'X_o)^{-1}X_o'e_n$ $\lambda = 0$, und sowohl die bedingte als auch die unbedingte Verteilung von $\frac{\sigma^2}{\tilde{\sigma}^2} \underline{\lambda}_{X,X_o}$ ist eine zentrale χ^2-Verteilung mit n-k+m-1 Freiheitsgraden.

Die weitere Vorgehensweise ist dieselbe wie in Abschnitt 3. Ersetzt man in (17) ϱ^2 durch $\varrho^2_{y|x_1...x_{k-m}}$ und $n-1$ durch $(f_1 + f_2)/2$, so gilt wiederum (23) mit $p = m/2$ und $q = (n-k-1)/2$. Man kann also den Stichprobenumfang mit Hilfe von Abb. 1 bestimmen.

Speziell ergibt sich der Stichprobenumfang für den Test eines einzelnen partiellen Regressionskoeffizienten für $f_1 = 1$, und es gilt die Approximationsformel (12) mit dem entsprechenden partiellen Regressionskoeffizienten anstelle von ρ_1.

LITERATUR

Bock J. (1977): "Sample size determination for the comparison of the regression coefficient with a constant in the case of simple linear regression model II" Biom. Journ., Vol.19, No.1, pp. 23-29

Cohen J. (1969): "Statistical Power Analysis for the Behavioural Sciences" Academic Press, New York

Cramer H. (1951): "Mathematical Methods of Statistics" Princeton University Press, Princeton

Fisher R.A. (1928): "The general sampling distribution of the multiple correlation coefficient" Proceed. of the Royal Stat.Soc., London, A, 121, p.654

Odeh R.E. and M. Fox (1975): "Sample Size Choice: Charts for Experiments with Linear Models" Marcel Dekker Inc., New York

Rasch D., Herrendörfer G., Bock J. und K.Busch (1978): "Verfahrensbibliothek Versuchsplanung und -auswertung" VEB Deutscher Landwirtschaftsverlag, Berlin

Yoong-Sin Lee (1971): "Some Results on the Sampling Distribution of the Multiple Correlation Coefficient" Journ.Roy.Stat.Soc.B. 33, No. 1, pp. 117-136

Akademie der Landwirtschaftswissenschaften
der Deutschen Demokratischen Republik
Abt. Biomathematik und EDV

Forschungszentrum für Tierproduktion
2551 Dummerstorf, Kreis Rostock

MARRIAGE BETWEEN THE SUPPLEMENTARY VARIABLE TECHNIQUE AND THE IMBEDDED MARKOV CHAIN TECHNIQUE-I

M L Chaudhry

Kingston

ABSTRACT

Although the main purpose of this paper is to derive relations between the probability generating functions of the number in the system for the queueing process $M/G^k/1$ in which the size of service batch is fixed at three epochs of time - random, just after departure and just before arrival, the method discussed is general and can be applied to several other queueing processes in which the size of arrival groups or of service groups is variable. In addition, usefulness of the mathematical expressions is illustrated by indicating as to how some of the parameters of the process under consideration may be evaluated.

INTRODUCTION

When the service time distribution or the interarrival time distribution is arbitrary, the steady-state solutions to queueing processes (in continuous time) have been obtained by using, among other techniques, the technique of supplementary variable or renewal theoretic arguments. To eliminate the non-Markovian aspect of stochastic processes occurring in the theory of queues, Kendal (1951, 1953) suggested and used the imbedded Markov chain technique. Since then the technique has been used by several other researchers.

In the theory and application of queueing processes one, sometimes, needs to know the probabilities at three epochs of time - random, just after departure and just before arrival. For example, an arriving customer may be interested more in the distribution of the number in the system (or in the queue) at his arrival epoch than in the other two distributions. This distribution, besides giving other information, is not only useful for determining the average number in the system at an arrival epoch, but also is used to determine his waiting time distribution or its average. Foster and Perera (1964), using renewal theoretic arguments, establish relations between steady-state probability generating functions (p.g.f.'s) of the number in the system at three epochs of time mentioned earlier for the queueing processes $M/G^k/1$. In the system $M/G^k/1$ customers following a Poisson process arrive singly and are served in groups of exactly size k by a single server whose service time distribution is arbitrary. It appears difficult to obtain similar relations by the procedure used by Foster and Perera for the more general queueing processes in which either the size of arrival groups or of service groups is variable.

Although some new results for the more general systems mentioned above have been obtained by using the procedure discussed below, the main purpose of this paper is not only to give another application of the procedure used by Chaudhry and Templeton (1978) for finding the relation between the p.g.f.'s of the distributions P_n, P_n^+ for the queueing process $M/G^B/1$ in which server's capacity is B, but to discuss some other interesting questions. In addition, the usefulness of the mathematical expressions is illustrated by indicating as to how some of the parameters of the process under consideration may be evaluated. It is hoped that these relations should prove useful to the practitioners of queueing theory. To discuss the procedure and other pertinent questions, we wish to consider here the queueing process $M/G^k/1$ for which some results are known.

THE QUEUEING PROCESS $M/G^k/1$

$P(z) = \sum_{n=0}^{\infty} P_n z^n$

Let $N(t)$ be a random variable (r.v.) representing the number in the system and $X(t)$ the elapsed service time of the group under service at time t. In the limiting case as $t \to \infty$, $N(t)$ shall be replaced by N. The same shall apply to the r.v.'s to be introduced later. Further, let the customers arrive at epochs $0 = \sigma_0'$, $\sigma_1', \ldots, \sigma_r', \ldots$. The interarrival times, $\sigma_n' - \sigma_{n-1}'$, $n = 1,2,\ldots$, are independently identically distributed (i.i.d.) exponential r.v.'s with mean $1/\lambda$. Let $\sigma_1, \sigma_2, \ldots, \sigma_n, \ldots$ be the epochs of departures. The service times, $\sigma_{n+1} - \sigma_n$, $n = 1,2,\ldots$, of groups are i.i.d. r.v.'s with common density $b(v)$ such that

$$1/\mu = \int_0^{\infty} v\, b(v)\, dv .$$

To get $P(z)$, the p.g.f. of the number in the system in the steady-state case, we introduce the following notations and probabilities.

(a) $\eta(x)$ is the conditional service rate so that the service time density and distribution functions are, respectively, given by

$$b(x) = \eta(x) \exp [- \int_0^x \eta(t)dt]$$

and

$$B(x) = 1 - \exp [- \int_0^x \eta(t)dt]$$

(b) $$P_n(x,t) = P[N(t) = n , x<X(t)\leq x+dx] , \quad n\geq k$$

(c) $$P_n(t) = P[N(t) = n , \text{server idle}] , \quad n<k$$

Besides, let

(i) $$P_n^- = \lim_{n\to\infty} P[N(\sigma_n'-0) = n] .$$

This means P_n^- is the limiting probability (as $n\to\infty$) of n being in the system just before an arrival epoch.

(ii) $$P_n^+ = \lim_{n\to\infty} P[N(\sigma_n+0) = n] .$$

This means P_j^+ is the limiting probability (as $n\to\infty$) of n being in the system just after a departure epoch.

Assuming that the various limiting probabilities exist which they do when $\rho = \lambda/k\mu < 1$, we first find the probability generating function (p.g.f.), $P(z) = \sum_{n=0}^{\infty} P_n z^n$, of P_n and from it the p.g.f. of P_n^+. That the p.g.f. of P_n^- equals the p.g.f. of P_n follows from intuitive arguments.

Adapting the arguments as given in Cox (1955) or Chaudhry and Templeton (1978), we find the p.g.f., $P(z)$ of $N(t)$ in the limiting case as $t\to\infty$. Since the arguments are the same, we only give the partial differential-difference equations in the case when $t\to\infty$ with the notation

$$\lim_{t\to\infty} P_n(x,t) = P_n(x) , \quad \lim_{t\to\infty} P_n(t) = P_n .$$

Perhaps we could have abridged a few more steps but we are not doing so for the sake of completeness.

(1) $$0 = -\lambda P_n + \lambda P_{n-1} + \int_0^\infty P_{n+k}(x) \eta(x)dx , \quad 1\leq n<k$$

(2) $$0 = -\lambda P_0 + \int_0^\infty P_k(x) \eta(x)dx$$

(3) $$\frac{\partial P_n(x)}{\partial x} = -(\lambda+\eta(x)) P_n(x) + \lambda P_{n-1}(x) , \quad n\geq k$$

which are to be solved under the boundary conditions,

(4) $$P_n(0) = \int_0^\infty P_{n+k}(x)\ \eta(x)dx\ , \qquad n>k$$

(5) $$P_k(0) = \int_0^\infty P_{2k}(x)\ \eta(x)dx + \lambda\ P_{k-1}$$

It may be noted that since $P_n(x) = 0$, $0 \le n < k$, there will also be no contribution to the boundary conditions when $n<k$.

Define the p.g.f. of $P_n(x)$ by

(6) $$P(z;x) = \sum_{n=k}^{\infty} P_n(x)z^n$$

From (3) and (6), as usual, one gets the solution

$$P(z;x) = A(z)\ e^{-\lambda(1-z)x}\ (1-B(x))\ ,$$

where $A(z)$ is independent of x and is given by

$$P(z;0) = A(z)\ .$$

Consequently,

$$P(z;x) = P(z;0)\ e^{-\lambda(1-z)x}\ (1-B(x))\ .$$

Whence

(7) $$P(z) \equiv \sum_{n=0}^{\infty} P_n z^n = \int_0^\infty P(z;x)dx + \sum_{n=0}^{k-1} P_n z^n$$

$$= P(z;0)\ \frac{1-\bar{b}(\lambda-\lambda z)}{\lambda-\lambda z} + \sum_{n=0}^{k-1} P_n z^n\ .$$

To find $P(z;0)$, first use (6) when $x=0$, (4) and (5); then simplify the expression by using (1) and (2). The result is

(8) $$P(z;0) = \frac{\lambda\ z^k(z-1)\ \sum_{n=0}^{k-1} P_n z^n}{z^k - \bar{b}(\lambda-\lambda z)}\ .$$

Substituting (8) in (7) and after a bit of simplification, we finally get

(9) $$P(z) = \frac{(z^k-1)\ \sum_{n=0}^{k-1} P_n z^n}{[\{z^k/\bar{b}(\lambda-\lambda z)\}-1]}$$

<u>$P^+(z)$</u>

Now to get $P^+(z)$, we have

$$P_n^+ = C \int_0^\infty P_{n+k}(x)\ \eta(x)dx\ , \qquad n\ge 0$$

where C is a normalizing constant. $P^+(z)$ then is given by

(10) $$P^+(z) \equiv \sum_{n=0}^{\infty} P_n^+ z^n = C\,\lambda \frac{(z-1)}{z^k-1} P(z) ,$$

Now using the normalizing condition, $P^+(1-) = P(1-)$, we get from (10), $C\,\lambda = k$.

Finally,

(11) $$P^+(z) = k \frac{(z-1)}{(z^k-1)} P(z) ,$$

or when written explicitly, it is given by

(12) $$P^+(z) = \frac{k(z-1) \sum_{n=0}^{k-1} P_n z^n}{[\{z^k/\bar{b}(\lambda-\lambda z)\}-1]} .$$

It may be mentioned that the k unknown probabilities, $P_0, P_1, \ldots, P_{k-1}$, contained in $P^+(z)$ or $P(z)$ may be evaluated in a manner similar to the one suggested by Foster and Nyunt (1961).

Although (12) is not yet expressed in terms of the k unknown probabilities P_i^+ , $0 \le i \le k-1$, it can be done as follows. For, as from (11)

(13) $$P_n = (1/k) \sum_{i=0}^{n} P_i^+ , \quad n \ge 0$$

(14) $$\sum_{n=0}^{k-1} P_n z^n = (1/k) \sum_{n=0}^{k-1} \sum_{i=0}^{n} P_i^+ z^j$$

$$= \sum_{i=0}^{k-1} P_i^+(z^i - z^k)/k(1-z) .$$

Substituting (14) in (12) gives

(15) $$P^+(z) = \sum_{i=0}^{k+1} (z^k - z^i)\, P_i^+/[\{z^k/\bar{b}(\lambda-\lambda z)\}-1] .$$

Further, (15) may be expressed as

(16) $$P^+(z) = k(1-\rho)(z-1) \prod_{i=1}^{k-1} \frac{(z-z_i)}{(1-z_i)} \Big/ [\{z^k/\bar{b}(\lambda-\lambda z)\}-1] ,$$

where z_i , $1 \le i \le k-1$ are the roots inside the unit circle of the equation

(17) $$z^k = \bar{b}(\lambda-\lambda z) .$$

By applying the normalizing condition, one can see either from (9) or from (12) that

$$\sum_{n=0}^{k-1} P_n = \sum_{i=0}^{k-1} P_i^+ \{1-(i/k)\} = 1-\rho$$

where the first relation is obtained by using (14).

The results (15) and (16) have been obtained alternatively by Foster and Nyunt (1961) whereas (10) has been obtained by Foster and Perera (1964). The distributions of P_n, P_n^+ have been independently discussed by Takács (1961). It may, however, be pointed out again that the procedure discussed above in getting the relation (11) between the p.g.f.'s of P_n and P_n^+ is more general than the one followed by the above authors and gives similar relations for many other cases, e.g., see Chaudhry and Templeton (1978).

$P_q(z)$

To get the p.g.f. of the number in the queue (excluding those in service), we proceed as follows:

$$\text{Let } N_q = \begin{cases} \text{a r.v. representing the number in queue} \\ \text{at a random epoch in the steady-state case.} \end{cases}$$

Then as

$$N_q = \begin{cases} N - k, & \text{if } N \geq k \\ N, & \text{if } N < k \end{cases}$$

$$P_q(z) = E\{z^{N_q}\} = P(z)/\bar{b}(\lambda-\lambda z) .$$

or

$$P(z) = P_q(z)\ \bar{b}(\lambda-\lambda z) . \tag{18}$$

The result (18) for the process $M/G^k/1$ is new and has an interesting interpretation. It shows that the p.g.f. of the number in the system at a random instant is equal to the p.g.f. of the number in queue times the p.g.f. of the number that arrive during the service time of a group.

WAITING TIME (IN QUEUE) DISTRIBUTION

Let $V_{q\ell}(x)$ be the waiting time (in queue) of the customer who arrives last (and hence has the shortest waiting time) among those served in the nth batch. Further, let

$$W_{q\ell}(\tau) = \lim_{n\to\infty} P(V_{q\ell}(n)\leq\tau) .$$

Consequently, if we define

$$w_{q\ell}(\alpha) = \int_0^\infty e^{-\alpha\tau}\, d\, W_{q\ell}(\tau) ,$$

then, see Takács (1962),

$$P^+(z) = w_{q\ell}(\lambda-\lambda z)\ \bar{b}(\lambda-\lambda z)$$

which can be written as

$$\bar{w}_{q\ell}(\alpha) = P^+(1-(\alpha/\lambda))/\bar{b}(\alpha) , \tag{19}$$

where $P^{+}(z)$ is given by (12), (15) or (16).

It may be remarked here that $\bar{w}_q(\alpha)$ is the Laplace-Stieltjes' transform of the conditional waiting time (in queue) distribution, conditional on the fact that we are considering the waiting time of that customer who arrives last in a group.

MEASURES OF EFFICIENCY

Now we wish to find the relation between L, average number in the system at a random epoch and L^{+}, the average number in the system at a departure epoch, etc. From (11) and (18) one gets

(20) $$L = L^{+} + \{(k-1)/2\} = L_q + (\lambda/\mu) ,$$

where

$$L = E(N) = \left. \frac{d\,P(z)}{dz} \right|_{z=1} , \quad \text{etc.}$$

Similarly one can show, using (11), that

$$\text{Var}\,(N) = \text{Var}\,(N^{+}) + 2(k-1)L^{+} + \{(k-1)(k-11)/12\}$$

(21) $$= \text{Var}\,(N_q) + \{\lambda(1-4L_q)/\mu\} + (\lambda\sigma_s)^2 ,$$

where $\sigma_s{}^2$ is the variance of service time distribution and we have used (20) in the second equation of (21).

One may observe that (20) shows that on the average the queue looks greater to an observer than to a departing group. For $k=1$, $L=L^{+}$, as it should be. The case $k=1$ was first discussed by Khintchine (1932) using another method and later by several other authors.

Average waiting time, $W_{q\ell}$, and the variance of the waiting time, $\sigma_{q\ell}{}^2$, of the customer who arrives last in a group may be obtained from (19) and are given by

(22) $$\lambda\, W_{q\ell} = L^{+} - k\rho$$

and

(23) $$\sigma_{q\ell}{}^2 = [\text{Var}(N^{+}) - L^{+} - (\lambda\sigma_s)^2]/\lambda^2 .$$

NUMERICAL ANALYSIS

One can observe from the equations (20), (21), (22) and (23) that once L^{+} and $\text{Var}(N^{+})$ are known, it is not difficult to get the other parameters such as L, L_q, etc. Consequently, we proceed to discuss the numerical evaluation of L^{+}, $\text{Var}(N^{+})$. The numerical analysis of L^{+} and $\text{Var}(N^{+})$ where E_k is k-Erlang distribution, for the system $M/E_k^B/1$, has been carried out by Downton (1955, 1956). In view of this, one can use Downton's numerical results to get the numerical values of L, L_q, etc. We have prepared more detailed tables for L^{+} and $\text{Var}(N^{+})$ than those given in Downton (1955). Persons interested in the tables may write to the author.

CONCLUDING REMARKS, FURTHER WORK AND ACKNOWLEDGEMENTS

It is possible to obtain (15) directly without using (13) and (14) if we define the state of the system as in Chaudhry and Templeton (1978). In a sequel, it is planned to extend and unify the work due to Foster and Perera (1964) and Chaudhry and Templeton (1978) who have extended and unified the results due to Bailey (1954) and Jaiswal (1960). It may be remarked further that Bhat (1964) and Fraser (1978) while discussing different processes observe that the distribution of P_n^+ for the processes $M/G^B/1$ and $M/G^k/1$ is the same.

Part of this work was done under the financial support of the Defence Research Board grant number 3610-603.

REFERENCES

Bailey, N.T.T. (1954): On queueing processes with bulk service. J. Roy. Statist. Soc., B16 (1954), 80-87.

Bhat, U.N. (1964): Imbedded Markov chain analysis of single server bulk queue. J. Austral. Math. Soc. Vol. 4 (1964), 244-262.

Chaudhry, M.L. and Templeton, J.G.C. (1978): The queueing system $M/G^B/1$ in continuous time and its relation to the corresponding departure epochs. To appear in the J. Roy. Statist. Soc.

Cox, D.R. (1955): The analysis of non-Markovian stochastic processes by the inclusion of supplementary variables. Proc. Cambridge Phil. Soc. 51 (1955), 433-441.

Downton, F. (1955): Waiting time in bulk queues. J. Roy. Statist. Soc., B17 (1955), 80-87.

________ (1956): On limiting distributions arising in bulk service queues. J. Roy. Statist. Soc., B18 (1956), 256-274.

Foster, F.G. and Nyunt, K.M. (1961): Queues with batch departures I. Ann. Math. Statist. Vol. 32 (1961), 1324-1333.

Foster, F.G. and Perera, A.G.A.D. (1964): Queues with batch departures II. Ann. Math. Statist. Vol. 35 (1964), 1147-1156.

Fraser, H. (1978): Some problems in the theory of bulk queues, thesis submitted to the Royal Military College of Canada, Kingston, Ontario, Canada.

Jaiswal, N.K. (1960): Bulk service queueing problems. Opns. Res. 8 (1960), 139-143.

Kendal, D.G. (1951): Some problems in the theory of queues. J. Roy. Statist. Soc., B13 (1951), 151-187.

__________ (1953): Stochastic processes occurring in the theory of queues and their analysis by the method of imbedded Markov chain. Ann. Math. Statist. Vol. 24 (1953), 338-354.

Khintchine, A. (1932): Mathematical theory of a stationary queue. Mat. Sbornik 39 (1932), 73-84 (Russian).

Takács, L.: Introduction to the theory of queues. Oxford University Press, New York, 1962.

Department of Mathematics,
Royal Military College of Canada,
Kingston, Ontario, Canada,
K7L 2W3.

OPTIMAL PARAMETER ESTIMATION IN TWO-PLAYER ZERO-SUM DIFFERENTIAL GAMES

Jaroslav Doležal

Prague

ABSTRACT

In addition to the classical formulation of a two-player zero-sum differential game it is assumed that the minimizing player chooses the values of certain parameters to further decrease the pay-off functional at the expense of the maximizing player. To solve this problem in a best possible way for the minimizing player, a set of necessary optimality conditions is derived, which enables not only to determine the saddle-point strategies for both participating players, but also the optimal parameters. Based on these conditions an iterative numerical algorithm of the gradient type is suggested. As an illustration, two examples are solved applying the described algorithm.

INTRODUCTION

As it is known from the optimal control theory the optimal behaviour of any dynamical system is, as a rule, a function of various system parameters. These parameters can be sometimes used to further improve the performance of a system in question. In the theory of optimal control is such class of problems usually denoted as parameter optimization or optimal parameter estimation. Here we shall show that this procedure is applicable also to general two-player zero-sum differential games. In fact, the further described results can be viewed as the extensions of the existing methods in optimal control theory.

At present time there exist several papers which contain rather deep results dealing with parameter optimization for control systems. The fundamental results in this field are due to Hofer and Sagirow (1968) and Boltjanskij (1969). Later Ahmed and Georganas (1973) showed that the results of Boltjanskij (1969) could be easily obtained applying the general maximum principal of Gamkrelidze (1965). Recently Georganas (1975) presented imbedding techniques for optimal parameter estimation, provided that the optimal parameter could be expressed analytically. Also Lunderstädt (1976) describes necessary optimality conditions for parameter optimization using the maximum principle approach.

Doležal and Černý (1976a) used the calculus of variations to obtain the analogical necessary optimality conditions and proposed a first-order gradient algorithm based on the so-called influence functions for the iterative solution of parameter optimization problems. For optimal control problems this algorithm was described by Bryson and Ho (1969). Moreover, the gradient algorithm of Doležal and Černý (1976a) enables also treatment of control and parameter constrained problems using the projection technique, e.g., see Vasiljev (1974). An alternative gradient-restoration approach can be found in a survey paper of Miele (1975).

The aim of this contribution is to show that the recent results of Doležal and Černý (1976a) can be extended also to the case of general differential games. It is assumed that the minimizing player has the opportunity to choose the values of certain parameters before the game starts. The question is, what values of these parameters should he choose to further increase his own gain at the minimizing player's expense? In turn, the minimizing player has to solve the "worst-case" analysis problem, i.e., to determine the greatest lower bound for his expected pay-off. Otherwise speaking, the saddle-point of a game in question depends on certain parameters and the minimizing player can choose the most favourable saddle--point due to his aims.

For the case of linear quadratic differential games such problem was studied by Leondes and Siu (1977). They suggested several numerical methods to solve the iterative parameter optimization. However, their approach was limited only to system parameters (elements of the transition matrix). Moreover, in the linear quadratic differential game it was possible to use a priori knowledge of the optimal strategies for both players.

In our approach both, optimal strategies (open-loop) and optimal values of parameters, are determined for a general nonlinear two-player zero-sum differential game during the course of calculations. The developed first-order gradient algorithm can handle also the problems with control and parameter constraints. This algorithm, not including parameter optimization, was used earlier by Doležal and Černý (1976b) to obtain a numerical solution of differential games.

In the following sections we first present problem formulation and necessary optimality conditions. Then the gradient algorithm mentioned above is described in detail. Finally, the solution of two concrete differential games with parameters is included which confirms the practical importance of the suggested algorithm.

During all computations the interactive simulation program, called SIMFOR, for the solution of two-point boudary-value problems developed by Černý (1975) is used. Such approach saves a lot of routine programmer's work and enables a direct use of the whole EAI PACER 600 computer system installed at the Institute of Information Theory and Automation.

PROBLEM FORMULATION

In this section we shall give a precise formulation of the studied problem. We assume that all vectors are the column vectors except of the gradients of various functions, which are always treated as row-vectors. Further, all further defined functions are supposed to be continuously differentiable. As E^n will be denoted n-dimensional Euclidean space. For the sake of simplicity only the problems with fixed final time will be studied, i.e., without any loss of generality we may assume that the independent variable (time) $t \in [0,1]$. The generalization of the next results to the case of free final time can be done according to the scheme of Doležal and Černý (1976b).

Now consider a two-player zero-sum differential game

$$\text{(1)} \qquad \dot{x} = f(x,u,v,a,t), \qquad x(0) = x_0, \qquad t \in [0,1]$$

where $x(t) \in E^n$ denotes the state and $u(t) \in E^m$, resp.

$v(t) \in E^q$, the control variable of the minimizing, resp. maximizing, player at the time t and $a \in E^r$ the parameter; function $f : E^n \times E^m \times E^q \times E^r \times E^1 \to E^n$. The aim of the minimizing player is to choose a strategy $u(t)$ and a parameter a to minimize the cost functional (payoff)

(2) $$J(u,v;a) = [\varphi(x,a)]_1 + \int_0^1 L(x,u,v,a,t)\,dt ,$$

while the aim of the maximazing player is to maximize J using a strategy $v(t)$. Here $\varphi : E^n \times E^r \to E^1$ and $L : E^n \times E^m \times E^q \times E^r \times E^1 \to E^1$. The lower indices 0 and 1 denote the evaluation of the corresponding expressions at $t = 0$ and $t = 1$.

Finally, the choices of the both players have to satisfy the following constraints

(3) $$u(t) \in U \subset E^m , \qquad v(t) \in V \subset E^q , \qquad t \in [0,1]$$

and

(4) $$a \in A \subset E^r .$$

The solution of the differential game (1) - (4) is the well--known saddle-point, i.e., such strategy pair (u^*, v^*) satisfying (3) for which

(5) $$J(u^*, v; a) \leqq J(u^*, v^*; a) \leqq J(u, v^*; a), \qquad a \in A .$$

Here u and v are any strategies satisfying the constraints (3). Further we assume that both, $u(t)$ and $v(t)$ are piecewice continuous functions of t. Now

(6) $$J^*(a) = J(u^*, v^*; a)$$

is the so-called value of the differential game (1) - (4) depending on the particular choice of $a \in A$ by the minimizing player. Assume the existence of the saddle-point for each $a \in A$. Then the minimizing player clearly chooses a^* such that (if it exists)

(7) $$J^* = J^*(a^*) = J(u^*, v^*; a^*) \leqq J(u^*, v^*; a), \quad a \in A .$$

As long as the value a^* is principally known to both players when the game starts, the value of this parametrized differential game is J^* provided that both players act optimally. On the other hand, J^* is clearly the expected pay-off for the maximizing player, i.e., the greatest lower bound.

<u>Remark.</u> In this formulation of a differential game we do not consider the possible constraints on x and a at $t = 1$, e.g.,

see Doležal and Černý (1976b). The only reason is, similarly as in the case of free final time, to avoid notational complexity without any substantial gain.

NECESSARY OPTIMALITY CONDITIONS

It is obvious and well-known fact that to find the optimal parameter a^* , an augmented state approach can be applied. Namely, the r elements of a are considered as additional state variables which are free at both ends. For a moment let us neglect constraints (3) - (4).

Applying now the calculus of variations it is simple exercise to show that if $\lambda(t)$ satisfies the differential equation (T denotes the transposition and subscripts stand for the corresponding partial derivatives)

(8) $$\dot{\lambda} = -f_x^T \lambda - L_x, \qquad t \in [0,1],$$

(9) $$[\lambda - \varphi_x^T]_1 = 0 ,$$

then an optimal solution $u^*(t)$, $v^*(t)$, a^* of the differential game with parameters (1) - (2) satisfy the relations

(10) $$f_u^T \lambda + L_u^T = 0, \qquad t \in [0,1],$$

(11) $$f_v^T \lambda + L_v^T = 0, \qquad t \in [0,1],$$

(12) $$\int_0^1 (f_a^T \lambda + L_a^T) dt + [\varphi_a^T]_1 = 0$$

Equations (8) - (9) define the unknown multipliers $\lambda(t)$, $t \in [0,1]$, and equations (10) - (11) give the so-called saddle-point conditions. Finally, equation (12) determines the optimal a^* . Combining (1) - (2) and (8) - (12) one can easily see that, in fact, a nonlinear two-point boundary-value problem for the system of $2n$ differential equations has to be solved. However, such problem cannot be generally solved analytically and thus some iterative numerical methods must be applied.

If constraints (3) - (4) are present, then equations (10)-(12) have the following form

(13) $$H(u^*, v^*) = \min_{u \in U} H(u, v^*) = \max_{v \in V} H(u^*, v) ,$$

where

(14) $$H(u,v) = L(x,u,v,a,t) + \lambda^T f(x,u,v,a,t) , \quad t \in [0,1],$$

and

$$\left\{ \int_0^1 (\lambda^T f_a + L_a)\,dt + [\varphi_a]_1 \right\} \delta\bar{a} \geq 0, \tag{15}$$

where $\delta\bar{a}$ is any feasible parameter change. For further details see Ahmed and Georganas (1973).

FIRST-ORDER GRADIENT ALGORITHM

The numerical approach to the studied problem is based on the first-order gradient algorithm originally described by Bryson and Ho (1969) for optimal control problems. Its applicability to differential games was demonstrated by Doležal and Černý (1976b). Let us only point out the fact that control and parameter constraints are treated applying the idea of a projection.

The derivation of the algorithm is omitted, because it can be done rather easily having in mind the just mentioned references. The resulting algorithm then consists of the following steps.

Step 1. Select the feasible initial solution estimate, i.e., strategies $u(t)$, $v(t)$, $t \in [0,1]$, and the value of a parameter a not violating the constraints (13) - (14).

Step 2. Integrate system (1) in the sense of the increasing time (forward run) using the given initial condition x_0, on applying the values estimated in step 1. Record the histories $x(t)$, $u(t)$ and $v(t)$, $t \in [0,1]$, and the values $[\varphi_x]_1$ and $[\varphi_a]_1$.

Step 3. Integrate in the sense of the decreasing time (backward run) n-dimensional influence function $p(t)$ according to formulas

$$\dot{p} = -f_x^T p, \qquad p(1) = [\varphi_x^T]$$

Step 4. Compute the following integral (q-dimensional row-vector)

$$I_a = \int_0^1 (p^T f_a + L_a)^T dt + [\varphi_a^T]_1 .$$

Step 5. The existing estimates of $u(t)$, $v(t)$ and a are updated by adding the corrections

$$\left.\begin{aligned} \delta u(t) &= -W_p (p^T f_u + L_u)^T, \\ \delta v(t) &= W_E (p^T f_v + L_v)^T, \end{aligned}\right\} \quad t \in [0,1]$$

$$\delta a = -W_a I_a ,$$

where W_P, W_E and W_a are positively definite weighting matrices of the appropriate dimensions.

Step 6. Check, if the resulting new estimates

$$\left.\begin{array}{lll} u(t) & \hat{=} & u(t) + \delta u(t) \\ v(t) & \hat{=} & v(t) + \delta v(t) \\ a & \hat{=} & a + \delta a \end{array}\right\} \quad t \in [0,1],$$

satisfy the constraints (3) - (4). If this is not the case, perform the projection according to formulas

$$\left.\begin{array}{lll} u(t) & \hat{=} & \mathrm{proj}\,[u(t) \mid U] \\ v(t) & \hat{=} & \mathrm{proj}\,[v(t) \mid V] \end{array}\right\} \quad t \in [0,1],$$
$$a \;\hat{=}\; \mathrm{proj}\,[a \mid A],$$

where for $\gamma \in E^n$ and $Q \subset E^n$ we define

$$\mathrm{proj}\,[\gamma \mid Q] = \{\tilde{\gamma} \in Q \mid \|\gamma - \bar{\gamma}\| \leq \|\tilde{\gamma} - \gamma_0\|,\ \gamma_0 \in Q\},$$

i.e., under the projection of a point γ we understand the nearest point $\tilde{\gamma} \in Q$.

Step 7. Using the projected values compute the corresponding feasible changes $\delta\bar{u}(t)$, $\delta\bar{v}(t)$, $t \in [0,1]$, and $\delta\bar{a}$ and evaluate the relation

$$E = \int_0^1 (\delta\bar{u}^T W_P^{-1} \delta\bar{u} + \delta\bar{v}^T W_E^{-1} \delta\bar{v})\, dt + \delta\bar{a}^T W_a^{-1} \delta\bar{a}.$$

If $E < \varepsilon$ (ε is the permitted error) then stop the computations; else go to step 2.

The weighting matrices can be roughly determined using the comparison with the so-called predicted values. Anyhow, it is usually sufficient to set these values only at the beginning of computations. For further details see Doležal and Černý (1976a,b).

It is also evident that the in step 6 required projection can cause certain difficulties, because we are not always able to compute it in an analytical way. However, in a number of practically important cases, where the constraining sets are given as parallelepipeds, spheres, balls, etc., the desired projection is easily determined. Thus this approach can represent an interesting numerical tool for the studied parameter estimation problem.

ILLUSTRATIVE EXAMPLES

To illustrate the practical importance of the developed algorithm we solve in this section two illustrative examples. Both examples were solved using the SIMFOR simulation program of Černý (1975), in the connection with EAI PACER 600 computer (digital part).

During this section all variables will be scalars. Both examples are directly given in the form of (1) - (4), i.e., the duration of the game in question is normalized to the interval $t \in [0,1]$. As a stopping condition the value $\varepsilon = 10^{-10}$ is used. As before, a parameter a is used by the minimizing player (control u) to decrease the value of the studied differential game. Let us also note that the all figures are direct prints of the computer display using the Hard Copy Unit and all integrations were done using the 3rd order variable step Runga-Kutta method with overall permitted error $e_{max} = 10^{-4}$.

First example is the differential game of Leondes and Siu (1977). They used the explicite analytic form for the optimal strategies to optimize only the unknown parameter. As a result of the algorihm described in the previous section we obtain both, optimal strategies (as functions of t) and parameters.

Example 1. Given the system equations

$$\dot{x}_1 = -0.9x_1 + 0.9ax_2 + 0.9v, \quad x_1(0) = 2.0, \tag{16}$$

$$\dot{x}_2 = 0.45(x_1 + x_2) + 1.35u, \quad x_1(0) = 2.0,$$

and the cost functional

$$J = 0.5[x_1^2 + x_2^2]_1 + 0.45\int_0^1 (u^2 - v^2)\,dt. \tag{17}$$

As initial estimates $u(t) = v(t) = 0$, $t \in [0,1]$, and $a = -0.5$ were used. The values of $W_p = 0.4$, $W_E = 0.6$ and $W_a = 0.4$ were found satisfactory. The desired accuracy was then reached in 30 iterations of the algorithm. The convergence histories for the parameter a, the cost functional J, and also the corresponding values of the error E are given in Table 1, where N denotes the iteration number. It can be seen that practically only 15 iterations were needed to reach J^*. The converged time histories of all variables are depicted in Fig. 1.

The obtained solution is almost identical with the optimal values $a^* = -1.338$ and $J^* = 1.484$ as reported by Leondes and Siu (1977). It is necessary to have in mind that not only optimal value of the parameter, but also the corresponding strategies are computed. Thus we can expect the higher number of iterations than

TABLE 1

N	Example 1			Example 2		
	a	J	E	a	J	E
0	-0.5000	6.7619	1.6E+01	0.0000	1.12603	1.9E+00
1	-1.1309	2.9270	4.5E+00	0.8978	0.20320	4.4E-01
2	-1.1610	1.8928	1.3E+00	1.1648	0.25951	1.7E-02
3	-1.2687	1.5921	3.8E-01	1.0582	0.24942	2.3E-03
4	-1.3267	1.5121	1.3E-01	1.0349	0.24969	5.6E-04
5	-1.3572	1.4909	3.8E-02	1.0325	0.24944	1.8E-04
10	-1.3886	1.4839	2.4E-04	1.0289	0.24941	6.7E-07
15	-1.3671	1.4838	3.6E-06	1.0286	0.24941	6.6E-09
20	-1.3655	1.4838	9.2E-08	1.0285	0.24941	7.8E-11
30	-1.3657	1.4838	5.9E-11	-	-	-

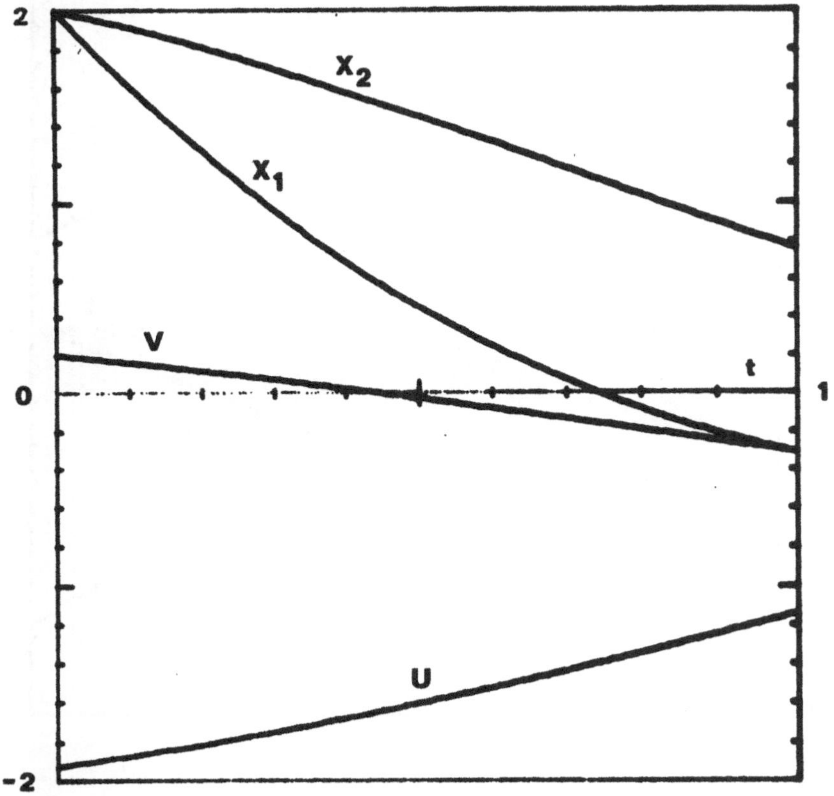

FIG. 1

in the case when the strategies are known a priori as in the just mentioned reference. This hypothesis was also confirmed by an experiment, when seeking the optimal strategies for the fixed a ; e.g. for $a' = -0.5$ the accuracy $\varepsilon = 10^{-10}$ was reached in 22 iterations and the resulting $J' = 1.8658$. This means that a considerable decrease of the value of this game is achieved. Also for other initial solution estimates the algorithm showed similar behaviour.

Example 2. Consider the problem of Quintana and Davison (1972) studied also by Doležal and Černý (1976b) with the initial state (first component) as a parameter. The system equations are as follows:

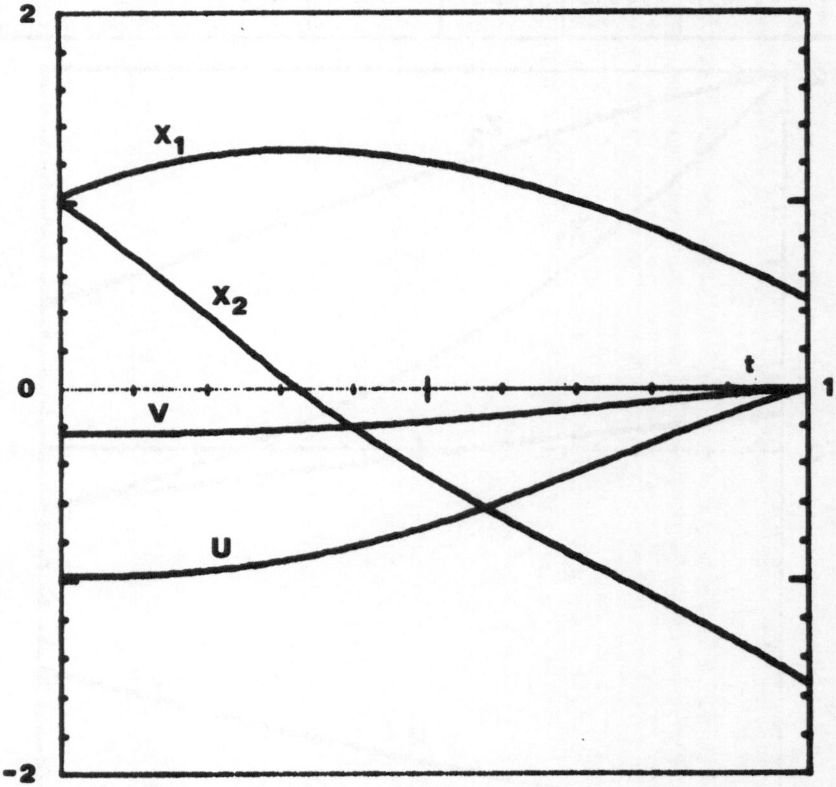

FIG. 2

$$(19)\quad \dot{x}_1 = 1.5\,x_2, \qquad x_1(0) = a$$
$$\dot{x}_2 = 1.5[-x_1 + u - v + (1 - x_1^2)\,x_2], \qquad x_2(0) = 1.$$

The cost functional

$$(20)\quad J = 0.5\,[x_1^2]_1 + 0.75 \int_0^1 (0.5\,u^2 - 2.0\,v^2)\,dt .$$

To obtain the problem with fixed initial state, as required in (1), let us perform the substitution

$$(21)\quad \begin{aligned} y_1 &= x_1 - a, \\ y_2 &= x_2 . \end{aligned}$$

In this way we obtain the following differential game with parameter

$$(22)\quad \begin{aligned} \dot{y}_1 &= 1.5\,y_2, & y_1(0) = 0, \\ \dot{y}_2 &= 1.5\,[-(y_1 + a) + u - v + y_2 - (y_1 + a)^2\,y_2], & y_2(0) = 1, \end{aligned}$$

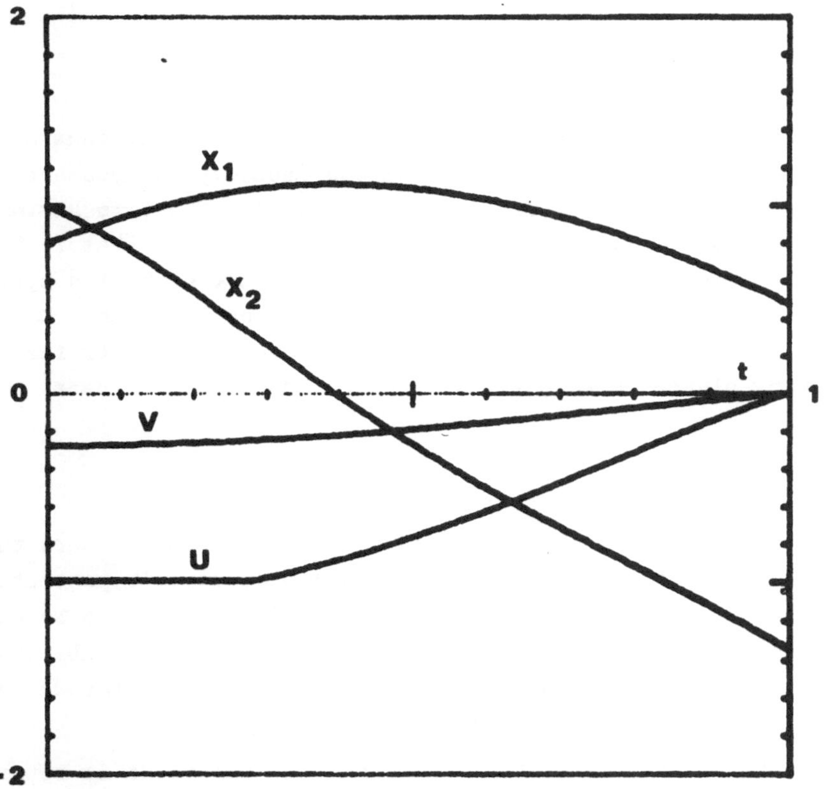

FIG. 3

$$(23)\quad J = 0.5\,[(y_1 + a)^2]_1 + 0.75 \int_0^1 (0.5u^2 - 2.0v^2)\,dt .$$

The nominal solution estimates were $u(t) = v(t) = 0$, $t \in [0,1]$ and $a = 0$. The desired accuracy was achieved in 20 iterations using the weighting constants $W_P = W_E = 0.5$ and $W_a = 1.0$. The details concerning the convergence can be again found in Table 1. The time histories of all variables are depicted in Fig. 2. If we now compare $J^* = 0.24941$ with the converged value $J = 0.54896$ obtained for $a = 0$ being fixed, i.e., with the problem studied by Doležal and Černý (1976b), we again observe a considerable decrease of the value of the game in question. Also in this case the choice of initial solution estimates is not crucial and the convergence is easily achieved for other initial estimates.

Now let us additionally assume the constraints of the type (4) - (5), i.e.,

$$(24)\quad \begin{array}{l} |u(t)| \leq 1, \quad |v(t)| \leq 1, \quad t \in [0,1] \\ 0 \leq a \leq 0.8 . \end{array}$$

Applying the same nominal estimates and weighting constants as in the unconstrained case the desired accuracy was reached in 19 iterations with $a^* = 0.8$ and $J^* = 0.26886$. The time histories for this case are depicted in Fig. 3. Let us observe that in this case the given constraints on $v(t)$ are inactive along the optimal solution. On the other hand, the constraints on $u(t)$ and a are active, which fact, in turn, causes the increase of costs for the minimizing player in comparison with the unconstrained case.

CONCLUSIONS

For a general class of two-player zero-sum games, where the minimizing player has the additional choice of certain parameters, we obtained the necessary optimality conditions using the calculus of variations. Based on these conditions, a previously published gradient algorithm of Doležal and Černý (1976b) was extended to handle also problems of this type.

Practical experience with the suggested algorithm is reported and two illustrative examples are solved in detail. As long as here only half of the parameter optimization problem in differential games is solved, a question of the optimal choice of certain parameters also for the maximizing player immediately arise. This problem is a subject of the present study and the results will be published elsewhere.

Finally, the author would like to acknowledge the support of his colleagues from the Hybrid Computer Laboratory of the Institute. Especially the continuous help of Ing. P. Černý is gratefully acknowledged.

REFERENCES

Ahmed N.U. and Georganas N.D. (1973): On Optimal Parameter Selection. IEEE Trans. on Automatic Control AC-18 (1973), No. 3, 313-314.

Boltjanskij V.G. (1969): Mathematical Methods of Optimal Control. Nauka, Moscow 1969 (in Russian).

Bryson A.E. and Ho Y.C. (1969): Applied Optimal Control. Ginn and Company, Waltham, Massachusetts 1969.

Černý P. (1975): Digital Simulation Program SIMFOR for the Solution of Two-Point Boundary-Value Problems. Research Report ÚTIA ČSAV, No. 639, Prague 1975 (in Czech).

Doležal J. and Černý P. (1976a): The Application of Optimal Control Methods to the Determination of Multifunctional Catalyst. 23rd CHISA Conference, Mariánské Lázně 1976 (in Czech).

Doležal J. and Černý P. (1976b): On the Numerical Solution of Differential Games. Cybernetics Conference, Prague 1976 (in Czech).

Gamkrelidze R.V. (1965): On Some Extremal Problems in the Theory of Differential Equations with Applications to the Theory of Optimal Control. SIAM J. Control 3 (1965), No. 1, 106-128.

Georganas N.D. (1975): Optimal Parameter Selection by Imbedding Techniques. IEEE Trans. on Automatic Control AC-20 (1975), No. 1, 166-167.

Hofer E. and Sagirow P. (1968): Optimal Systems Depending on

Parameters. AIAA J. 6 (1968), No. 5, 953-956.

Leondes C.T. and Siu T.K. (1977): Parameter Optimization for Linear Quadratic Differential Games. Trans. ASME 99 (1977), Ser. G, No. 1, 58-62.

Lunderstädt R. (1976): Transportation Systems with Optimal Parameters. Problems of Control and Information Theory 5 (1975), No. 2, 109-116.

Miele A. (1975): Recent Advances in Gradient Algorithms for Optimal Control Problems. J. Optimiz. Theory Applics.17 (1975), Nos. 5-6, 361-430.

Quintana D.H. and Davison E.J. (1972): Two Numerical Techniques to Solve Differential Game Problems. Int. J. Control 16 (1972), No. 3, 465-474.

Czechoslovak Academy of Sciences
Institute of Information Theory and Automation

Pod vodárenskou věží 4
182 08 Praha 8 - Libeň
Czechoslovakia

REMARK ON THE LAWS OF LARGE NUMBERS AND THE CENTRAL LIMIT THEOREMS ON A LOGIC

Anatolij Dvurečenskij

Bratislava

ABSTRACT

In this communication the notion of the independence of observables in a state on a logic, as it was introduced Gudder(1967), will be studied. Some generalized forms of the weak law of large numbers and the central limit theorems for observables of a logic will be proved. The used methods are similar to those of the conventional probability theory.

1. PRELIMINARY DEFINITIONS AND RESULTS

Let us suppose that L be a poset with the first and the last elements 0 and 1, respectively, and an orthocomplementation $\perp : a \mapsto a^\perp$ which satisfies (i) $(a^\perp)^\perp = a$ for all $a \in L$; (ii) if $a < b$, then $b^\perp < a^\perp$ for $a,b \in L$; (iii) $a \vee a^\perp = 1$ for all $a \in L$. We say that a,b are orthogonal and write $a \perp b$ if $a < b^\perp$. We further assume that if $a < b$, then $b = a \vee (b \wedge a^\perp)$ and if $\{a_i\}$ is a sequence of mutually orthogonal elements of L, then $\bigvee_i a_i \in L$. A poset L satisfying the above axioms will be called a logic (Varadarajan (1962)).

A state is a map m from L into $\langle 0,1 \rangle$ such that $m(1) = 1$ and $m(\bigvee_i a_i) = \sum_i m(a_i)$ if $a_i \perp a_j$ for $i \neq j$. A system M of states of a logic L is called a quite full system if the statement $m(b) = 1$ whenever $m(a) = 1$, $m \in M$, implies $a < b$.

An observable is a map x from the Borel sets $B(R_1)$ of R_1 into L such that (i) $x(R_1) = 1$; (ii) $x(E) \perp x(F)$ if $E \cap F = \emptyset$, $E,F \in B(R_1)$; (iii) $x(\bigcup_i E_i) = \bigvee_i x(E_i)$ if $E_i \cap E_j = \emptyset$, $i \neq j$, $E_i \in B(R_1)$. If f is a Borel function on R_1 and x an observable, then $f \circ x: E \mapsto x(f^{-1}(E))$, $E \in B(R_1)$, is an observable. We say that an observable x is bounded if there is a compact set C such that $x(C) = 1$. The mean value of x in the state m is $m(x) = \int_{R_1} t \, dm_x(t)$ if the integral on the right-hand side exists and is finite, where m_x is a measure on $B(R_1)$: $m_x(E) = m(x(E))$, $E \in B(R_1)$.

Let $x_1,\ldots,x_n$ be observables (x_i may be unbounded for some i = 1, ..., n) of a logic L. If there is quite full system M of states on L and a unique observable z such that $m(x_1),\ldots,m(x_n)$ exist and are finite and $m(z) = m(x_1) + \ldots + m(x_n)$ for every $m \in M$, then z is called the sum of $x_1,\ldots,x_n$ and is written $z = x_1 + \ldots + x_n$.

If there is a quite full system M of states on L such that for any two bounded observables x,y there is a unique sum z = x + y, then L is called a sum logic. In Gudder (1966) it is shown, that a sum logic is a lattice. From this moment we shall suppose that L is a sum logic.

We say that observables $x_1,\ldots,x_n$ have a joint distribution in the state m if there is a measure m_n on $B(R_n)$ such that

$$m_n(E_1 \times \ldots \times E_n) = m\left(\bigwedge_{j=1}^{n} x_j(E_j)\right) \text{ for all } E_j \in B(R_1),\ j=1,\ldots,n.$$

A system of obsefvables $x_1,\ldots,x_n$ is independent in the state m if

$$m(x_1(E_1) \wedge \ldots \wedge x_n(E_n)) = m(x_1(E_1)) \ldots m(x_n(E_n)),\ E_j \in B(R_1),\ j=1,\ldots,n.$$

A system of observables $\{x_t : t \in T\}$ is independent in m if any finite subsystem is independent,

According to Gudder (1967) we introduce the notion of a strong independence in a state, which converts into the notion of an independence in a state in the conventional probability theory. This notion enables us to study properties of a characteristic function of the sum $z = x_1 + \ldots + x_n$ by means of $\varphi_{x_1},\ldots,\varphi_{x_n}$, where $\varphi_x(u) = \int_{R_1} e^{iut} \cdot dm_x(t)$. Thus, we say that observables $x_1,\ldots,x_n$ are strongly independent in the state m if for any n Borel function $f_1,\ldots,f_n$, for which $f_1 \circ x_1 + \ldots + f_n \circ x_n$ has a sense, we have

$$m_{f_1 \circ x_1 + \ldots + f_n \circ x_n} = m_{f_1 \circ x_1} * \ldots * m_{f_n \circ x_n},$$

where the * denotes the convolution.

As usualy, a system $\{x_t : t \in T\}$ of observables is strongly independent in the state m if every finite subsystem is strongly independent in the state m.

Let a be real, we define an observable I_a by $I_a(\{a\}) = 1$ and it may be shown that $aI_1 = I_a$, where on the left-hand side we have the constant function $a(t) = a$, $t \in R_1$ and $a \circ I_1 = a\, I_1$.

If $m(x^2)$ is finite, then m(x) is finite, too, and we define the variance $V_m(x)$ of an observable x in the state m by

$$V_m(x) = m((x - m(x)\, I_1)^2),$$

and in this case the Chebychev inequality $m((x - m(x)I_1)((-\varepsilon,\varepsilon)^c)) \leq \leq V_m(x)\varepsilon^{-2}$ holds.

Observables $x_1,\dots,x_n$ are uncorrelated in the state m if

$$V_m(x_1 +\dots+ x_n) = V_m(x_1) +\dots+ V_m(x_n),$$

if the sum $x_1 +\dots+ x_n$ exists. A system $\{x_t : t \in T\}$ of observables is uncorrelated in m if every finite subsystem is uncorrelated in m.

2. CONVERGENCE OF OBSERVABLES

For purposes of the last sections we need to introduce some types of convergences of observables with respect to a state. These forms of convergences are equivalent to corresponding convergences in the conventional probability theory (Gudder(1967)). If x, $x_1, x_2, \dots$ are observables and m is a state, the we say that (i) x_n converges in the state m to, and write $x_n \xrightarrow{m} x$ if $\lim_{n\to\infty} m((x-x_n)((-\varepsilon,\varepsilon)^c)) = 0$ for every $\varepsilon > 0$; (ii) converges in square mean (m) to x if $\lim_{n\to\infty} m((x-x_n)^2) = 0$.

A distribution function of x in the state m is a function $F_x(t) = m(x(-\infty,t))$. We say that a sequence of distribution functions $\{F_n\}$ of of a sequence x_n converges weakly to F_x and write $F_n \xrightarrow{w} F_x$ if $F_n(t) \to F_x(t)$ at each continuity point of F_x. Due to the direct and inverse limit theorems (Gnedenko (1965)) $F_n \xrightarrow{w} F_x$ iff $\varphi_n \to \varphi_x$. This note will be used in the following.

LEMMA 2.1. There holds $x_n \xrightarrow{m} I_a$ iff $F_n \xrightarrow{w} F_a$, where F_n, F_a are distribution functions of x_n and I_a, respectively, in the state m.

Proof. It is easily to see that x_n and I_a are compatible observables for every n. If we take into account the calcul for compatible observables (Varadarajan (1968), theorem 6.17) , then the proof is finished. Q.E.D.

From this moment we suppose that $\{x_i\}$ is such a sequence of observables and m is a state such that (i) for all n =1,2,... the sums $x_1+\dots+x_n$ exist; (ii) if $m(x_1),\dots,m(x_n)$ exist and they are finite, n=1,2,..., then $m(x_1 +\dots+ x_n) = m(x_1) +\dots+ m(x_n)$.

Let $\{x_i\}$ be a sequence of observables and m be a state, we shall investigate convergence of $s_n = n^{-1}(\sum_{i=1}^{n}(x_i - m(x_i)\, I_1))$, if $m(x_i)$, i= =1,2,... is finite.

We say that for a sequence $\{x_i\}$

(i) the weak law of large numbers (w.l.l.n.) in the state m holds if $s_n \xrightarrow{m} 0$;

(ii) the squaremean law of large numbers (s.m.l.l.n.) in the state m holds if $m(s_n^2) \to 0$;

(iii) the central limit theorem (c.l.t.) in the state m holds if distribution functions of ns_n/B_n converges to $F(t) = \frac{1}{\sqrt{2\pi}}\int_{-\infty}^{t} \exp(-\frac{u^2}{2})\,du$ at each $t \in R_1$, where $B_n^2 = V_m(x_1 + \ldots + x_n)$.

3. Weak laws of large numbers

THEOREM 3.1. (Gudder (1967))(Chebychev) If $\{x_i\}$ is a sequence of uncorrelated observables in the state m and if there is a constant K such that $m(x_i^2) \leq K$, i=1,2,..., then for $\{x_i\}$ the w.l.l.n. and the s.m.l.l.n. hold.

Proof. Due to the finiteness and boundedness of $m(x_i^2)$, i=1,2,..., and by the Chebychev inequality, we have $m(s_n((-\varepsilon,\varepsilon)^c)) \leq \varepsilon^{-2} V_m(n^{-1}(x_1 + \ldots + x_n)) = \varepsilon^{-2}n^{-2}\sum_{i=1}^{n} V_m(x_i) \leq \varepsilon^{-2}n^{-1}K \longrightarrow 0$.

Similarly, $m(s_n^2) = V_m(n^{-1}(x_1 + \ldots + x_n)) \longrightarrow 0$. Q.E.D.

THEOREM 3.2. (Khinchin) Let $\{x_i\}$ be a sequence of strongly independent observables, which have the same distribution function in the state m and let $m(x_1) = a$, then

$$n^{-1}(x_1 + \ldots + x_n) \xrightarrow{m} I_a.$$

Proof. If we change $x_i - I_a$ instead of x_i, we may assume that $a = 0$. According to Lemma 2.1. it suffices to show that a sequence of distribution functions of $y_n = n^{-1}(x_1 + \ldots + x_n)$ converges weakly to the distribution function of I_o, but due to the known theorems of the probability theory it is sufficient to examine a convergence of corresponding functions φ_n of y_n in the state m to $\varphi_o(u) \equiv 1$, $u \in R_1$.

If φ is a characteristic function of x_1, then the strong independence of $\{x_i\}$ in m implies $\varphi_n(u) = \varphi^n(\frac{u}{n})$ for all $u \in R_1$, n=1,2,... . Since φ' exists, $\varphi'(0) = 0$, and φ' is continuous, then, by Taylor's theorem, we have $\varphi(\frac{u}{n}) = 1 + O(\frac{u}{n})$, where $\lim_{n\to\infty} O(\frac{u}{n}) = 0$, $u \in R_1$. Therefore $\varphi^n(\frac{u}{n}) = (1 + O(\frac{u}{n}))^n \to 1$ for every $u \in R_1$, and the proof is finished. Q.E.D.

THEOREM 3.3. (Markov) If a sequence of observables satisfies in the state m the Markov condition

$$n^{-2} V_m(x_1 + \ldots + x_n) \to 0, \tag{4.1}$$

then for $\{x_i\}$ the w.l.l.n. and the s.m.l.l.n. hold in the state m.

Proof. Using the Chebychev inequality we have $m(s_n((-\varepsilon,\varepsilon)^c)) \leq \varepsilon^{-2}n^{-2} V_m(x_1 + \ldots + x_n) \to 0$. Q.E.D.

REMARK 1. If $\{x_i\}$ is a sequence of uncorrelated observables in m, then the Markov condition (3.1) has the following form $n^{-2}\sum_{i=1}^{n} V_m(x_i) \longrightarrow 0$, therefore the Chebychev theorem follows from the Markov theorem.

If x is an observable, then $y = x^2/(1+x^2)$ is an observable if we put $y = f\circ x$, where $f(t) = t^2/(1+t^2)$.

THEOREM 3.4. In order to the w.l.ln. to hold in the state m for a sequence $\{x_i\}$ it is necessary and sufficient

$$\lim_{n\to\infty} m\left[\frac{(\sum_{i=1}^{n}(x_i - m(x_i)I_1))^2}{n^2 + (\sum_{i=1}^{n} x_i - m(x_i)I_1))^2}\right] = 0. \tag{3.2}$$

Proof. The necessity follows from this calcul

$m(s_n((-\varepsilon,\varepsilon)^c)) = \int_{|t|\geq\varepsilon} dms_n(t) \geq \int_{|t|\geq\varepsilon} t^2/(1+t^2)\, dm_{s_n}(t) = \int t^2/(1+t^2)\, dm_{s_n}(t) - \int_{|t|<\varepsilon} t^2/(1+t^2)\, dm_{s_n}(t) \geq \int t^2/(1+t^2)\, dm_{s_n}(t) - \varepsilon^2 = m(s_n^2/(1+s_n^2)) - \varepsilon^2.$

Therefore $0 \leq m(s_n^2/(1+s_n^2)) \leq \varepsilon^2 + m(s_n((-\varepsilon,\varepsilon)^c))$ and the(3.2) holds.

The sufficiency. Let (3.2) hold, then for each $\varepsilon > 0$ we have

$m(s_n((-\varepsilon,\varepsilon)^c)) = \int_{|t|\geq\varepsilon} dm_{s_n}(t) \leq (1+\varepsilon^2)\varepsilon^{-2} \int_{|t|\geq\varepsilon} t^2/(1+t^2)\, dm_{s_n}(t) \leq (1+\varepsilon^2)\varepsilon^{-2} \int t^2/(1+t^2)\, dm_{s_n}(t) = (1+\varepsilon^2)\varepsilon^{-2}\, m(s_n^2/(1+s_n^2)) \to 0.$ Q.E.D.

REMARK 2. We may show that Theorems 3.1. --3.3 follow from Theorem 3.4. For example, if the Markov condition (3.1) is satisfied, then $m(s_n^2/(1+s_n^2)) = \int t^2/(1+t^2)\, dm_{s_n}(t) \leq \int t^2\, dm_{s_n}(t) = n^{-2} V_m(x_1 + \ldots + x_n) \to 0$.

THEOREM 3.5. If a sequence of observables $\{x_i\}$ satisfies the following condition in the state m

$$\lim_{n\to\infty} n^{-(1+\delta)} m(|\sum_{i=1}^{n}(x_i - m(x_i)I_1)|^{1+\delta}) = 0 \tag{3.3}$$

for $0 < \delta \leq 1$, then the w.l.ln. holds.

Proof. We claim to verify fulfilling of (3.2). Indeed,

$m(s_n^2/(1+s_n^2)) = \int t^2/(1+t^2)\, dm_{s_n}(t) \leq \int |t|^{1+\delta}\, dm_{s_n}(t) = n^{-(1+\delta)} m(|\sum_{i=1}^{n}(x_i - m(x_i)I_1)|^{1+\delta}) \to 0.$ Q.E.D.

REMARK 3. If in (3.3) we put $\delta = 1$, then the Markov condition is satisfied.

4. CENTRAL LIMIT THEOREMS

The role of the gaussian distributions in the conventional probability theory and its applications is, doubtless, very important. Moreover, in the quantum field theory, which motivates the theory of logic, they give a satisfactory description of the radiation field of coherent sources, as it was remarked in Holevo (1973),p.372 . We now show the role of the gaussian distribution from a purely probabilistic viewpoint for the logic theory. If a distribution function F_x of an observable x in the state m has the form $F_x(t) =$

$= \frac{1}{\sqrt{2\pi}} \int_{-\infty}^{t} \exp(-\frac{u^2}{2})\, du$, then F_x is called the gaussian distribution.

THEOREM 4.1. (Lindeberg - Levy) Let $\{x_i\}$ be a sequence of strongly independent observables which have the same distribution function in the state m and let $m(x_1^2)$ be finite, then for $\{x_i\}$ the c.l.t. holds in m.

Proof. The proof is analogical to that of the conventional probability theory. Let φ be a characteristic function of $(x_1 - m(x_1) I_1)/V_m(x_1)$, then $\varphi(0) = 1$, $\varphi'(0) = 0$, $\varphi''(0) = -1$ and φ'' is continuous. By Taylor's theorem we have $\varphi(u) = 1 - \frac{u^2}{2} + O(u^2)$, where $\lim_{u \to 0} \frac{O(u^2)}{u^2} = 0$. If φ_n is a characteristic function of $y_n = B_n^{-1} \sum_{i=1}^{n} (x_i - m(x_i) I_1)$ in m, then by the strong independence

$$\varphi_n(u) = (\varphi(\frac{u}{n}))^n = (1 - \frac{u^2}{2n} + O(\frac{u^2}{n}))^n \longrightarrow \exp(-\frac{u^2}{2}),\ u \in R_1,$$

and this property implies a weak convergence of corresponding distribution functions. Q.E.D.

We say that a sequence $\{x_i\}$ of observables of a logic satisfies the Lindeberg condition in the state m if for every $\tau > 0$

$$\lim_{n \to \infty} B_n^{-2} \sum_{i=1}^{n} \int_{|t - m(x_i)| > \tau B_n} (t - m(x_i))^2\, dm_{x_i}(t) = 0. \qquad (4.1)$$

THEOREM 4.2. (Lindeberg) If a sequence $\{x_i\}$ of strongly independent observables satisfies the Lindeberg condition (4.1) in the state m, then the c. l.t. holds for $\{x_i\}$ in m.

Proof. Since a sequence of coordinate functions $\{\pi_i\}$ is independent on $(R_\infty, B(R_\infty), \mu)$, $\mu = m_{x_1} \times m_{x_2} \times \ldots$, and they satisfy the Lindeberg condition, then, by the known result of the probability theory, it follows that the distribution functions F_n of $\sum_{i=1}^{n} (\pi_i - E(\pi_i)) / \sigma(\sum_{i=1}^{n} \pi_i)$ converge weakly to the distribution function of N(0,1). Because of the equality of F_n and the distribution function of y_n, respectively, the remaining part of the proof is shown. Q.E.D.

REMARK 4. Theorem 4.1 follows from Theorem 4.2. Indeed, let $\tau > 0$, then $B_n = \sigma\sqrt{n}$, where $\sigma^2 = V_m(x_1)$. If $m(x_1) = a$, then

$$\sum_{i=1}^{n} B_n^{-2} \int_{|t - m(x_i)| > \tau B_n} (t - m(x_i))^2\, dm_{x_i}(t) = \sigma^{-2} \int_{|t - a| > \tau\sigma\sqrt{n}} (t - a)^2\, dm_{x_1}(t) \longrightarrow 0.$$

THEOREM 4.3. (Ljapunov) If for a sequence $\{x_i\}$ of strongly independent observables in the state m we may choose an $\delta > 0$ such that

$$\lim_{n \to \infty} B_n^{-(2+\delta)} \sum_{i=1}^{n} \int |t - m(x_i)|^{2+\delta}\, dm_{x_i}(t) = 0, \qquad (4.2)$$

then for $\{x_i\}$ the c. l.t. holds in m.

Proof. It suffices to verify that the condition (4.2) implies the validity of (4.1). Indeed,

$$B_n^{-2}\sum_{i=1}^{n}\int_{|t-m(x_i)|>\tau B_n}(t-m(x_i))^2\,dm_{x_i}(t)\le$$

$$\le B_n^{-2}(\tau B_n)^{-\delta}\cdot\sum_{i=1}^{n}\int_{|t-m(x_i)|>\tau B_n}|t-m(x_i)|^{2+\delta}\,dm_{x_i}(t)\le$$

$$\le\tau^{-\delta}B_n^{-(2+\delta)}\sum_{i=1}^{n}\int|t-m(x_i)|^{2+\delta}\,dm_{x_i}(t)\to 0. \qquad \text{Q.E.D.}$$

CORROLARY 4.3.1. If a sequence of uniformly bounded observables $\{x_i\}$ is strongly independent in the state m and if $B_n\to\infty$, then the c.l.t. holds in m.

Proof. If $\|x_i\|\le K$, $i=1,2,\ldots$, for some K, then $\int|t-m(x_i)|^{2+\delta}\,dm_{x_i}(t)\le(2K)^{2+\delta}/B_n^{2+\delta}\to 0$ for $n\to\infty$. Q.E.D.

REFERENCES

GNEDENKO, B.V. (1965): Kurs teorii verojatnostej, Moskva.

GUDDER, S.P. (1966): Uniqueness and existence properties of bounded observables. In: Pac.J.Math.Analys. and Appl., 19,1,(1966) 81--93.

———— (1967): Hilbert space, independence, and generalized probability. In. J. Math. Analysis and Appl. 20,1, (1967) 48--61.

HALMOS, P.R. (1953): Teorija mery, Moskva (1953).

HOLEVO, A.S. (1973): Statistical decision theory for quantum systems. In: J. Multivar. Anal., 3,4 (1973) 337 -- 394.

VARADARAJAN, V.S. (1962): Probability in physics and a theorem on simultaneous observability. In: Comm. Pure Appl.Math. 15, (1962) 189--217.

———— (1968) Geometry of quantum theory. New York (1968).

Slovak Academy of Sciences
Institute of Measurement and
Measuring Technic
Dúbravska cesta
885 27 Bratislava

ПЕРЕДАЧА ИНФОРМАЦИЙ СВЕРТОЧНЫМИ КОДАМИ С ОБНАРУЖЕНИЕМ ОШИБОК

Владимир Н. Дынькин, Станислав И. Жегалов

Москва

АННОТАЦИЯ

Получено выражение вероятности необнаружения ошибки для сверточного кода при передаче по двоичному симметричному каналу без памяти (ДСК) через спектр двойственного кода. Получены верхняя и нижняя границы этой вероятности. Построена стратегия передачи по ДСК с обратной связью с обнаружением ошибок сверточными кодами, при которой эффективная скорость всегда выше некоторой фиксированной константы, а вероятность ошибки асимптотически стремится (экспоненциально) к нулю.

ВЕРОЯТНОСТЬ НЕОБНАРУЖЕНИЯ ОШИБКИ ДЛЯ ДВОИЧНОГО СВЕРТОЧНОГО КОДА

Одним из распространенных методов передачи информации является передача с обнаружением ошибок и обратной связью. В работах Коржика (1965), Леонтьева (1972) и Левенштейна (1977) рассматривается двоичное блочное кодирование применительно к этому методу для случая передачи по ДСК.

В настоящей работе рассматривается двоичное сверточное кодирование при передаче по ДСК с обнаружением ошибок и обратной связью.

Двоичный сверточный систематический (n_o, k_o) - код задается полубесконечной порождающей матрицей

$$G = \begin{Vmatrix} IG_1 & OG_2 & \ldots & OG_m & & & \\ & IG_1 & \ldots & OG_{m-1} & OG_m & & \\ & & \ddots & & & & \\ & & & IG_1 & OG_2 & \ldots & OG_m \ldots \end{Vmatrix}$$

где I - $(K_0 \times K_0)$ - единичная матрица, O - $(K_0 \times K_0)$ - нулевая матрица, $G_z = [g_{is}^{(z)}]$ - $(K_0 \times \tau_0)$ - двоичные матрицы, $z = \overline{1,m}$, $\tau_0 = n_0 - K_0$. Величину m будем называть задержкой кодирования. Скорость равна

$$R = K_0 / n_0$$

Полубесконечная проверочная матрица сверточного кода имеет вид

$$H = \begin{Vmatrix} G_1^T I' & & & \\ G_2^T O' & G_1^T I' & & \\ \vdots & \vdots & \ddots & \\ G_m^T O' & G_{m-1}^T O' \ldots & G_1^T I' & \\ & G_m^T O' \ldots & G_2^T O' & \ddots \\ & & \vdots & \end{Vmatrix}$$

где I' - $(\tau_0 \times \tau_0)$ - единичная матрица, O' - $(\tau_0 \times \tau_0)$ - нулевая матрица.

Пусть $C = [C_1, C_2, \ldots]$ - переданный кодовый вектор, $Y = C \oplus E$ - принятый вектор, $E = [E_1, E_2, \ldots]$ - вектор ошибок (E_i - i-й подблок, $E_i = e_1^{(i)}, e_2^{(i)}, \ldots, e_{n_0}^{(i)}$). Вектор синдрома равен

$$S = [S_1, S_2, \ldots] = Y \cdot H^T = E \cdot H^T$$

Вероятностью необнаружения ошибки для сверточного кода будем называть вероятность

$$p_H(\ell) = p(E_1, E_2, \ldots, E_\ell \text{ - кодовое слово}, E_1 \neq 0) = \\ = p(S_1 = 0, S_2 = 0, \ldots, S_\ell = 0, E_1 \neq 0) = p(S^{(\ell)} = 0, E_1 \neq 0),$$

где величину ℓ будем называть задержкой декодирования.

Двойственный сверточный (n_0, τ_0) - код (обозначим его код D) имеет порождающую матрицу вида

$$G_D = \begin{Vmatrix} G_1^T I' & G_2^T O' & \ldots & G_m^T O' & & & \\ & G_1^T I' & \ldots & G_{m-1}^T O' & G_m^T O' & & \\ & & \ddots & & & & \\ & & & G_1^T I' & G_2^T O' & \ldots & G_m^T O' \ldots \end{Vmatrix}$$

Вероятность необнаружения ошибки (n_0,K_0) - сверточного кода выражается через весовой спектр (n_0,τ_0) кода D следующим образом

$$p_H^{\ell}(\ell)=2^{-\ell\tau_0}\Big[\sum_{W=0}^{(\ell-1)n_0} B_W^{(\ell)}(0)(1-2p)^W(1-2q^{n_0})+\sum_{W=0}^{\ell n_0} B_W^{(\ell)}(1)(1-2p)^W\Big],$$

где p - переходная вероятность , $q=1-p$, $B_W^{(\ell)}(0)$ - число кодовых слов кода D длины ℓn_0 с весом W, первые K_0 информационных символов которых нулевые, $B_W^{(\ell)}(1)$ - не все нулевые.

ГРАНИЦЫ ВЕРОЯТНОСТИ НЕОБНАРУЖЕНИЯ ОШИБКИ

а) Граница случайного кодирования.

Пусть при кодировании после поступления на вход кодера очередного информационного подблока (соответственно, выхода кодового подблока) величины $g_{is}^{(z)}$ $(z=\overline{1,m};\ i=\overline{1,K_0};\ s=\overline{1,\tau_0})$ выбираются заново случайно и независимо с вероятностями

$$p(g_{is}^{(z)}=1)=p(g_{is}^{(z)}=0)=1/2$$

Таким образом мы получим ансамбль меняющихся во времени систематических сверточных кодов. Средняя по этому ансамблю вероятность необнаружения ошибки $\overline{p_H}(\ell)$ равна $(\tau_0=1)$

(2) $$\overline{p_H}(\ell)=2^{-\ell}(1-q^{K_0}),\quad \ell=m$$

и

$$p_H(\ell)<2^{-\ell}(1-q^{n_0})\{1-q^{K_0(m-1)}+q^{K_0(m-1)}[1+(1-2p)q^{K_0}]^{\ell-m}\},\ \ell>m$$

Из (3) следует, что при любой фиксированной задержке кодирования средняя по ансамблю меняющихся во времени сверточных кодов вероятность необнаружения ошибки экспоненциально стремится к нулю с ростом ℓ.

б) Граница минимального расстояния

$$p_H(\ell)\geqslant p^{d(\ell)}q^{\ell n_0-d(\ell)}=(p/q)^{d(\ell)}q^{\ell n_0}$$

где $d(\ell)$ - минимальное расстояние сверточного кода на длине ℓ.

ПЕРЕДАЧА ИНФОРМАЦИИ ПО ДСК С ОБРАТНОЙ СВЯЗЬЮ

Рассмотрим систему передачи информации, изображенную на рисунке

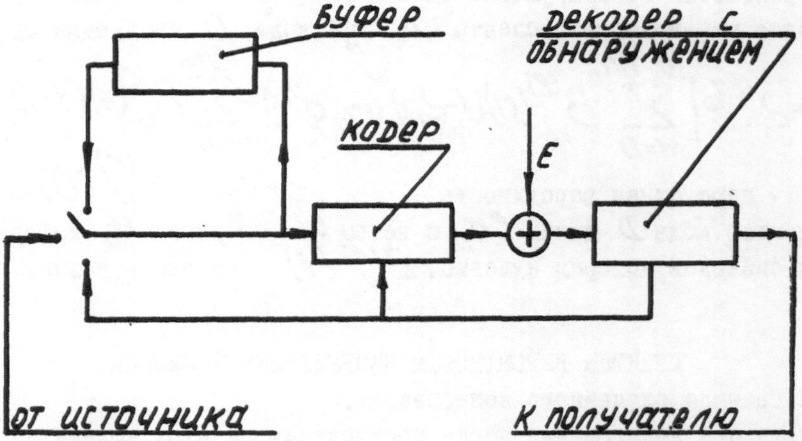

Порождаемые в последовательные моменты времени независимо и равновероятно двоичные символы источника кодируются сверточным кодом и посылаются в канал связи. Одновременно символы источника поступают в буфер. После приема из канала очередных n_0 символов декодер вычисляет очередное значение синдрома и, если последний не равен нулю, содержимое регистра декодера сдвигается на два подблока назад, с помощью обратной связи (она предполагается бесшумной) производится сдвиг на два подблока назад регистра кодера и, когда необходимо, переключение переключателя. Нас будут интересовать параметры этой системы: эффективная скорость передачи $R_{эф}$ - скорость передачи с выхода декодера, и остаточная вероятность ошибки $\overline{p_0}(\ell)$ - вероятность того, что получателю выдается неверный подблок. Для этой системы

$$(4) \qquad R_{эф} > R\,\frac{2q^{n_0}-1}{q^{n_0}}, \quad 2q^{n_0} > 1,$$

а средняя по ансамблю меняющихся во времени сверточных кодов остаточная вероятность ошибки

$$(5) \qquad \overline{p_0}(m) < (1-q^{n_0})\,2^{-m}\left(1+\frac{1-q^{k_0}}{2q^{n_0}-1}\,q^{n_0}\right)$$

Правая часть в (4) от m не зависит, т.е., при $2q^{n_0}>1$ эффективная скорость $R_{эф}$ всегда выше константы, равной

$$k_0(2q^{n_0}-1)/(n_0 q^{n_0})$$

а

$$\overline{p_0}(m) \xrightarrow[m\to\infty]{} 0$$

Сравним полученные результаты со случаем блочного кодирования. Левенштейн (1977) показал, что при передаче по ДСК с мгновенной обратной связью с использованием для обнаруженияошибок блочных кодов обе величины: и эффективная скорость передачи и остаточная вероятность ошибки стремятся к нулю одновременно при увеличении длины блока. Таким образом, сверточные коды позволяют добиться принципиально лучших результатов, чем блочные.

При наличии задержек в прямом и обратном каналах (обозначим их T_1 и T_2 соответственно, в качестве единицы измерения возьмем время передачи n_0 символов)

$$R_{эф} > R/\left(\frac{1-q^{n_0}}{2q^{n_0}-1} + \frac{T(1-q^{n_0})}{q^{n_0}} + 1\right), 2q^{n_0} > 1,$$

где $T = T_1 + T_2$

ЛИТЕРАТУРА

Коржик В.И. (1965): Границы по вероятности необнаружения ошибок и оптимальные групповые коды в канале с обратной связью. Радиотехника, 1965,20,1, 27-33.

Леонтьев В.К. (1972) Кодирование с обнаружением ошибок. Проблемы передачи информации, 1972, 8,2, 6-14.

Левенштейн В.И. (1977) О границах вероятности необнаружения ошибки. Проблемы передачи информации, 1977,13,1, 3-18

Институт проблем управления,
Москва, Профсоюзная ул., 81
СССР

THE EXTINCTION OF GENERATIONS IN GENERATION-DEPENDENT BELLMAN-HARRIS BRANCHING PROCESSES WITH EXPONENTIAL LIFESPAN

Lutz Edler

Mainz

ABSTRACT

If $V^{\langle k\rangle}$ is the time when in a Bellman-Harris branching model the k-th generation disappears out of the population, and if all individuals have exponentially distributed lifespans, the asymptotic behavior of the tail of the distribution of the extinction time $V^{\langle k\rangle}$, $P(V^{\langle k\rangle} > t)$, is obtained, even if the distributions of the lifespans and the offspring sizes vary generation-dependent. Furthermore the times of extinction of several successive generations can be specified for the generation-independent case of the Markov branching model in continuous time. If the initial number of individuals and the absolute time grow up appropriately linked, a Poisson limit theorem for generation sizes will be given.

1. INTRODUCTION AND SUMMARY

The Bellman-Harris process is a well-known branching population model (Cf. Athreya-Ney (1972)),where each individual has a random lifespan and produces at death a random number of offsprings, which belong to the next generation and behave themselves in the same manner but independent of each other and of previous born individuals (branching property). We call such a Bellman-Harris process *generation-dependent*,if an individual of generation k has a random life-length with distribution function G_k and a random number of offsprings given by its probability generating function h_k, both depending on the generation index k , $k = 0,1,2, \dots$.

Let $Z(t)$ be the *number of individuals alive at time* t ,and let $Z^{\langle k\rangle}(t)$ be the *number of individuals of the k-th population-generation alive at time* t , if the population is initiated by one ancestor of age 0 at time 0 . We recall that this single ancestor forms the 0-th population-generation, his children the first, their children the second, and so on.

Such a generation-dependent Bellman-Harris process ($Z(t)$, $t\geq 0$) studied by Fildes (1971) , Fearn (1976) and also in a more general context by Edler (1976) and (1977) can serve as a stochastic population model, where a deterministic varying envivironment effects only the generations.

In this paper we shall study $V^{\langle k\rangle}$, *the time of extinction of the* k-*th population-generation*. We follow an advice of Bühler (1974) who stated a result for the generation-independent Markov branching process in continuous time, when the number of offsprings is bounded. This extinction time is the same as the time of the *last death* of an individual of the k-th population-generation and corresponds in some sense to the problem of the *first birth* of an individual of the k-th population-generation treated by Bühler (1972) and (1974) for the Bellman-Harris process. These problems have also, but in another way, been attacked by Biggins (1976) even for the more general Crump-Mode-Jagers process.

In the following, we stay in the neighbourhood of the Markov branching process in continuous time and suppose throughout the subsequent

Assumption 1. *For each generation* r , $r = 0,1, \ldots$, *let the lifespan be exponentially distributed with distribution function*

$$G_r(t) = \begin{cases} 0 & \text{if } t<0 \\ 1 - \exp(-\lambda_r t) & \text{if } t\geq 0 \end{cases}$$

and $0<\lambda_r \leq \bar{\lambda} < \infty$, *and the offspring distribution be given by its probability generating function* $h_r(s)$, $0\leq s\leq 1$, *satisfying* $h_r''(1) \leq \bar{m} < \infty$.

Then we define

$$m_k = h_k'(1) \tag{1.1}$$

as the mean number of offsprings of an individual in generation k ,

$$T_k = \left| \left\{ j\in\{0,1, \ldots , k-1\} : \lambda_j = \lambda_k \right\} \right| \tag{1.2}$$

as the number of λ's in $\{\lambda_0, \ldots ,\lambda_{k-1}\}$ equal to λ_k and

$$D_k = \Big(\prod_{j=o}^{k-1} m_j\lambda_j\Big)\Big(T_k! \prod_{j=o,\ldots,k-1,\ \lambda_j\neq\lambda_k} (\lambda_j - \lambda_k) \Big)^{-1} \tag{1.3}$$

for $k\geq 1$ and $T_0 = 0$, $D_0 = 1$. Our main result concerning the tail of the distribution of $V^{\langle k\rangle}$ is then

Theorem 1.1. *If* $\lambda_k = \min\{\lambda_0,\ldots,\lambda_k\}$, *then as* $t\rightarrow\infty$

$$P(V^{\langle k\rangle}> t) = D_k t^{T_k}\exp(-\lambda_k t) + o(t^{T_k}\exp(-\lambda_k t)) \quad \text{for } k = 0,1, \ldots . \tag{1.4}$$

After some preliminary results for the one- and two-dimensional probability generating functions of $Z^{\langle k\rangle}(t)$ in section 2 we infer in section 3 this theorem at once from an asymptotic representation of the probability generating function of $Z^{\langle k\rangle}(t)$ as t tends to infinity, given in Theorem 3.1. For the generation-independent Markov branching process we get some more detailed results. The rather technical proofs will be given in section 5. In section 4 we show weak convergence of the size of a population-generation at time t to a Poisson distribution, if both t and the number of initial individuals N tend to infinity, linked by relation (4.1). Furthermore, in this asymptotic the size of a population-generation decreases ultimately in independent jumps.

Throughout this paper $|A|$ denotes the cardinal number of the set A and $a(t) = o(b(t))$ means $a(t)/b(t) \longrightarrow 0$ as $t \to \infty$, and we shall use the convention, that empty products will be defined as 1.

2. PRELIMINARY RESULTS

To study $Z(t)$ or $Z^{\langle k\rangle}(t)$ in the generation-dependent case, we have to consider subpopulations initiated by individuals living in any generation i, $i \in \mathbb{N}_0$, if $\mathbb{N}_0 = \{0,1,2,\dots\}$. In this slightly more general case, if the population is initiated by one ancestor of age 0 at time 0 in generation i, we define $Z_i(t)$ as the *number of individuals alive at time* t, $Z_i^{\langle k\rangle}(t)$ as the *number of individuals in the* k-*th population-generation alive at time* t and $V_i^{\langle k\rangle}$ as the *time of extinction of the k-th population-generation, with one ancestor in generation* i. We call $(Z_i(t), t \geq 0)$ the *generation-dependent Bellman-Harris process with one ancestor of generation* i.

Then this single ancestor of generation i makes up the 0-th population-generation. He dies at some time L_i with distribution function G_i and produces at his death all his offsprings, say, N_i with a distribution given by h_i. Therefore, the number of his descendants alive at time t in the k-th population-generation is the same as the number of descendants after k-1 generations of his children living the time $t - L_i$ later beginning from the birth-time L_i. From this we conclude the *representations*

$$(2.1)\qquad \begin{aligned} Z_i^{\langle 0\rangle}(t) &= \delta(L_i - t) \\ Z_i^{\langle k\rangle}(t) &= \sum_{j=1}^{N_i} Z_{i+1}^{\langle k-1\rangle}(t - L_i, j) \end{aligned}$$

for $k \in \mathbb{N}$, if $\mathbb{N} = \{1,2,\dots\}$, and for all $i \in \mathbb{N}_0$, where $\delta(x) = 0$ if $x \leq 0$ and $\delta(x) = 1$ if $x > 0$ and where $Z_{i+1}^{\langle k-1\rangle}(t - L_i, j)$ for $j = 1, \dots, N_i$ are independent random variables, which have, given $L_i = u$, the same distribution as $Z_{i+1}^{\langle k-1\rangle}(t-u)$.

Now, we introduce the probability generating functions

$$F_i^{\langle k\rangle}(s,t) = E\left[s^{Z_i^{\langle k\rangle}(t)}\right],$$

(2.2)
$$F_i^{\langle k\rangle}(s,\sigma;t,\tau) = E\left[s^{Z_i^{\langle k\rangle}(t)}\sigma^{Z_i^{\langle k\rangle}(t+\tau)}\right]$$

for $i,k \in \mathbb{N}_0$, $t,\tau \geq 0$ and $|s|, |\sigma| \leq 1$.

For a generation-dependent Bellman-Harris process with data (G_r, h_r), $r \geq i$, one deduces easily from (2.1) with standard conditioning a system of integral equations for these generating functions, cf. Edler ((1976), section 7.2 and Lemma 8.1). Under the assumption 1 these systems reduce to

Lemma 2.1. *Suppose* Assumption 1 *for* $r \geq i$, $i \in \mathbb{N}_0$. *Then for* $t,\tau \geq 0$, $|s|, |\sigma| \leq 1$,

(2.3) $$F_i^{\langle 0\rangle}(s,t) = 1 - (1-s)\exp(-\lambda_i t),$$

(2.4) $$F_i^{\langle n\rangle}(s,t) = \exp(-\lambda_i t) + \int_0^t h_i(F_{i+1}^{\langle n-1\rangle}(s,t-u))\lambda_i \exp(-\lambda_i u)du,$$

(2.5) $$F_i^{\langle 0\rangle}(s,\sigma;t,\tau) = 1 - \exp(-\lambda_i t)\left[s(1-\sigma)\exp(-\lambda_i \tau) + 1-s\right],$$

(2.6) $$F_i^{\langle n\rangle}(s,\sigma;t,\tau) = \exp(-\lambda_i(t+\tau)) + \int_t^{t+\tau} h_i(F_{i+1}^{\langle n-1\rangle}(\sigma,t+\tau-u))\lambda_i \exp(-\lambda_i u)du$$
$$+ \int_0^t h_i(F_{i+1}^{\langle n-1\rangle}(s,\sigma;t-u,\tau))\lambda_i \exp(-\lambda_i u)\, du$$

for $n \in \mathbb{N}$.

Obviously the distribution of $Z_i^{\langle k\rangle}(t)$ is determined by the offspring distributions of the generation indices $i, i+1, \ldots, i+k-1$ and by the life-length distributions of the indices $i, i+1, \ldots, i+k$.

3. THE ASYMPTOTIC DISTRIBUTION OF THE SIZE AND OF THE EXTINCTION TIME OF A POPULATION-GENERATION

For the generation-dependent Bellman-Harris process with exponential lifespan special results about $V_i^{\langle k\rangle}$ and $Z_i^{\langle k\rangle}(t)$ will be derived from the following

Theorem 3.1. *If* $(Z_i(t), t \geq 0)$ *is a generation-dependent Bellman-Harris process under* Assumption 1 *and* $\lambda_{i+k} = \min\{\lambda_i, \ldots, \lambda_{i+k}\}$ *for* $i,k \in \mathbb{N}_0$ *fixed, then for* $s \in [0,1]$ *and* $t \geq 0$

(3.1) $$F_i^{\langle k\rangle}(s,t) = 1 - (1-s)D_{i,k}t^{T_{i,k}}\exp(-\lambda_{i+k}t)(1 + \rho_{i,k}(s,t)),$$

with

(3.2) $$T_{i,k} = \left|\left\{ j \in \{i,i+1,\ldots,i+k-1\} : \lambda_j = \lambda_{i+k}\right\}\right|,$$

(3.3) $$D_{i,k} = \left(\prod_{j=i}^{i+k-1} m_j\lambda_j\right)\left(T_{i,k}! \prod_{j=i,\ldots,i+k-1,\lambda_j \neq \lambda_{i+k}} (\lambda_j - \lambda_{i+k})\right)^{-1}$$

for $k \geq 1$, $T_{i,0} = 0$, $D_{i,0} = 1$ *and* $\rho_{i,k}(s,t) \rightarrow 0$ *as* $t \rightarrow \infty$.

The *proof* deferred to section 5 is based on Lemma 5.1 and Lemma 5.2 showing how to get $F_j^{\langle n\rangle}$ from $F_{j+1}^{\langle n-1\rangle}$. This reduction can be used to derive $F_i^{\langle k\rangle}$ by steps from the trivially known $F_{i+k}^{\langle o\rangle}$.

Remark 3.1. The condition $\lambda_{i+k} = \min\{\lambda_i,\dots,\lambda_{i+k}\}$ imports, that the "last" mean lifespan $1/\lambda_{i+k}$ is not smaller than all preceding ones.

If $\lambda_i = \min\{\lambda_i,\dots,\lambda_{i+k}\}$ and $k \geq 1$, then as $t \to \infty$

$$F_i^{\langle k\rangle}(s,t) = 1 - \exp(-\lambda_i t)(\, p_1(1-s,t) + \sigma_1(s,t)\,) \tag{3.4}$$

with a polynomial p_1 of a degree greater than 1 in $1-s$ and of a degree not greater than $|\{ j \in \{i+1,\dots,i+k\} : \lambda_j = \lambda_i\}|$ in t. The degree of $1-s$ in p_1 depends on the offspring distribution and can be made arbitrarily large.

If $\lambda_j = \min\{\lambda_i,\dots,\lambda_{i+k}\}$, $k \geq 2$ and $j \neq i,i+k$, then as $t \to \infty$

$$F_i^{\langle k\rangle}(s,t) = 1 - \exp(-\lambda_j t)(\, p_2(1-s,t) + \sigma_2(s,t)\,) \tag{3.5}$$

with a polynomial p_2 of a degree greater then 1 in $1-s$. The functions $\sigma_i(s,t)$, $i = 1,2$, are bounded in absolute value with respect to t and tend to 0 as t tends to infinity for all $s \in [0,1]$. An explicit computation of the polynomials p_1 and p_2 becomes very complicated even for simple offspring distributions and small k, as can be seen in an example given by Edler ((1976), pp. 114-115).

Therefore, a decreasing mean lifespan at two successive generations implies, that roughly speaking "all individuals of the latter population-generation vanish asymptotically as t tends to infinity of the same order of magnitude". This is in contrast to the situation of Theorem 3.1 from which we deduce the

Corollary 3.2 *Under the assumptions of* Theorem 3.1 *holds as* $t \to \infty$

$$P(\, Z_i^{\langle k\rangle}(t) = 1\,) = D_{i,k}\, t^{T_{i,k}} \exp(-\lambda_{i+k} t) + o(t^{T_{i,k}} \exp(-\lambda_{i+k} t)) \,,$$

$$P(\, Z_i^{\langle k\rangle}(t) \geq 2\,) = o(t^{T_{i,k}} \exp(-\lambda_{i+k} t)) \,.$$

Interpretation. The probability that two or more individuals of an arbitrary population-generation live at time t tends in higher order to 0 than the probability that exactly one individual is alive at time t, if t tends to infinity. Therefore, asymptotically speaking, in a population-generation lives at last exactly one individual and with its death the whole population-generation is extinguished.

Since $P(\, V_i^{\langle k\rangle} \leq t\,) = P(\, Z_i^{\langle k\rangle}(t) = 0\,)$ we obtain from Theorem 3.1 at once the following Theorem 3.2 which reduces for $i = 0$ to Theorem 1.1.

Theorem 3.2. *Under the assumptions of* Theorem 3.1 *holds as* $t \to \infty$

$$P(V_i^{\langle k\rangle} > t) = D_{i,k} t^{T_{i,k}} \exp(-\lambda_{i+k} t) + o(t^{T_{i,k}} \exp(-\lambda_{i+k} t)) \quad . \tag{3.6}$$

Remark 3.2. If $\lambda_{i+k} < \lambda_j$ for $j=i, \dots, i+k-1$, then this reduces to

$$P(V_i^{\langle k\rangle} > t) = \prod_{j=i}^{i+k-1} (m_j \lambda_j / (\lambda_j - \lambda_{i+k})) \exp(-\lambda_{i+k} t) + o(\exp(-\lambda_{i+k} t)). \tag{3.7}$$

If only the offspring distributions are generation-dependent and if $0 < 1/\lambda < \infty$ is the mean life-length, then

$$P(V_i^{\langle k\rangle} > t) = (\lambda^k / k!) (\prod_{j=i}^{i+k-1} m_j) t^k \exp(-\lambda t) + o(t^k \exp(-\lambda t)) \quad . \tag{3.8}$$

Finally, in the complete generation-dependent case we get back the result of Bühler (1974) for the tail of the distribution of the extinction time:

Corollary 3.2. *If* $(Z(t), t \geq 0)$ *is a Markov branching process in continuous time with* $0 < 1/\lambda < \infty$ *as mean life-length and* $0 < m < \infty$ *as mean number of offsprings and if the second moment of the offspring distribution exists, then for* $k \in \mathbb{N}_0$

$$P(V^{\langle k\rangle} > t) = (k!)^{-1} (m\lambda t)^k \exp(-\lambda t) + o(t^k \exp(-\lambda t)) \quad \text{as } t \to \infty . \tag{3.9}$$

For the generation-independent Markov branching process the assertion of Corollary 3.2 can be strengthened if the number of offsprings is bounded:

Theorem 3.3. *If* $(Z(t), t \geq 0)$ *is a Markov branching process in continuous time with mean lifespan* $\lambda = 1$ *and an offspring distribution given by the generating function* $h(s) = \sum_{i=0}^{n} p_i s^i$ *with* $n \geq 1$ *and* $p_n > 0$, *then as* $t \to \infty$ *for* $k \in \mathbb{N}$ *and* $s \in [0,1]$

$$F^{\langle k\rangle}(s,t) = 1 - (1-s)(m^k/k!) t^k e^{-t} + \sum_{i=1}^{k} c_i^{\langle k\rangle}(s) t^{k-i} e^{-t} + o(e^{-t}) \tag{3.10}$$

where for $k \in \mathbb{N}$

$$c_1^{\langle k\rangle}(s) = \sum_{j=2}^{n} (1-s)^j (-1)^j h^{(j)}(1) m^{k-1} \left[(k-1)! j! (j-1)\right]^{-1} \tag{3.11}$$

and for $i = 2, \dots, k$ *and* $k \geq 2$

$$c_i^{\langle k\rangle}(s) = \sum_{j=0}^{n^i} d_{i,j}^{\langle k\rangle} s^j \tag{3.12}$$

with not further specified real numbers $d_{i,j}^{\langle k\rangle}$ *for* $j = 0, \dots, n^i$. ($h^{(j)}$ denotes the j-th derivative of h(s) with respect to s.)

We defer an outline of the *proof* to section 5. Yet, if A_i and $A_{i,j}$ are some real numbers depending on the offspring distribution, we deduce from Theorem 3.3 at once if t tends to infinity the following:

$$P(Z^{\langle k\rangle}(t)=j) = \begin{cases} (m^k/k!)t^k e^{-t} + o(t^k e^{-t}) & \text{for } j=1 \\ A_j t^{k-1} e^{-t} + o(t^{k-1}e^{-t}) & \text{for } j=2,\dots,n \\ A_{i,j} t^{k-i} e^{-t} + o(t^{k-i}e^{-t}) & \text{for } j=n^{i-1}+1,\dots,n^i \ , \ i=2,\dots,k \ . \end{cases}$$

Interpretation. The probability, that the k-th population-generation consists at time t of exactly one individual, is of the same order of magnitude as the probability, that in the (k+i)-th population-generation live at time t $n^{i-1}+1$ to n^i individuals, $i = 1, \dots, k$, and we can say that asymptotically with highest probability there exists exactly one line of the population.

A corresponding result for non-bounded numbers of offsprings has been only obtained under strong conditions on $h(s)$, see Edler ((1976), pp. 126-127).

The crucial point in the expansion of the probability generating functions of Theorem 3.1 and 3.3 is the integrability of the tails $1 - G_i(t) = \exp(-\lambda_i t)$ of the lifespan distributions. Merely for some classes of distribution functions G characterized by suitable conditions on the tails $1 - G(t)$ the asymptotic behavior of $P(V^{\langle k\rangle} > t)$ for $t \to \infty$ could be determined. To this one expands $F^{\langle k\rangle}(s,t)$ directly in $s = 1$ and uses the asymptotic behavior of the first two moments of $Z_i^{\langle k\rangle}(t)$.

A simple generalization is attainable for *multi-phase exponentially distributed lifespans* :

Let $(Z(t), t \geq 0)$ be a generation-dependent Bellman-Harris process where

$$G_r(t) = \begin{cases} 0 & \text{if } t < 0 \\ 1 - \exp(-\lambda_r t) \sum_{j=0}^{n_r} (\lambda_r t)^j / j! & \text{if } t \geq 0 \end{cases}$$

are Erlang-distribution functions for the life-lengths with $\lambda_r > 0$ and $n_r \in \mathbb{N}_0$ for $r \in \mathbb{N}_0$; and where the second moments of the offspring distributions have an uniform upper bound. An Erlang-distributed life-length with parameter (λ_r, n_r) and an offspring distribution given by h_r corresponds to a life-length distribution which is the independent sum of n_r+1 phases, where each phase has an exponentially distributed length with parameter λ_r, and at the end of the first n_r phases there will be generated exactly one offspring, and at the end of the last phase there will be generated the usual number of offsprings given by h_r. Hence, to the given data $(\lambda_r, n_r; h_r(s))$, for $r \in \mathbb{N}_0$, we can construct a corresponding generation-dependent Bellman-Harris process, which satisfies Assumption 1 and has the data $(\tilde{\lambda}_j, \tilde{h}_j(s))$ for $j \in \mathbb{N}_0$:

If $v_{-1} = 0$ and $v_r = \sum_{l=0}^{r} n_l + r + 1$ for $r = 0, 1, \dots$, then $\tilde{\lambda}_j = \lambda_r$ if $j = v_{r-1}, \dots, v_r - 1$ and $\tilde{h}_j(s) = s$ if $j = v_{r-1}, \dots, v_r - 2$ and $\tilde{h}_j(s) = h_r(s)$ if $j = v_r - 1$. The distribution of $V_0^{\langle k\rangle}$ is then the same as that of $\tilde{V}_0^{\langle v_k - 1\rangle}$, if $\tilde{V}_0^{\langle n\rangle}$ is the time of extinction of the n-th population-generation of this corresponding process with the data $(\tilde{\lambda}_j, \tilde{h}_j(s))$.

4. A LIMIT THEOREM

Throughout this section we assume without loss of generality $i = 0$ and consider a sequence of generation-dependent Bellman-Harris processes $\{(Z(t;N), t \geq 0)$, $N = 1,2, \dots \}$, where $Z(0;N) = N$ is the number of ancestors of age 0 at time 0 in the 0-th population-generation. Further, we suppose Assumption 1 for $r \in \mathbb{N}_0$, $\{\lambda_r; r = 0,1, \dots\}$ nonincreasing and $\lambda_r, m_r > 0$ for $r \in \mathbb{N}_0$.

Then we study $Z^{\langle k \rangle}(t;N)$ the *number of individuals of the* k*-th population-generation alive at time* t *, if* N *ancestors initiate the population.* We set

$$F^{\langle k \rangle}(s,t;N) = E\left[s^{Z^{\langle k \rangle}(t;N)}\right],$$

$$F^{\langle k \rangle}(s,\sigma;t,\tau;N) = E\left[s^{Z^{\langle k \rangle}(t;N)}\sigma^{Z^{\langle k \rangle}(t+\tau;N)}\right].$$

Because of the branching property $Z^{\langle k \rangle}(t;N)$ is distributed as the independent sum of N random variables distributed as $Z^{\langle k \rangle}(t)$ in a population with one ancestor. Hence Theorem 3.1 yields for each $k \in \mathbb{N}_0$

Theorem 4.1. *If* $\{t_{N,k}$, $N=1,2,..\}$ *is a sequence of real numbers and* v_k *a positive real number such that*

$$N\, t_{N,k}^{T_k} \exp(-\lambda_k t_{N,k}) \xrightarrow[N \to \infty]{} v_k , \tag{4.1}$$

then for $s \in [0,1]$

$$F^{\langle k \rangle}(s,t_{N,k};N) \xrightarrow[N \to \infty]{} \exp(-D_k(1-s)v_k) ; \tag{4.2}$$

furthermore, if we set $u_k = v_k \exp(-\lambda_k \tau)$ *for* $\tau \geq 0$ *, then for* $s,\sigma \in [0,1]$

$$F^{\langle k \rangle}(s,\sigma;t,\tau;N) \xrightarrow[N \to \infty]{} \exp(-D_k\left[(1-s\sigma)u_k + (v_k - u_k)(1-s)\right]) . \tag{4.3}$$

Proof. Because of the above mentioned branching property and Theorem 3.1

$$F^{\langle k \rangle}(s,t_{N,k};N) = (F^{\langle k \rangle}(s,t_{N,k}))^N = (1 - (1-s)x_{N,k}/N)^N$$

with

$$x_{N,k} = D_k\, N\, t_{N,k}^{T_k} \exp(-\lambda_k t_{N,k})(1 + \rho_k(s,t_{N,k})) ,$$

where $\rho_k = \rho_{0,k}$ from Theorem 3.1. (4.1) implies $t_{N,k} \to \infty$ and therefore $\rho_k(s,t_{N,k}) \to 0$ as $N \to \infty$. Hence $x_{N,k} \longrightarrow v_k D_k$ and then the first assertion follows from the well-known fact, that $y_N \to y$ implies $(1 - y_N/N)^N \longrightarrow e^{-y}$ as N tends to ∞.

For the second assertion of Theorem 4.1 first of all we need like in Theorem 3.1 an asymptotic representation of the two-dimensional generating function $F_i^{\langle k \rangle}(s,\sigma;t,\tau)$ defined in (2.2). In analogy to Theorem 3.1 one obtains from the system (2.5) and (2.6) , see Edler (1976), for $k \in \mathbb{N}_0$, $s, \sigma \in [0,1]$, $t,\tau \geq 0$, the following representation

(4.4) $F^{\langle k\rangle}(s,\sigma;t,\tau) = 1 - D_k[s(1-\sigma)\exp(-\lambda_k\tau) + 1-s]\exp(-\lambda_k t)t^{T_k}(1 + \rho_k(s,\sigma;t,\tau))$

with $\rho_k(s,\sigma;t,\tau) \longrightarrow 0$ as $t \to \infty$. Observing $[s(1-\sigma)\exp(-\lambda_k\tau) + 1-s]\, v_k =$
$= (s-s\sigma)u_k - (1-s)v_k = (1-s\sigma)u_k + (v_k - u_k')(1-s)$, we then reach (4.3) as in the first part of this proof.

Corollary 4.1. *If* t *and* $N \longrightarrow \infty$ *so that*

(4.5) $$N\, t^{T_k} \exp(-\lambda_k t) \longrightarrow v_k > 0 ,$$

then

a) $Z^{\langle k\rangle}(t;N)$ *is asymptotically* Poisson *distributed with parameter* $D_k v_k$;

b) *the decrements* $Z^{\langle k\rangle}(t;N) - Z^{\langle k\rangle}(t+\tau;N)$, $Z^{\langle k\rangle}(t+\tau;N)$ *are asymptotically independent.*

Proof. a) follows at once from (4.2). For b) it is easily shown, that in the limit as t , $N \to \infty$ according to (4.5) the common characteristic function of the random variables $Z^{\langle k\rangle}(t;N) - Z^{\langle k\rangle}(t+\tau;N)$ and $Z^{\langle k\rangle}(t+\tau;N)$ is equal to the product of the limits of each single characteristic function of these two random variables. To that end, one has merely to replace in (4.2) and (4.3) s and σ by e^{ix} and e^{iy}, see Edler (1976), since these expansions hold also for $|s|$, $|\sigma| \leq 1$.

Remark. Corollary 4.1 justifies the following intuitive argument:
$p = D_k t^{T_k} \exp(-\lambda_k t) + o(t^{T_k}\exp(-\lambda_k t))$ is because of Corollary 3.1 the probability, that there is exactly one individual of the k-th population-generation alive at time t . In the case of N ancestors the last individuals of a generation act independently of each other. Now, if N and $t \to \infty$ so that $Np \to \alpha$, a Poisson limit theorem for binomial distributions yields, that the number of individuals of this population-generation is asymptotically Poisson distributed with parameter α .

For the special cases discussed in Remark 3.2 we obtain

Corollary 4.2. *For* $k \in \mathbb{N}$ $Z^{\langle k\rangle}(t;N)$ *is for* $N,t \to \infty$ *asymptotically* Poisson *distributed*

a) *with parameter* $v_k \cdot \prod_{j=0}^{k-1} [m_j\lambda_j/(\lambda_j - \lambda_k)]$, *if* $\{\lambda_k\}$ *is strictly decreasing and* $N\exp(-\lambda_k t) \longrightarrow v_k > 0$ as $N,t \to \infty$;

b) *with parameter* $v\cdot(\lambda^k/k!)\cdot \prod_{j=0}^{k-1} m_j$, *if* $\lambda_n = \lambda$ *for all* n *and* $Nt^k\exp(-\lambda t) \longrightarrow v > 0$;

c) *with parameter* $v\cdot(m^k/k!)$, *if* $\{(Z(t;N), t \geq 0) : N = 1,2,\ldots\}$ *is a sequence of Markov branching processes with* $\lambda = 1$ *and* $Nt^k\exp(-t) \longrightarrow v > 0$ *as* $N,t \to \infty$.

5. PROOFS FOR SECTION 3

In the following, we frequently use the integration formula

$$(5.1)\qquad \int_0^t (t-y)^n e^{a(t-y)} dy = \int_0^t y^n e^{ay} dy = e^{at}\sum_{j=o}^{n} \frac{n!t^j(-1)^{n-j}}{j!a^{n-j+1}} + (-1)^{n+1} n!/a^{n+1}$$

$$= n!(-a)^{-n-1}\left\{1 - e^{at}\sum_{j=o}^{n} t^j(-a)^j/j!\right\}, \text{ if } a \neq 0,\ n\in\mathbb{N}_o .$$

In the proofs below A_j, B_j, $A_{i,j}$, $B_{i,j}$ will be some not further specified non-negative real numbers independent of t, and $\sigma_j(t)$, $\sigma_j(s,t)$ and $\bar{\sigma}_j(t)$ will be some real valued functions with bounded absolute value in t and converging to 0 as $t \to \infty$.

The following two lemmas are crucial for the results in section 3.

Lemma 5.1. *Suppose* Assumption 1 *and let* $n\in\mathbb{N}$, $j,p\in\mathbb{N}_o$ *and* $0<\lambda<\lambda_j$. *If for* $s\in[0,1]$, $t\geq 0$ *with some non-negative real numbers* a_r, $r = 0,1, \ldots , p$,

$$(5.2)\qquad F_{j+1}^{\langle n-1\rangle}(s,t) = 1 - \exp(-\lambda t)\left[(1-s)\sum_{r=o}^{p} a_r t^r + \sigma_1(s,t)\right],$$

then for $s\in[0,1]$, $t\geq 0$

$$(5.3)\qquad F_j^{\langle n\rangle}(s,t) = 1 - \exp(-\lambda t)\left[(1-s)\sum_{q=o}^{p} \bar{a}_q t^q + \sigma_2(s,t)\right]$$

with

$$(5.4)\qquad \bar{a}_q = m_j\lambda_j \sum_{h=q}^{p} a_h b_{hq} \quad \textit{for } q = 0,1, \ldots , p$$

and

$$(5.5)\qquad b_{hq} = h!(-1)^{h-q}/\left[q!(\lambda_j-\lambda)^{h-q+1}\right] \quad \textit{for } q = 0,1, \ldots , h \textit{ and } h = 0,1,\ldots,p.$$

Proof. For $j\in\mathbb{N}_o$, due to Sewastjanow ((1975), p.6), we have for $s\in[0,1]$

$$(5.6)\qquad h_j(s) = 1 - (1-s)m_j + \tfrac{1}{2}(1-s)^2 R_j(s)$$

with $R_j(s)$ nondecreasing, $0\leq R_j(s)\leq \delta_j$ and $R_j(s) \to \delta_j$ as $s \to 1$, if we set $\delta_j = h_j''(1)$. Now, since being a generating function, $F_{j+1}^{\langle n-1\rangle}(s,t-u)\in[0,1]$ for $s\in[0,1]$ and $0\leq u\leq t$ and putting (5.2) into (5.6), we obtain

$$(5.7)\qquad h_j(F_{j+1}^{\langle n-1\rangle}(s,t-u)) = 1 - m_j\exp(-\lambda(t-u))\left\{(1-s)\sum_{r=0}^{p} a_r(t-u)^r + \sigma_1(s,t-u)\right\}$$

$$+ \tfrac{1}{2}\exp(-2\lambda(t-u))\left\{(1-s)\sum_{r=0}^{p} a_r(t-u)^r + \sigma_1(s,t-u)\right\}^2 + R_j(F_{j+1}^{\langle n-1\rangle})$$

with $R_j(F_{j+1}^{\langle n-1\rangle}) = R_j(F_{j+1}^{\langle n-1\rangle}(s,t-u))\in[0,\delta_j]$. Inserting (5.7) in (2.4) we get

$$(5.8)\qquad F_j^{\langle n\rangle}(s,t) = 1 - (1-s)m_j\lambda_j\int_0^t \sum_{r=0}^{p} a_r(t-u)^r \exp(-\lambda(t-u))\exp(-\lambda_j u)du + J + K,$$

if we define

$$(5.9)\qquad J = J(s,t) = -A_1\int_0^t \sigma_1(s,t-u)\exp(-\lambda(t-u))\exp(-\lambda_j u)du$$

$$(5.10)\qquad K = K(s,t) = A_2 \int_0^t \exp(-2\lambda(t-u))\exp(-\lambda_j u)\cdot \cdot\left\{(1-s)\sum_{r=0}^{p} a_r(t-u)^r + \sigma_1(s,t-u)\right\}^2 R_j(F_{j+1}^{\langle n-1\rangle})du$$

For the remaining part of the proof we have to treat the three integrals in (5.8). The first one will be evaluated using (5.1) :

$$(5.11)\qquad \int_0^t (t-u)^r \exp(-\lambda(t-u))\exp(-\lambda_j u)du = \exp(-\lambda_j t)\int_0^t y^r \exp((\lambda_j - \lambda)y)dy$$

$$= \exp(-\lambda t)\left\{\sum_{w=0}^{r} \frac{r!(-1)^{r-w}}{w!(\lambda_j-\lambda)^{r-w+1}} t^w + (-1)^{r+1}\frac{r!\exp(-(\lambda_j-\lambda)t)}{(\lambda_j-\lambda)^{r+1}}\right\}$$

$$= \exp(-\lambda t)\left\{\sum_{w=0}^{r} b_{rw}\, t^w + \sigma_r(t)\right\} ,$$

with b_{rw} from (5.5) and $\sigma_r(t)$ as arranged.

Since $\lambda_j > \lambda$, $|\sigma_1(s,t)|$ bounded for $t \geq 0$ and $\sigma_1(s,t) \to 0$ as $t \to \infty$, Lemma 1 of Smith (1954) yields $\int_0^t \sigma_1(s,t-u)\exp(-(\lambda_j - \lambda)u)du \longrightarrow 0$ as $t \to \infty$ and hence

$$(5.12)\qquad J(s,t) = \exp(-\lambda t)\,\sigma_3(s,t) .$$

Next, one easily shows with the aid of (5.1) that

$$\int_0^t \exp(-2\lambda(t-u))\exp(-\lambda_j u)du = \exp(-\lambda t)\,\bar{\sigma}_1(t)$$

$$\int_0^t (t-u)^r \exp(-2\lambda(t-u))\exp(-\lambda_j u)du = \exp(-\lambda t)\,\bar{\sigma}_2(t)$$

and applying these two formulas to K from (5.10) we obtain

$$(5.13)\qquad |K(s,t)| \lesssim A_3 \int_0^t \exp(-2\lambda(t-u))\exp(-\lambda_j u)\,\sigma_1(s,t-u)^2 du + \sum_{r=0}^{p} B_r \int_0^t |\sigma_1(s,t-u)|\,(t-u)^r \exp(-2\lambda(t-u))\exp(-\lambda_j u)du + \sum_{r,q=0}^{p} B_{r,q}\int_0^t (t-u)^{r+q}\exp(-2\lambda(t-u))\exp(-\lambda_j u)du$$

$$\lesssim \exp(-\lambda t)\,\sigma_4(s,t) .$$

Finally, we sum up $(1-s)\sum_{r=0}^{p}\sigma_r(t)$, $\sigma_3(s,t)$ and $\sigma_4(s,t)$ to $\sigma_2(s,t)$, exchange in $\sum_{r=0}^{p} a_r \sum_{w=0}^{r} b_{rw}\, t^w$ the order of summation and then conclude (5.3) from (5.8) together with (5.11) - (5.13) .

Corollary 5.1. *Under the assumptions of* Lemma 5.1 *holds*

a) $$F_j^{\langle n\rangle}(s,t) = 1 - (1-s)\bar{a}_p t^p \exp(-\lambda t) + \beta(s,t) \quad \textit{for} \quad p \geq 1$$

with $\bar{a}_p = a_p m_j \lambda_j/(\lambda_j-\lambda)$ *and* $|\beta(s,t)| \leq A_6 t^{p-1}\exp(-\lambda t) + B_6 \exp(-\lambda t)$;

b) *if* $|\sigma_1(s,t)| \leq \exp(-bt)B(s,t)$ *for a real* $b>0$ *and a function* B *bounded with respect to* t *on* $[0,\infty)$ *for all* $s\in[0,1]$, *then* $|\sigma_2(s,t)| \leq \exp(-\bar{b}t)\bar{B}(s,t)$ *with* $\bar{b}$ *and* $\bar{B}$ *having the same properties as* b *and* B .

Lemma 5.2. *Suppose* Assumption 1 *and let* $n\in\mathbb{N}$, $j,p\in\mathbb{N}_0$, $0<\lambda=\lambda_j$. *If for* $s\in[0,1]$, $t\geq 0$, *with some nonnegative real numbers* a *and* b

$$F_{j+1}^{\langle n-1\rangle}(s,t) = 1 - (1-s)at^p\exp(-\lambda t) + \bar{\sigma}_1(s,t) \tag{5.14}$$

with

$$|\bar{\sigma}_1(s,t)| \leq \begin{cases} A_1\exp(-\lambda t)t^{p-1} + B_1\exp(-\lambda t) & \textit{if} \quad p\geq 1 \\ A_2\exp(-\lambda t - bt) & \textit{if} \quad p=0 \end{cases} \tag{5.15}$$

then for $s\in[0,1]$, $t\geq 0$

$$F_j^{\langle n\rangle}(s,t) = 1 - (1-s)m_j\lambda a(p+1)^{-1}t^{p+1}\exp(-\lambda t) + \bar{\sigma}_2(s,t) \tag{5.16}$$

with

$$|\sigma_2(s,t)| \leq A_3\exp(-\lambda t)t^p + B_2\exp(-\lambda t) \ . \tag{5.17}$$

Proof. Proceeding as in the forestanding proof of Lemma 5.1 we obtain now from (5.14)

$$\begin{aligned} F_j^{\langle n\rangle}(s,t) &= 1 - (1-s)m_j\lambda a \int_0^t (t-u)\exp(-\lambda(t-u))\exp(-\lambda u)du + J(s,t) + K(s,t) \\ &= 1 - (1-s)\, m_j\lambda(p+1)^{-1}at^{p+1}\exp(-\lambda t) + J(s,t) + K(s,t) \ , \end{aligned} \tag{5.18}$$

with

$$J(s,t) = m_j\lambda \int_0^t \bar{\sigma}_1(s,t-u)\exp(-\lambda u)du$$

and

$$K(s,t) = \frac{1}{2}\lambda \int_0^t \left\{(1-s)a\exp(-\lambda(t-u))(t-u)^p - \bar{\sigma}_1(s,t-u)\right\}^2 R_j(F_{j+1}^{\langle n-1\rangle})\exp(-\lambda u)du.$$

Then, using similiar methods as in the proof of Lemma 5.1 one can show, see Edler ((1976), pp. 136-137), that

$$|J(s,t)| \leq A_4\exp(-\lambda t)t^p + A_5\exp(-\lambda t) \quad \text{if} \quad p\geq 0 \ ,$$

$$|K(s,t)| \leq \begin{cases} B_3\exp(-\lambda t)t^p + B_4\exp(-\lambda t) & \text{if} \quad p\geq 1 \\ B_5\exp(-\lambda t) & \text{if} \quad p=0 \end{cases}$$

Hence $\bar{\sigma}_2(s,t) = J(s,t) + K(s,t)$ satisfies (5.17) and the assertion follows from (5.18).

Proof of Theorem 3.1. For $k=0$ (3.1) follows from (2.3) with $\rho_{i,o} = 0$, since $T_{i,o} = 0$ and $D_{i,o} = 1$.

Let k be greater 0. Now we derive $F_i^{\langle k\rangle}$ from $F_{i+k}^{\langle o\rangle}$:
If $\lambda_{i+k-1} > \lambda_{i+k}$, (2.3), Lemma 5.1 and Corollary 5.1 b) yield for the first step $F_{i+k}^{\langle o\rangle} \longrightarrow F_{i+k-1}^{\langle 1\rangle}$, abbreviating $i+k-1 = n$,

$$F_n^{\langle 1\rangle}(s,t) = 1 - (1-s)\exp(-\lambda_{i+k}t)m_n\lambda_n/(\lambda_n - \lambda_{i+k}) + \rho_{n,1}(s,t)\exp(-\lambda_{i+k}t)$$

with $|\rho_{n,1}(s,t)| \leq \exp(-b_n t)\bar{\rho}_n(s,t)$, where $b_n > 0$ and $\bar{\rho}_n(s,t)$ is bounded with respect to t.

If $\lambda_{i+k-1} = \lambda_{i+k}$, Lemma 5.2 yields in the case $p = 0$, again setting $i+k-1 = n$,

$$F_n^{\langle 1\rangle}(s,t) = 1 - (1-s)m_n\lambda_n\exp(-\lambda_{i+k}t)\,t + \rho'_{n,1}(s,t)$$

with $|\rho'_{n,1}(s,t)| \leq A_{n,1}\exp(-\lambda_{i+k}t)$.

Generally for the l-th step $F_{i+k-l+1}^{\langle l-1\rangle} \longrightarrow F_{i+k-l}^{\langle l\rangle}$, abbreviating $i+k-l = j$, we argue as follows:

If $\lambda_j > \lambda_{i+k}$, according to Corollary 5.1 a) the highest exponent of t will be maintained and multiplied by the factor $m_j\lambda_j/(\lambda_j - \lambda_{i+k})$; and the absolute value of the remainder term can be bounded from above of the same order of magnitude.

If $\lambda_j = \lambda_{i+k}$, then according to Lemma 5.2 the highest exponent, say p, of t will be increased by 1 and multiplied by the factor $m_j\lambda_j/(p+1)$. The absolute upper bound of the remainder term also has an order of magnitude increased by one t-exponent.

This works for $l = 1,2, \ldots, k$. Deriving $F_i^{\langle k\rangle}$ from $F_{i+k}^{\langle o\rangle}$, at the transitions $F_{j+1}^{\langle l-1\rangle} \longrightarrow F_j^{\langle l\rangle}$ for $l = 1,\ldots,k$, $j = i+k-l$, one has only to attend, if λ_j is greater or equal to λ_{i+k} and then to proceed as indicated above. $T_{i,k}$ then counts the number of transitions where λ_i is equal to λ_{i+k} and $D_{i,k}$ collects the multiplying factors of all transitions.

Proof of Theorem 3.3. We shall give here only an outline of the proof and refer for the computations to Edler ((1976), pp. 120 - 125).

If, as assumed, $\lambda = 1$ and $h(s) = \sum_{i=0}^{n} p_i s^i$, we get from Lemma 2.1 for $k \geq 1$

$$F^{\langle k\rangle}(s,t) = e^{-t} + \int_0^t \sum_{i=0}^{n} p_i (F^{\langle k-1\rangle}(s,t-u))^i e^{-u} du$$

$$(5.19) \qquad = e^{-t} + \int_0^t \sum_{w=0}^{n} a_w (1 - F^{\langle k-1\rangle}(s,t-u))^w e^{-u} du$$

$$= e^{-t} + \sum_{w=0}^{n} a_w \int_0^t (1 - F^{\langle k-1\rangle}(s,t-u))^w e^{-u} du ,$$

if we expand h in $s = 1$, set $a_o = 1$ and $a_w = f_w(-1)^w/w!$ for $w = 1,\ldots,n$, where f_w is the w-th factorial moment of the offspring distribution given by h.

If $n = 1$ $Z^{\langle k\rangle}(t)$ is for all k 0 or 1 and with an induction argument we deduce from Lemma 2.1

$$F^{\langle k\rangle}(s,t) = 1 - (1-s)(m^k/k!)t^k e^{-t}$$

and, setting all $c_i^{\langle k\rangle}(s)$ and the o-term equal to 0, (3.7) follows.

If $n \geq 2$ it is suitable to proof first the slightly stronger

Lemma. *If we suppose the assumptions of* Theorem 3.3 *and set*

$$N(k) = \{ (x,y) \in \mathbb{N}_o^2 : 2 \leq x \leq n^k , 0 \leq y \leq kn^{k-1} \}$$

then for $s \in [0,1]$, $t \geq 0$ *and* $k \in \mathbb{N}$

$$(5.20) \quad F^{\langle k\rangle}(s,t) = 1 - (1-s)\frac{m^k}{k!}t^k e^{-t} + \sum_{i=1}^{k} c_i^{\langle k\rangle}(s)t^{k-i}e^{-t} + \sum_{(a,b)\in N(k)} c_{a,b}^{\langle k\rangle}(s)t^a e^{-bt}$$

with

$$o_1^{\langle k\rangle}(s) = \sum_{j=2}^{n} (1-s)^j m^{k-1}(-1)^j f_j [(k-1)!j!(j-1)]^{-1}$$

and some polynomials $c_i^{\langle k\rangle}(s)$ *of a degree not greater than* n^i *for* $i = 2, \ldots, k$ *and* $c_{a,b}^{\langle k\rangle}(s)$ *of a degree not greater than* n^k *in* s.

The *proof* using again an induction argument is based on (5.19) and an explicit computation and will be omitted here.

Hence, to complete the proof of Theorem 3.3 in the case $n \geq 2$ we only remark, that the last term on the right of (5.20) is $o(\exp(-t))$ since $(a,b)\in N(k)$ implies $a \geq 2$. An explicit formula for the coefficients of $c_i^{\langle k\rangle}(s)$ for $i = 2, \ldots, k$ seems very difficult to obtain.

ACKNOWLEDGEMENTS

This paper comprises a part of my doctoral thesis supported in part by Stiftung Volkswagenwerk and supervised by Professor W. J Bühler. I whish to thank Professor Bühler for his helpful advices and encouragement.

REFERENCES

Athreya, K. B. , Ney, P. *Branching processes*. Springer, Berlin-Heidelberg-New-York, 1972.

Biggins, J. D. (1976) The first- and last-birth problems for a multitype age-dependent branching process. *Adv. Appl. Prob.* 8 , 1976, 446-459.

Bühler, W. J. (1972) The distribution of generations and other aspects of the family structure of branching processes. *Proc. 6-th Berkeley Symp. Math. Statist. Prob.* Vol. III, 1972, 463-480.

Bühler, W. J. (1974) On the family structure of populations. *Adv. Appl. Prob. 6*, 1974, 192-193.

Edler, L. (1976) *Generationsabhängige Verzweigungsprozesse unter besonderer Berücksichtigung des Aussteibeverhaltens bei Bellman-Harris Prozessen mit exponentiellen Lebensdauern.* Dissertation, 1976, Mainz.

Edler, L. (1977) Strict supercritical generation-dependent Crump-Mode-Jagers branching processes. Submitted to *J. Appl. Prob.*

Fearn, D. H. (1976) Supercritical age dependent branching processes with generation dependence. *Ann. Prob. 4*, 1976, 27-37.

Fildes, R. (1971) An age dependent branching process with variable lifetime distributions. *Adv. Appl. Prob. 4*, 1971, 453-474.

Jagers, P. *Branching processes with biological applications*. Wiley, London-New York-Sydney-Toronto, 1975.

Savage, I. R. , Shimi, I. N. (1969) A branching process without rebranching. *Ann. Math. Statist. 40* , 1969, 1850-1851.

Sewastjanow, B. A. *Verzweigungsprozesse*. Oldenbourg, München-Wien, 1975.

Smith, W. L. (1954) Asymptotic renewal theorems. *Proc. Roy. Soc. Edinburgh Sect. A. 64*, 1954, 9-48.

Stratton, H. H. Jr. , Tucker, H. G. (1964) Limit distributions of a branching stochastic process. *Ann. Math. Statist. 35*, 1964, 557-565.

Johannes Gutenberg-Universität Mainz
Fachbereich 17 Mathematik

Saarstrasse 21 , Postfach 3980
D-6500 Mainz
W. Germany

РАНГОВЫЕ ПРОЦЕДУРЫ ИЗМЕРЕНИЯ И КЛАССИФИКАЦИИ БЕЗ ЭТАЛОНА

Алексей Ефимов, Вячеслав Кутеев

Харьков

ВВЕДЕНИЕ

Процедура измерения или классификации предполагает необходимым сравнение случайного параметра объекта с системой привнесенных извне эталонов.

Здесь, в предположении, что закон распределения параметра, характеризующего совокупность объектов, известен, предлагается способ оценивания значений их параметра, требующий, вместо процедуры сравнения объекта с эталоном, лишь упорядочивания выборки из множества объектов. Это означает возможность измерять и классифицировать в ситуации, когда система эталонов отсутствует, но объекты допускают сравнение между собой, приводящее к суждению "больше-меньше".

БЕЗЭТАЛОННОЕ ИЗМЕРЕНИЕ

Для того, чтобы произвести измерение значения x непрерывной случайной величины X , необходимо располагать системой эталонов (разновески, шкала и т.д.) и устройством сравнения (компаратором), способным выносить суждения "больше-меньше" относительно измеряемого значения и эталона. Измерительная процедура заключается в том, что значение измеряемой величины путем последовательных сравнений локализуется в сужающихся областях $\Delta x_1, \Delta x_2, \ldots, \Delta x_K$.

Наблюдателю, таким образом, указывается, что измеряемое значение локализовано в области Δx_K и определяется положением этой области на шкале. Далее наблюдатель может выбрать в качестве точечной оценки любую определенную точку отрезка Δx_K - его середину, левый или правый конец и т.д. Если при этом известен и закон распределения X , может (в принципе) быть построена оптимальная шкала, что позволяет при этом же числе сравнений несколько улучшить разрешающую способность.

Разумеется, знание закона распределения измеряемой величины никак не может компенсировать отсутствие измерительного прибора. Действительно, если прибора нет, а оценить значение x все-таки нужно, лучшее, что может быть сделано - это выбор в качестве оценки x математического ожидания $E[X]$. Тогда значение ошибки $\xi = X - E[X]$, а ее дисперсия равна дисперсии $D[X]$ измеряемой величины и улучшена быть не может.

Однако, знание закона распределения может быть использовано гораздо более эффективно, если привлечь результаты теории порядковых статистик. Покажем, каким образом, зная закон распределения $F(x)$ объекта измерения X и не располагая системой эталонов, имея вместо измерительного прибора один лишь компаратор, можно оценить значения x .

Пусть $\mathcal{X}$ - генеральная совокупность некоторых образцов $\varkappa$, X - генеральная совокупность характеризующих их значений, подчиненная закону $F(x)$. Требуется, имея компаратор для сравнения образцов, оценить значение x , присущее отдельному образцу $\varkappa$.

Образуем выборку из n образцов $\varkappa_1, \varkappa_2, \ldots, \varkappa_n$ и, используя компаратор, упорядочим ее путем ранжирования:

$$\varkappa_{(1)} < \varkappa_{(2)} < \ldots < \varkappa_{(n)}.$$

Теперь и неизвестные значения x_i , соответствующие образцам $\varkappa_i$, образуют вариационный ряд $x_{(1)}, x_{(2)}, \ldots, x_{(n)}$. Обратим внимание на то, что элементы вариационного ряда представляют собой значения соответствующих порядковых статистик. Плотность распределения i -ой порядковой статистики в выборке объема n Крамер (1948) имеет вид:

$$f_{j,n}(x) = \frac{n!}{(j-1)!\,(n-j)!}\left[F(x)\right]^{j-1}\left[1-F(x)\right]^{n-j} f(x),$$

где $f(x)$ - плотность распределения X .

Естественно принять в качестве точечной оценки $\hat{x}_{(i)}$ неизвестного значения, занимающего i -е место в вариационном ряду, некоторую функцию от параметра расположения $\mu_{i,n}$ распределения i -й порядковой статистики -

$$\hat{x}_{(i)} = S(\mu_{i,n}),$$

где i и n - параметры распределения.

Допускаемая при этом ошибка также зависит от номера порядковой статистики i и объема выборки n -

$$\xi_{(i)} = X_{(i)} - S(\mu_{i,n}).$$

Если критерием точности выбрать дисперсию ошибки оценивания, то в качестве оценки следует брать математическое ожидание порядковой статистики

(0)
$$\hat{x}_{(i)} = E_n[X_{(i)}].$$

Дисперсия ошибки оценивания определяется в этом случае дисперсией i -ой порядковой статистики

$$D[\xi_{(i)}] = D_n[X_{(i)}].$$

Математические ожидания $E_n[X_{(i)}]$, однозначно определяемые для данного объема выборки законом распределения генеральной совокупности, образуют на оси x систему n детерминированных точек.

Увеличение объема выборки ($n \to \infty$) приводит к асимптотическим распределениям порядковых статистик, которые исследованы в Гнеденко (I94I), Крамер (I948), Гумбель (I965). При этом, естественно, изменяются и стремятся к своим асимптотическим значениям и их средние и дисперсии.

Анализ асимптотических распределений показывает, что существует широкий класс распределений, для которых дисперсия i -ой порядковой статистики $D_n[X_{(i)}]$ уменьшается по сравнению с дисперсией исходного распределения $D[X]$ и стремится к нулю с ростом объема выборки n. К этому классу относятся распределения ограниченных случайных величин. Например, при равномерном распределении на отрезке $[0;I]$ дисперсия i -ой порядковой статистики определяется выражением

$$D_n[X_{(i)}] = \frac{i(n-i+1)}{(n+1)^2(n+2)}$$

и стремится к нулю при $n \to \infty$ независимо от номера i.

Аналогично ведет себя дисперсия порядковых статистик при выборках из нормального распределения и из некоторых других. Следует отметить, что дисперсия не уменьшается только для крайних порядковых статистик некоторых распределений (например, для наибольшей порядковой статистики гамма-распределения). Однако, если рассматривать центральные порядковые, то в этом случае $D_n[X_{(i)}] \to 0$ при $n \to \infty$ для любого непрерывного распределения.

При сравнительно небольших объемах выборок дисперсии порядковых статистик могут оказаться в десятки раз меньше, чем дисперсия исходной совокупности. Это значит, что, не сравнивая образец с эталонами, а лишь упорядочивая выборку образцов $\{x_i\}$ достаточного объема, можно, пользуясь оценкой (1), оценить значения $\{x_i\}$ элементов выборки сколь угодно точно в смысле дисперсии ошибки. Роль системы эталонов, роль шкалы при этом играет множество средних $\{E_n[X_{(i)}]\}$. При ограниченном диапазоне распределения измеряемой величины и неограниченном росте объема выборки n расстояние между точками этой "шкалы" (ее "разрешающая способность") неограниченно уменьшается. Точность при таком способе измерений будет ограничиваться чувствительностью компаратора.

ИДЕНТИФИКАЦИЯ ОБЪЕКТА С НЕНАБЛЮДАЕМЫМ ВХОДОМ

Предложенный метод оценивания позволяет поставить новую задачу идентификации: определение неизвестной статической характеристики звена с ненаблюдаемым входом.

Пусть значения входной величины X с известным законом $F(x)$ не могут быть измерены. Выходная величина $Y=\varphi(X)$ наблюдаема, но зависимость $\varphi(x)$ неизвестна. Известно лишь, что она монотонна.

Измерив n значений Y_i и ранжировав выборку $\{Y_i\}$, имеем возможность ранжировать и выборку $\{X_i\}$. Предположим, что зависимость $y=\varphi(x)$ параметризована, линейна по параметрам и имеет вид:

$$y=\theta_1\varphi_1(x)+\theta_2 ,$$

где $\varphi_1(x)$ - известная функция. Тогда, имея набор упорядоченных наблюдений $Y_{(1)}, Y_{(2)}, \ldots, Y_{(n)}$, получаем:

$$Y_{(i)}=\theta_1\varphi_1(X_{(i)})+\theta_2 .$$

В этом случае математические ожидания фактических наблюдений $Y_{(i)}$ представляют собой линейные функции искомых параметров θ_1 и θ_2 :

$$E_n[Y_{(i)}]=\theta_1 E_n[\varphi_1(X_{(i)})]+\theta_2=\theta_1 e_i+\theta_2 ,$$

а ковариации наблюдений - известны с точностью до постоянного множителя θ_1^2 :

$$cov_n[Y_{(i)},Y_{(j)}]=\theta_1^2 cov_n[\varphi_1(X_{(i)}),\varphi_1(X_{(j)})]=\theta_1^2 v_{ij}$$

Выражения для оценок θ_1 и θ_2 определяются из системы нормальных уравнений обобщенным методом наименьших квадратов и имеют вид Lloyd (1952):

(I) $$\hat{\theta}_1 = 1^T \Gamma Y, \qquad \hat{\theta}_2 = -e^T \Gamma Y$$

где

$$\Gamma = V_n^{-1}(1e^T - e1^T)V_n^{-1}/b, \quad b = (1^T V_n^{-1} 1)(e^T V_n^{-1} e) - (1^T V_n^{-1} e)^2,$$

1 - единичный вектор размерности $n \times 1$;

Y - вектор упорядоченных наблюдений размерности $n \times 1$;

e - вектор математических ожиданий $E_n[\varphi_1(X_{(i)})]$ размерности $n \times 1$;

V_n - ковариационная матрица размерности $n \times n$.

Дисперсии и ковариации коэффициентов определяются выражениями:

(2) $$D_n[\hat{\theta}_1] = \theta_1^2(1^T V_n^{-1} 1)/b, \quad D_n[\hat{\theta}_2] = \theta_1^2(e^T V_n^{-1} e)/b,$$

(3) $$cov_n[\hat{\theta}_1, \hat{\theta}_2] = -\theta_1^2(1^T V_n^{-1} e)/b.$$

Так как оценки $\hat{\theta}_1$ и $\hat{\theta}_2$ получены методом наименьших квадратов, то они оказываются несмещенными и эффективными в классе линейных. Это обстоятельство показывает, что оценки $E_n[X_{(i)}]$ "не хуже", чем прямые измерения величины X в том смысле, что могут служить основой для применения статистических методов интерпретации.

Рассмотрим пример.

Предположим, что распределена равномерно на отрезке $[0;I]$. Градуировочная характеристика прибора представляет собой прямую $y = \theta_1 x + \theta_2$ с неизвестными коэффициентами θ_1 и θ_2.

Для их определения применим описанный выше метод. В этом случае вектор математических ожиданий и ковариационная матрица V_n порядковых статистик определяются элементами

$$e_i = \frac{i}{n+1}, \quad v_{ij} = \frac{i(n-j+1)}{(n+1)^2(n+2)}$$

соответственно, а обратная матрица V_n^{-1} - элементами

$$v_{ij}^{-1} = (n+1)(n+2)\begin{cases} -1, & \text{при } i = j \pm 1, \\ 2, & \text{при } i = j, \\ 0, & \text{в остальных случаях.} \end{cases}$$

Используя (I), получаем оценки $\hat{\theta}_1$ и $\hat{\theta}_2$ неизвестных коэффициентов

$$\hat{\theta}_1 = \frac{n+1}{n-1}\left(Y_{(n)} - Y_{(1)}\right), \quad \hat{\theta}_2 = \frac{nY_{(1)} - Y_{(n)}}{n-1}.$$

Оценки определяются лишь крайними элементами выборки. Операция ранжирования в этом случае не нужна - достаточно выбрать наибольший и наименьший элементы.

Дисперсии и ковариация оценок, определяемые из (2) и (3), имеют вид:

$$D_n[\hat{\theta}_1] = \frac{2\theta_1^2}{(n+1)(n+2)}, \quad D_n[\hat{\theta}_2] = \frac{n\theta_1^2}{(n^2-1)(n+2)},$$

$$cov_n[\hat{\theta}_1, \hat{\theta}_2] = -\frac{\theta_1^2}{(n-1)(n+2)}.$$

В таблице приведены значения оценок и параметров модели $y = \theta_1 x + \theta_2$, полученных при помощи статистического моделирования для выборок объемов n = 5,20,50 из равномерной совокупности. Истинные значения коэффициентов $\theta_1 = 2$ и $\theta_2 = $ I. Полученные результаты иллюстрирует

ТАБЛИЦА I

Оценка Параметров Линейной Функции

n	θ_1	θ_2	$Y_{(1)}$	$Y_{(n)}$	$\hat{\theta}_1$	$\hat{\theta}_2$	$D_n[\hat{\theta}_1]$	$D_n[\hat{\theta}_2]$
5	2	I	I,I72	2,772	2,4	0,772	0,I9	0,II9
20	2	I	I,0224	2,9547	2,I36	0,92I	0,0I73	0,0I82
50	2	I	I,0664	2,9922	2,004	I,027	0,003	0,0036

РАНГОВАЯ ПРОЦЕДУРА БЕЗЭТАЛОННОЙ КЛАССИФИКАЦИИ

Пусть имеется совокупность объектов $\mathcal{K}$, параметром которых является случайная величина X с законом распределения $F(x)$. Система эталонов, обеспечивающая измерение параметра предъявляемого объекта, отсутствует. Объекты совокупности $\mathcal{K}$ допускают лишь сравнение между собой по параметру X. Сравнение осуществляется с помощью соответствующего компараторного устройства.

Отсутствие эталонов не дает возможности построить вектор значений параметра X, однако наличие компаратора позволяет получить вектор их рангов. В этой ситуации оказывается возможным разбить совокупность объектов $\mathcal{K}$ на L непересекающихся классов $\{\pi_i\}_1^L$.

Подчеркнем, что значения эталонов $\{a_j\}_1^K$, $K \geq L-1$ заданы, но они не реализуются с помощью конкретных приборов, а присутствуют в качестве объективных требований, предъявляемых к процессу классификации или разбиения совокупности $\mathcal{K}$ на классы $\{\pi_i\}_1^L$.

Воспользуемся тем, что любая выборка значений случайной величины $\{X_i\}_1^n$, будучи упорядочена, порождает соответствующий ранговый вектор $R = \{R_i\}_1^n$. При этом ранг R является монотонной функцией значения x (большему значению соответствует больший ранг). Это позволяет перейти от процедур, основанных на операциях с выборками значений, к процедурам с операциями на выборках их рангов.

Пусть $\{\varkappa_i\}_1^n$ – выборка объектов из совокупности $\mathcal{K}$ объема n с неизвестными значениями параметра X. Совокупность эталонов задана множеством их значений $\{a_j\}_1^K$. Значения эталонов $\{a_j\}_1^K$, а следовательно, и упорядоченные значения $\{a_{(j)}\}_1^K$, принадлежат области изменения $\mathcal{E}$ случайной величины X. Поэтому $a_{(j)}$ можно интерпретировать как квантиль распределения $F(x)$ уровня $F(a_{(j)})$. Конкретный эталон можно представить как некоторый эталонный объект, т. е. объект с эталонным значением параметра a_j. Объединяя выборки значений $\{x_i\}_1^n$ и $\{a_j\}_1^K$, получаем новую выборку объемом $N=n+K$:

(4)
$$\{x'_\ell\}_1^N=\{x_i\}_1^n\cup\{a_j\}_1^K .$$

С помощью компаратора упорядочим объекты выборки $\{\varkappa_i\}_1^n$. При этом неизвестные значения их параметра образуют вариационный ряд $\{x_{(i)}\}_1^n$, определяемый ранговым вектором $R=\{R_1^n, R_2^n, \dots, R_n^n\}$, где R_i^n указывает место объекта $\varkappa_i$ в упорядоченной выборке $\{\varkappa_{(i)}\}_1^n$ (или место x_i в $\{x_{(i)}\}_1^n$).

Построим теперь множество рангов $\{r_j^N\}_1^K$ значений эталонов $\{a_j\}_1^K$ в выборке (4). Как известно Крамер (1948), состоятельной оценкой квантиля распределения $F(x)$ является порядковая статистика соответствующего уровня. Кроме того, существует асимптотическое соотношение

(5)
$$F(x_{(j)})=\frac{j}{n+1} ,$$

где $j=R_i^n$ – ранг x_i в выборке объема n, выполняющееся при $n\to\infty$, $i\to\infty$, $i/n\to p=const\ (0<p<1)$. Так как в нашем случае объем выборки объектов n ограничен, K фиксировано, а значения эталонов $\{a_j\}_1^K$ известны, то интерпретируя их как квантили распределения $F(x)$, из (5) получаем оценки рангов

(6)
$$\hat{r}_j^N=(N+1)\,F(a_{(j)}),\quad j=\overline{1,K},$$

где $a_{(j)}$ – элемент упорядоченной выборки $\{a_{(j)}\}_1^K$.

Теперь найдем множество рангов $\{R_i^N\}_1^n$ значений $\{x_i\}_1^n$ в выборке (4). Так как множества рангов $\{R_i^n\}_1^n$ и $\{\hat{r}_j^N\}_1^K$ определены, то ранг неизвестного значения x_i в выборке $\{x'_\ell\}_1^N$ можно определить с помощью следующего итерационного соотношения:

(7) $$R_{i,j}^{N} = R_{i,j-1}^{N} + z\left(R_{i,j-1}^{N} - \hat{\tau}_j^{N}\right), \quad j=\overline{1,K},$$

где $R_{i,0}^{N} = R_i^n$, $R_{i,K}^{N} = \hat{R}_i^{N}$, $z(t)=0$ при $t<0$ и $z(t)=1$ при $t \geqslant 0$. Изменяя в (7) j от I до K для каждого $i=\overline{1,n}$, мы получаем множество рангов $\{\hat{R}_i^N\}_1^n$.

Таким образом, ранги $\{R_i^N\}_1^n$ и $\{\tau_j^N\}_1^K$ определяют множество рангов всех элементов выборки (4), т.е.

$$\{\hat{R}_e'\}_1^N = \{\hat{R}_i^N\}_1^n + \{\hat{\tau}_j^N\}_1^K .$$

Значения эталонов $\{a_j\}_1^{K}$ осуществляют разбиение $T(a)$ пространства $\mathcal{E}$ на подпространства $\{Q_i\}_1^L$, соответствующие классам $\{\pi_i\}_1^L$. Так как значения рангов монотонно зависят от значений параметров объектов, то разбиению $T(a)$ соответствует разбиение $T'(\hat{\tau})$ множества рангов $\{\hat{R}_e'\}_1^N$ на L непересекающихся подмножеств $\{\mathcal{R}_e'\}_1^L$, которое осуществляется значениями рангов эталонов $\{\hat{\tau}_j^N\}_1^K$. В этом случае в подмножества $\{\mathcal{R}_e'\}_1^L$ входят только элементы множества $\{\hat{R}_i^N\}_1^n$, а $\{\hat{\tau}_j^N\}_1^K$ являются точками разбиения и не включаются ни в какое из подмножеств $\{\mathcal{R}_e'\}_1^L$. Полученное разбиение $T'(\hat{\tau})$ однозначно задает разбиение $T(\hat{\tau})$ множества $\{R_i^n\}_1^N$ на непересекающиеся подмножества $\{\mathcal{R}_e\}_1^L$, так как ранги $\hat{R}_i^N$ и R_i^n связаны соотношением (7).

Таким образом, правило классификации заключается в следующем: объект $\varkappa$ из выборки $\{\varkappa_i\}_1^n$ относится к классу π_j, если его ранг в выборке (4) принадлежит подмножеству $\mathcal{R}_j$.

Отметим, что предложенная безэталонная процедура классификации корректна только в том случае, если соотношение (6) позволяет получать целые значения рангов эталонов $\{\hat{\tau}_j^N\}_1^K$. Поэтому на объем выборки n при заданном числе эталонов K следует наложить соответствующее условие.

ТОЧНОСТЬ РАНГОВОЙ КЛАССИФИКАЦИИ

Обсудим точность данного метода классификации. Так как разбиение $T(\hat{\tau})$ осуществляется с помощью оценок рангов эталонов (6), а выборка значений параметра объектов случайна, то возможны ситуации, когда объект $\varkappa_i$ не будет включен в класс π_j $(R_i \in \mathcal{R}_j)$, в то время как $x_i \in Q_j$. Либо наоборот - объект $\varkappa_i$ включается в класс π_j $(R_i \in \mathcal{R}_j)$, при этом значения его параметра $x_i \bar{\in} Q_j$.

В этом случае допускаются ошибки классификации, соответствующие ошибкам первого и второго рода.

Вероятность этих ошибок можно найти, определив совместную плотность распределения рангового вектора R и случайной величины X. Тогда критерием точности классификации может служить сумма этих вероятностей.

Однако на практике более важен критерий, построенный на сумме апостериорных вероятностей неправильной классификации, так как он характеризует относительное число неправильно классифицированных объектов. Этот критерий непосредственно связан с критерием суммы вероятностей ошибок первого и второго рода и зависит от них. Однако для ранговой процедуры его использование позволяет обойти трудности, связанные с их вычислением.

Рассмотрим этот критерий для обычной эталонной классификации. В этом случае ошибки первого и второго рода возникают из-за погрешности измерения ξ. Вероятности ошибок

$$\alpha_i = P\{X+\xi \bar{\in} Q_i | X \in Q_i\}, \quad \beta_i = P\{X+\xi \in Q_i | X \bar{\in} Q_i\}$$

можно определить, если известно распределение погрешности ξ. Тогда плотности распределения параметра X объектов, включенных в класс π_i и не включенных в него в результате классификации (определение по формуле Байеса), имеют вид:

$$(8) \quad f(x|\pi_i) = \begin{cases} \dfrac{(1-\alpha_i)f(x)}{c_i}, & x \in Q_i, \\ \dfrac{\beta_i f(x)}{c_i}, & x \bar{\in} Q_i, \end{cases} \qquad f(x|\bar{\pi}_i) = \begin{cases} \dfrac{(1-\beta_i)f(x)}{1-c_i}, & x \bar{\in} Q_i, \\ \dfrac{\alpha_i f(x)}{1-c_i}, & x \in Q_i, \end{cases}$$

где $f(x)$ - плотность распределения X, а

$$c_i = (1-\alpha_i)\int_{x \in Q_i} dF(x) + \beta_i \int_{x \bar{\in} Q_i} dF(x).$$

Из (8) следует, что относительные объемы неправильно классифицированных объектов при ошибках первого и второго рода характеризуются вероятностями:

$$(9) \qquad q_{1\pi_i}(\alpha_i, \beta_i) = \frac{\alpha_i}{1-c_i} \int_{x \in Q_i} dF(x), \quad q_{2\pi_i}(\alpha_i, \beta_i) = \frac{\beta_i}{c_i} \int_{x \bar{\in} Q_i} dF(x),$$

сумма которых и является критерием точности классификации.

В нашем случае вероятности (9) можно определить, если найти распределение параметра объектов, попавших в класс π_i, в результате ранговой процедуры.

Для этого воспользуемся тем, что распределение случайной величины X представляет собой смесь распределений порядковых статистик

(10) $$f(x)=\frac{1}{n}\sum_{j=1}^{n} f_{j,n}(x).$$

Тогда плотность распределения параметра объектов, включенных в класс π_i, имеет вид

(11) $$f^*(x|\pi_i)=\frac{1}{\rho_i}\sum_{j\in R_i} f_{j,n}(x),$$

где ρ_i - число элементов множества R_i. Из (10) и (11) получаем плотность распределения $f^*(x|\bar{\pi}_i)$:

$$f^*(x|\bar{\pi}_i)=\frac{1}{n-\rho_i}\left[n f(x)-\rho_i f^*(x|\pi_i)\right].$$

Следовательно, относительные объемы неправильно классифицированных объектов при ранговой процедуре характеризуются вероятностями:

$$q_{1\pi_i}(n)=\int_{x\in Q_i} f^*(x|\bar{\pi}_i)dx,\quad q_{2\pi_i}(n)=\int_{x\bar{\in} Q_i} f^*(x|\pi_i)dx$$

и зависят от объема исходной выборки объектов n.

При увеличении n (а следовательно и N) объемы неправильно классифицированных объектов стремятся к нулю. Объясняется это сходимостью порядковых статистик к соответствующим квантилям исходного распределения при $F(x)$. Именно тогда соотношения (5) и (6) позволяют точно определить ранги значений эталонов $\{a_j\}_1^K$, что дает возможность провести безошибочную классификацию.

Отсюда следует, что при некотором $n \geqslant n_0$ безэталонная классификация может оказаться не хуже эталонных методов в смысле минимума вероятности $q_{\pi_i} = q_{1\pi_i} + q_{2\pi_i}$. При этом n_0 можно найти из соотношения:

$$n_0 = int[max(n_1, n_2)]+1,$$

где n_1 и n_2 - наибольшие действительные корни уравнений

$$q_{1\pi_i}(n)=q_{1\pi_i}(\alpha_i,\beta_i),\quad q_{2\pi_i}(n)=q_{2\pi_i}(\alpha_i,\beta_i)$$

ЛИТЕРАТУРА

Гнеденко Б.В. (I94I) : Предельные теоремы для максимального члена вариационного ряда. ДАН СССР, т.32, I94I, 7-9.

Гумбель Э. (I965) : Статистика экстремальных значений. "Мир", М., I965.

Крамер Г. (I948) : Математические методы статистики. ИЛ, М., I948.

Lloyd E.H. (I952) : Least-squares estimation of location and scale parameters using order statistics. Biometrika, v.32, 1952, 88-95.

Харьковский Институт Радиоэлектроники
Проспект Ленина I4, 3I0059, Харьков-59
СССР

SIMULTANEOUS INTERFERENCE IN SEQUENTIAL PREDICTION

András Farkas

Budapest

ABSTRACT

This paper contains a computer assisted psychological experiment. The concrete psychological phenomenon to be investigated is the simultaneous interference in the interaction of probability learning and sequential pattern learning. In the experiments a 50:0 type probability learning-task and a single alternation learning-task were interfered by human subjects. Subexperiments were performed on control groups, and a considerable interference could be observed. The basic experiment has demonstrated negative-recency effect.Classic no-memory stochastic models for learning developed by Estes (1955), Luce (1959), Bush and Mosteller (1955) and others are derived from the positive-recency principle, thus they are not applicable to this experimental situation. Our model is basically a 4 operators linear model with short-term memory, in which a so-called alternating tendency is involved by us. Our model ensures negative-recency and can be applied to the basic and subexperiments as well.

INTRODUCTION

Probability Theory and Mathematical Statistics have a considerable role in modern psychological research. Mathematical Psychology became a separate area of scientific research already at the beginning of the fifties. Mathematical Learning Theory has an important position in Mathematical Psychology, within which sequential prediction is a significant trend of investigations. The problem of sequential prediction arose in connection with examination of the effects of partial reinforcement. At the initial stage two types of sequential prediction experiments have appeared: Hunter (1920) in his experiments introduced time-maze, thus realizing sequential pattern-learning, while Brunswik (1939) and Humphreys (1939) introduced probability learning experiments in animal and human subjects. The best way to illustrate sequential prediction would be to present a classic two-choice probability learning experiment performed on rats in a simple T-maze by Brunswik. The experimental apparatus is shown in Fig. 1. In the experiment a rat is placed at the starting point s and then it runs to the choice point c. At point c the rat is in a two-choice situation; it can turn right to goal-box A_1 or left to goal-box A_2, where it may or may not receive an reinforcement finding food. Several tries are carried out during which the reward can be changed from A_1 to A_2 and vice versa. In Brunswik's experiment the reinforcement was scheduled to A_1 with probability π_1 and to A_2 with probability $1-\pi_1$. The optimal strategy would be always to choose the side with a reward of greater probability. The so-called probability matching phenomenon discovered by Brunswik indicates that the probability of choice approaches the probability of reward.

A similarly constructed experiment was presented by Humphreys on human subjects. In this case a simple alternative had to be predicted, after which each time the subject received a sign referring to the correctness of his choice..

A number of recent studies examined the effect of various random and non-random reinforcement schedules in decision behaviour. Accordingly, there are two types of sequential prediction: if the distribution of reward is random it is the case of probability learning, if the distribution of reward is regular it is the case of sequential pattern learning.

American mathematical psychologists developed stochastic models for probability learning,such as Estes (1955) , Luce (1959) , Bush and Mosteller (1955). The phenomenon of probability matching is described quite adequately by classic no-memory models. On base of these models simulation of decision behaviour could be carried out by computer. A great number of investigators established short-term memory models for sequential pattern learning, such as Restle (1961) Vitz and Todd (1967).

Development of current research tends to set up models adequate to describe both probability learning and sequential pattern learning. The model set up by us fits well this research paradigm. Based on the Monte Carlo method it is applicable to simulate the 4 sub-experiments of the simultaneous interference experiment.

PSYCHOLOGICAL BACKGROUND

The phenomenon of simultaneous interference belongs to the area of learning and behaviour psychology. Studies referring to this field of research are concerned with the investigation of the interaction among various learning processes. A previous learning process interacting with a new learning process is called transfer effect. The phenomenon in which an old learning process affects the new one is the proactive transfer. The situation reversed is the retroactive transfer. When two or more learning processes happen at the same time the term of simultaneous transfer can be introduced. If in the course of various learning processes one becomes handicapped by the other, this is the case of interference. According to this the negative interaction of simultaneous learning processes is to be entitled as simultaneous interference.

The psychological experiment carried out by us can be considered a methodical novelty, since such an experiment between the two types of sequential prediction has not been performed before, at least, none that we are aware of.

Investigating the basic characteristies of the learning processes the so-called positive-recency principle has been established, which denotes the fact that the most effective informations are those that are acquired most recently. Concerning the Brunswik experiment this means that in the T-maze the probability of choosing the recently rewarded goal-box will increase. As Overall and Brown (1956), Anderson and Whalen (1960), Friedman et al. (1968) have demonstrated in 50:50 type probability learning experiments on human subjects, negative-recency effect will occur. This means that the probability of choosing the alternative that is not reinforced increases. In our experiment significant negative-recency could be proved.

COMPUTER ASSISTED PSYCHOLOGICAL EXPERIMENT

Concerning simultaneous interference a basic experiment, two sub-experiments and an additional experiment were performed, each of them consisted of 80 predictions. The additional experiment was carried out to demonstrate that our model can handle single alternating experiment as opposed to the classic stochastic model for learning which we started from.

Subjects: 40 psychologically naive secondary school students were applied for the experiment. Each subject participated in 1 experiment only.

Apparatus: The experimental apparatus was basically a modified Humphreys' keyboard with the only difference that the feedback concerning the correctness of the choice was made through sound instead of light.

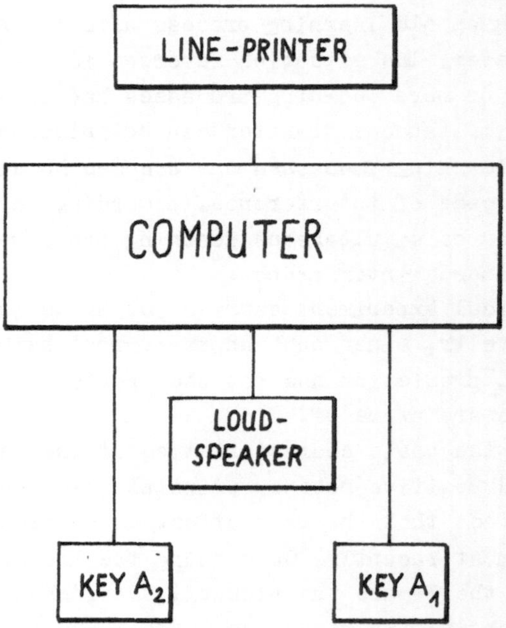

FIG. 1

As shown in Fig. 1 the apparatus contained 2 keys located on left and right, respectively, a loud-speaker, a computer and a line-printer.

Procedure: In every experiment the subjects operated keys A_1 or A_2 in arbitrary sequence 80 times. After each operation of key the computer responded by sound indicating the correctness of choice. A trill was heard when the prediction was correct. A monotonous sound was emitted if the non-rewarded key was operated. A period of 3 sec was available for each prediction. Exceeding this time limit the loud-speaker radiated a bass sound. The run of the experiment was controlled by the computer. The sounds applied were also generated by a computer program.

Experiments and results.

Subexperiment No. 1: Key A_1 was non-rewarded and the computer attached random reward of 50 % probability to key A_2. The distribution of random reward was derived from the article by Galanter and Bush (1959). Although this distribution is not very ideal it is still applied because subexperiment No. 1 was already used with another aim of research in the papers of the author published in 1976 and 1977.

The experimental subjects in the postexperimental interview gave a report on the fact that no reward was experienced on key Even so, they did not follow the optimum strategy and in the last block of 20 tries they chose the rewarded key in 65 % on the average. This value is somewhat lower than those referred to in literature.

Subexperiment No. 2: Key A_2 was non-rewarded and the computer attached alternatively reward and non-reward to key A_1. The situation of single-alternation was recognized by all the experimental subjects but one for the 25th try on the average. Subsequently, key A_1 was chosen in the overwhelming majority of the cases if that key was rewarded.

Basic experiment: This is the actual interference experiment in which the rewards applied in subexperiments Nos. 1 and 2 appeared simultaneously, that is, a probability learning task and a sequential pattern learning task were interfered. More specifically, a probability learning of type 50:0 and a single alternation were involved.

Of the 10 experimental subjects only 1 solved the problem and the reward on key A_1 was not obtained in 2 cases only. The others followed their instinct in the process of learning or set up a series of inadequate hypotheses which were tried and later refused. The results exhibit a very intensive interference effect in the interaction of the two types of sequential prediction.

Additional experiment: Reward alternatingly appeared on keys A_2 and A_1. Learning took place about twice as fast as in subexperiment No. 2. The problem was solved by all the subjects but one for the 12th try on the average. Similarly to subexperiment No. 2 the criterion of learning was 8 subsequent correct choices.

MODELLING

In the model applied for the simulation of experimental results we have started from the 4 operators linear model devised by Bush and Mosteller (1955) . The Bush-Mosteller model is introduced in the Brunswik experiment described earlier. The experimental apparatus is shown in Fig. 2.

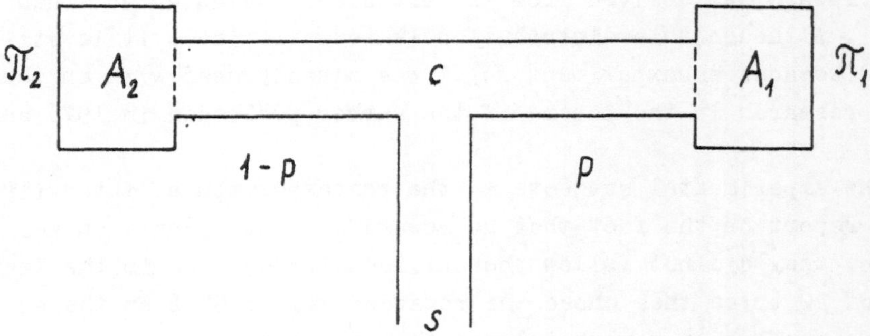

FIG. 2

The two alternatives A_2 and A_1 represent the choice between the left or right sides. The two outcomes O_1 and O_2 correspond to reward and non-reward. The possible events are shown in Table 1.

TABLE 1

Event	Alternative	Outcome	Identification
E_{11}	A_1	O_1	right turn + reward
E_{12}	A_1	O_2	right turn + non-reward
E_{21}	A_2	O_1	left turn + reward
E_{22}	A_2	O_2	left turn + non-reward

Denote the probability of choosing the right side by p. Probability p is transformed in each try. An operator Q_{ij} corresponds to each event E_{ij}.

$$Q_{11}p = \alpha_1 p + 1 - \alpha_1 \qquad Q_{21}p = \alpha_1 p$$

$$Q_{12}p = \alpha_2 p \qquad Q_{22}p = \alpha_2 p + 1 - \alpha_2$$

α_1 is the reward parameter and α_2 is the non-reward parameter. From the formulae it becomes clear that this model is derived from the positive-recency principle. Probability p increases with preceding events E_{11} or E_{22} and decreases with preceding events E_{12} or E_{21}.

Our experimental results have demonstrated negative-recency effects. Table 2 shows the data of the basic experiment.

TABLE 2

Recency Effects

	Negative recency		Positive recency	
	Alternation after reward	Repetition after non-reward	Alternation after non-reward	Repetition after reward
1	24	20	20	15
2	26	18	22	13
3	16	8	26	29
4	39	24	8	8
5	52	1	23	1
6	21	13	23	22
7	25	24	14	16
8	24	13	22	20
9	21	28	8	22
10	37	11	22	9

In subjects 1, 2, 4, 7, 9 and 10 negative-recency was obvious. In subjects 3, 6 and 8 the phenomenon of positive-recency can be observed. The behaviour of subject 5 could be interpreted as an alternation pressure. Assuming random behaviour the possibilities listed in Table 2 would occur 20 times on the average. The analysis of data has shown that if an alternation occurred frequently it continued. This observation has lead to the introduction of the so-called alternation tendency. If an experimental subject is reinforced in a spontaneously occurred alternation, there is a markedly growing probability that the alternation will be further continued. This phenomenon was also pointed out by Laming (1969) and (1973) in his choice -reaction time experiments. The alternation tendency can be established with the following complementation of the Bush-Mosteller model.

$$Q_{11}^{*}p = 1-(\alpha_1 p + 1 - \alpha_1) \quad \text{if} \quad p > 0.5 \wedge E(i-1) = E_{21} \wedge E(i) = E_{11}$$

$$Q_{11}^{**}p = \alpha_1 p \quad \text{if} \quad p < 0.5 \wedge E(i-1) = E_{21} \wedge E(i) = E_{11}$$

$$Q_{21}^{*}p = 1-\alpha_1 p \quad \text{if} \quad p < 0.5 \wedge E(i-1) = E_{11} \wedge E(i) = E_{21}$$

$$Q_{21}^{**}p = \alpha_1 p + 1 - \alpha_1 \quad \text{if} \quad p > 0.5 \wedge E(i-1) = E_{11} \wedge E(i) = E_{21}$$

If none of the systems of conditions are satisfied, proceed according to the specifications of the Bush-Mosteller model. $E(i)$ stands for the event taking place at the i'th try. In the transformation of the probability of choice the event $E(i-1)$ taking place at the directly preceding choice must also be remembered. This information is stored in the short-term memory.

SIMULATION

The mean learning curves were determined by computer for each experiment on the basis of the model. The data of each try are shown in the figures from left to right. The * symbols in the top and bottom lines, resp., indicate the place of rewards belonging to alternatives A_2 and A_1, resp. Symbols 0 show the proportion of choice A_2 with spacings of 0.1 made by 10 subjects in each try. Line of * starting from the centre is the mean learning curve. Fig. 3 shows the result of subexperiment No. 1. The parameter values applied for the mean learning curve are $\alpha_1 = 0.98$ and $\alpha_2 = 1$.

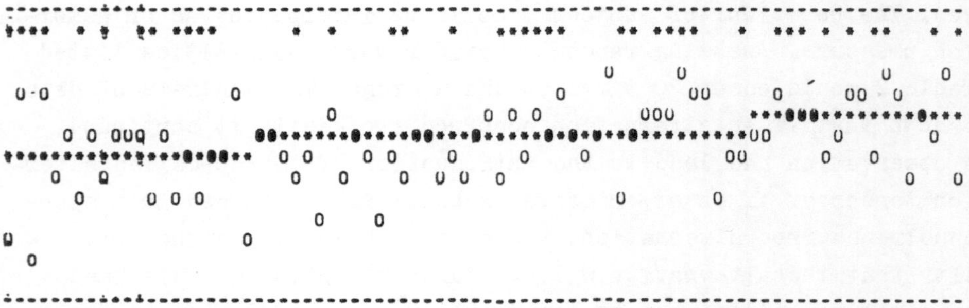

FIG. 3

Fig. 4 shows the result of subexperiment No. 2 in case of $\alpha_1 = 0.97$ and $\alpha_2 = 1$.

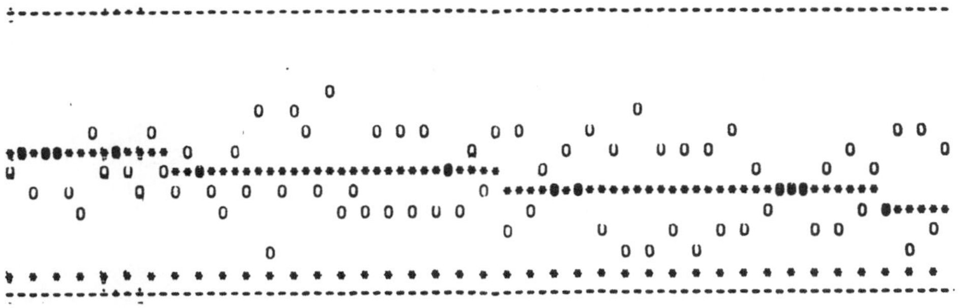

FIG. 4

In Fig. 5 the proportions of choice A_2 are indicated by ● where the reward was on the right side. The mean learning curve applies to these tries. $\alpha_1 = 0.91$ and $\alpha_2 = 1$.

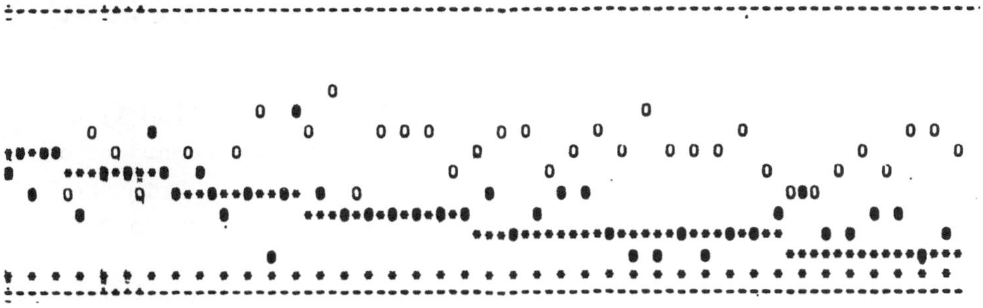

FIG. 5

Fig. 6 shows the data of the basic experiment. No significant displacement is observed in the mean learning curve in either direction A_1 or A_2.

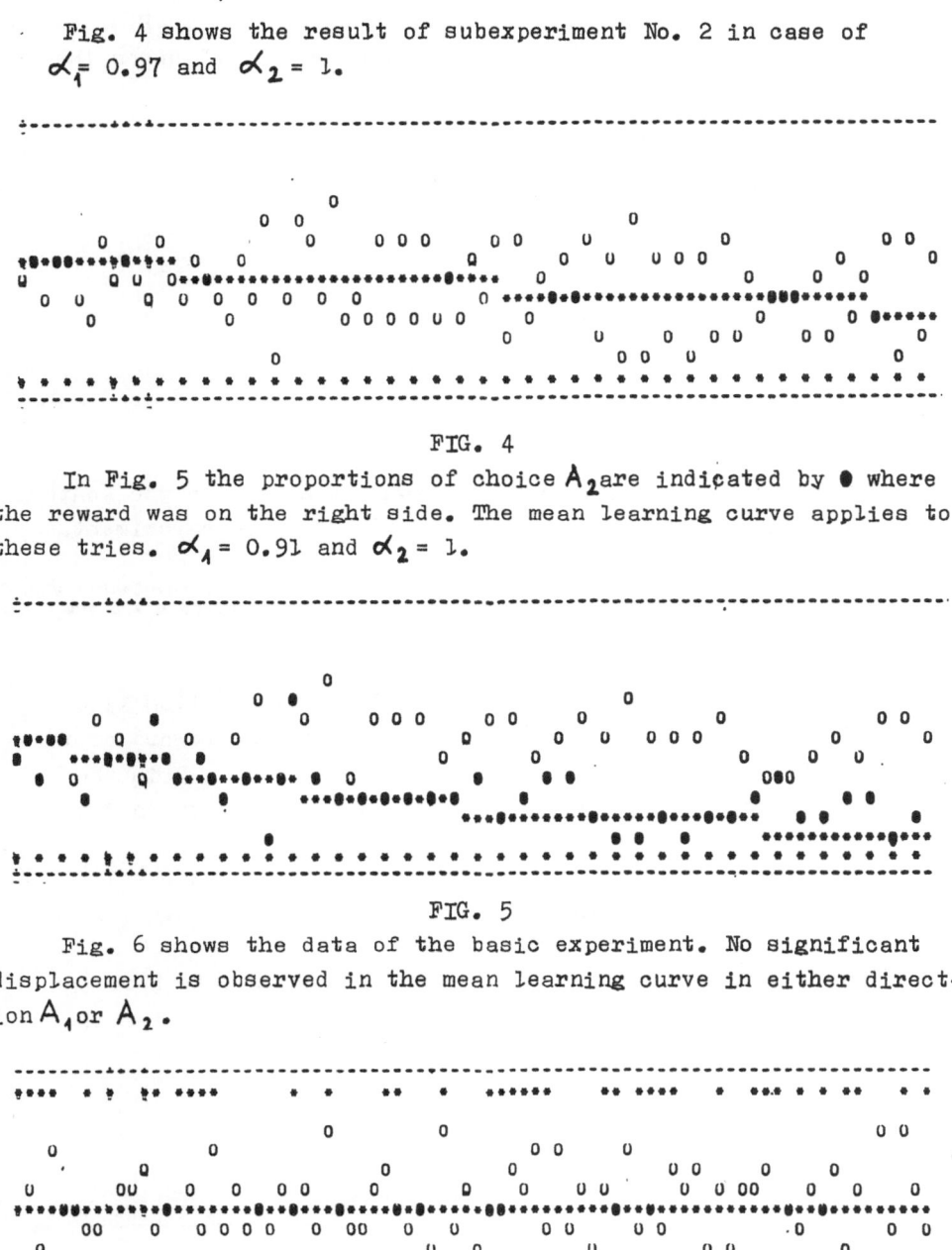

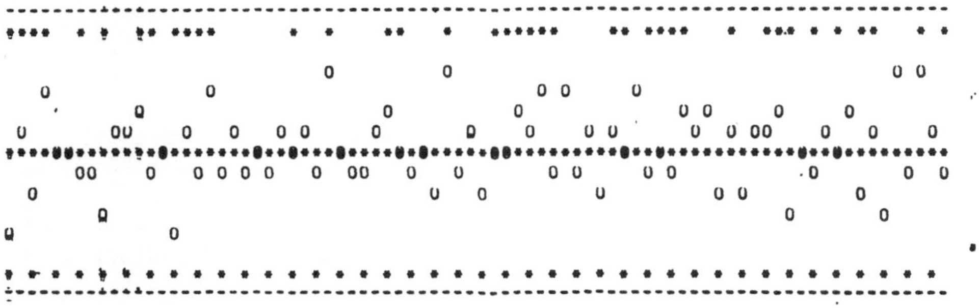

FIG. 6

The results of the additional experiment are shown in Fig. 7. In the last 2 cases the choice of α_1 and α_2 does not influence the mean learning curve for reasons of symmetry.

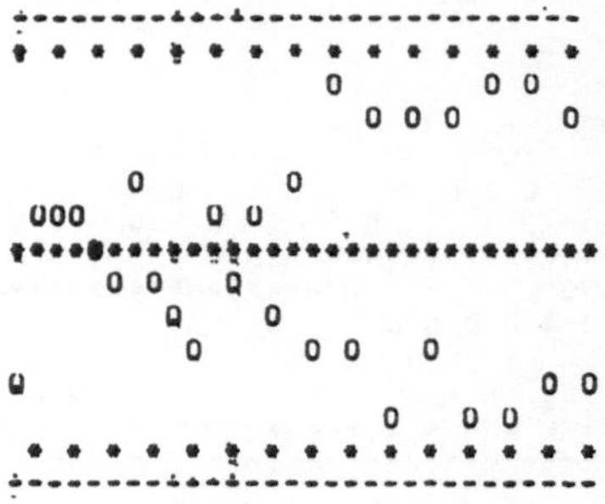

FIG. 7

On the basis of the model a simulation was made with the Monte-Carlo method. Fig. 8 shows the simulation of the basic experiment.

FIG. 8

The top line shows the data of left side and the bottom line shows the data of the right side while * indicates the reward of the left and right sides in the top and bottom lines, resp., 0 in the top line stands for choice A_2 and 0 in the bottom line stands for the choice of A_1.

The simulated simultaneous interference experimental results exhibit a high similarity with the majority of the data experienced in case of $\alpha_1 = 0.3$ and $\alpha_2 = 0.7$ values.

For reasons of brevity no detailed comparison is made here and only the negative recency of the simulated experiment is shown in Table 3.

TABLE 3

Recency in simulated data

Negative recency		Positive recency	
Alternation after reward	Repetition after non-reward	Alternation after non-reward	Repetition after reward
28	16	19	16

Fig. 9 shows the simulated result of the additional experiment with $\alpha_1 = 0.2$ and $\alpha_2 = 0.9$ values. From the result it is seen that our model is suitable for the simulation of the single-alternation task. The simulated result satisfies the learning criterion.

Subexperiments Nos. 1 and 2 may not contain rewarded alternation, thus they are actually simulated on the basis of the Bush-Mosteller model and the related results are not discussed here.

FIG. 9

Finally, it is noted that the experiments will be continued in connection with the simultaneous interference. We would like to determine the effect of frequency of the randomly distributed reward on the extent of interference.

REFERENCES

Anderson,N.H., and Whalen,R.E.(1960) : Likelihood judgments and sequential effects in a two-choice probability learning situation. Journal of Experimental Psychology, 60, 1960, 111-120.

Brunswik,E. (1939) : Probability as a determiner of rat behavior. Journal of Experimental Psychology, 25, 1939, 175-197.

Bush,R.R., and Mosteller,F. (1955) : Stochastic Models for Learning. Wiley, New York, 1955.

Estes,W.K. (1955) : Theory of elementary predictive behavior. In: Mathematical Models of Human Behavior-Proceedings of a Symposium. Dunlap and Associates, Stanford, 1955, 63-67.

Farkas,A. (1976) : Egyszerü és valószinüségi tanulási jelenségek. (Phenomena of Simple and probability learning.) In: 7. Neumann Kollokvium Kiadványa. (Transactions of the Seventh Neumann Conference.) 1976, University Press, Szeged, 243-267.

Farkas,A. (1977) : Az állat és az ember viselkedése analóg valószinüségi tanulási helyzetben. (Behaviour of animals and human subjects in analogous probability learning situation.) In: 8. Neumann Kollokvium Kiadványa. (Transactions of the Eighth Neumann Conference.) 1977, University Press, Szeged, 148-160.

Galanter,E., and Bush,R.R. (1959) : Some T-maze experiments. In: Studies in Mathematical Learning Theory. Stanford University Press, Stanford, 1959, 265-289.

Humphreys,L.G. (1939) : Acquisition and extinction of verbal expectations in a situation analogous to conditioning. Journal of Experimental Psychology, 25, 1939, 294-301.

Hunter,W.S. (1920) : The temporal maze and kinaesthetic sensory processes in the white rat. Psychobiology, 2, 1920, 1-17.

Laming,D.R.J. (1969): Subjective probability in choice-reaction experiments. Journal of Mathematical Psychology, 6, 1969, 81-120.

Laming,D.R.J. (1973) : Mathematical Psychology. Academic Press, London, 1973.

Luce,R.D. (1959) : Individual Choice Behavior. A Theoretical Analysis. Wiley, New York, 1959.

Overall,J.E., and Brown,W.L. (1959) : A comparison of the decision-behavior of rats and of human subjects. American Journal of Psychology, 80, 1967, 276-281.

Restle,F. (1961) : Psychology and the Judgment of Choice. A Theoretical Essay. Wiley, New York, 1961.

Vitz,P.C., and Todd,T.C. (1967) : A model of learning for simple repeating binary patterns. Journal of Experimental Psychology, 75, 1967, 108-117.

Hungarian Steel and Iron Association
Mathematical Department

1051 Budapest Október 6.u.7.
Hungary

SOME REMARKS ON THE ROLE OF INACCURACY IN SHANNON'S THEORY OF INFORMATION TRANSMISSION

Thomas R. M. Fischer

Dresden

ABSTRACT

The paper contains an approach to the problems of universal encoding and decoding which is based on the concept of inaccuracy. The areas of noiseless source coding, source coding for noisy channels and channel coding and decoding are covered. Another purpose of the paper is to get a deeper understanding of the role of inaccuracy in the theory of information transmission.

0. INTRODUCTION

The classical form of Shannon's inequality is

$$- \sum_{i=1}^{n} p_i \log p_i \leq - \sum_{i=1}^{n} p_i \log q_i \tag{1}$$

for all $n \geq 2$, where

$$\sum_{i=1}^{n} p_i = \sum_{i=1}^{n} q_i = 1, \quad p_i, q_i > 0,$$

with equality in (1) if and only if $p_i = q_i$ $(i = 1,\dots,n)$.

The most important application of Shannon's inequality is the theorem asserting that the average length of a codeword in an uniquely decipherable code cannot be smaller than the Shannon entropy (where the logarithm base is the number of symbols in the code alphabet). Moreover, it is possible to characterize the Shannon entropy in some sense by the validity of inequality (1), see P. Fischer (1972) and Aczél and Ostrowski (1973).

The expression on the right hand side of (1) is called inaccuracy or Bongard entropy. It was systematically investigated by Kerridge (1961) and Bongard (1963). Since that time it was considered by several authors, e. g. Nath (1968), Kannappan (1972), Ng (1974). Pötschke (1974) has proved a McMillan type theorem for inaccuracy and considered some relations between inaccuracy and hypothesis testing. But very little is known on the role of this function in the theory of information transmission. An important exception are Shannon's inequality and its application in noiseless codig theory.

However, recently some results concerning the "mismatch" that occurs when using a code, perhaps designed specifically for one source, on another source are obtained, where inaccuracy plays a crucial role, see Fischer (1977a). In the present paper we apply these and some other mismatch results to universal encoding and decoding problems.

Universal encoding and decoding problems has been treated in the information theoretic literature by various methods, e. g. Blackwell, Breiman and Thomasian (1959), Dobrushin (1959), Wolfowitz (1961), Stiglitz (1966), Goppa (1975), Fitingov (1966), Gilbert (1971), Davisson (1973), Sakrison (1969) and Ziv (1972). More recently Gray, Neuhoff and Shields(1975) obtained some results on the mismatch mentioned above by using a generalization of Ornstein's $\bar{d}$ distance. They took these results as a basis for proving universal source coding theorems, see also Neuhoff, Gray and Davisson (1975).

The aim of our paper is to develop an approach to the problems of universal encoding and decoding which is based on the concept of inaccuracy. Another purpose is to get a deeper understanding of the role of inaccuracy in Shannon's theory of information transmission. For this the results of Fischer (1977a) form the starting point of our considerations.

1. NOISELESS SOURCE CODING

For any nonempty and finite set A (called alphabet) and any positive integer n the symbol A^n denotes the set of all finite sequences (words) $w = a_1 \ldots a_n$ with $a_i \in A$, $i = 1, \ldots, n$. W(A) is the set of all finite sequences on A, that is $W(A) = \bigcup_n A^n \cup \{e\}$, where e is the empty word.

Let $[X,\pi]$ be a discrete memoryless source, that means X is a finite alphabet and $\pi = (\pi(x))_{x\in X}$ is a probability distribution (abbr. p. d.) on X. The probability that the source will produce the word $u = x_1 \dots x_n \in X^n$ is given by

$$\pi(u) = \prod_{i=1}^{n} \pi(x_i). \tag{2}$$

Let further a finite code alphabet Γ with at leǿast two letters be given. Let m denote the cardinal number of Γ. For any positive number n let $\gamma = \gamma(n)$ be an encoding function from X^n into $W(\Gamma)$.Finally, for any $u \in X^n$ let $l_\gamma(u)$ denote the length of the codeword $\gamma(u)$. Then the average codeword length of a given code $\gamma(X^n)$ is defined as

$$L_n(\pi,\gamma) =_{df} \frac{1}{n}\sum_{u\in X^n} \pi(u)\cdot l_\gamma(u).$$

Let us suppose that the encoder does not know the exact source probability distribution π. Suppose he has only some hypothesis σ on the true distribution π. For the following we assume that $\sigma(x) > 0$ for any $x \in X$ and that the hypothetic source $[X,\sigma]$ is memoryless too. Then the inaccuracy (or Bongard entropy) of the source $[X,\pi]$ with respect to σ is defined by

$$H(\pi,\sigma) =_{df} -\sum_{x\in X} \pi(x)\cdot \log_m \sigma(x).$$

For the Shannon entropy $H(\pi) =_{df} H(\pi,\pi)$ we get from (1)

$$H(\pi) = \inf_{\sigma} H(\pi,\sigma).$$

It is well-known that for any p. d. σ and any positive integer n thre is always an uniquely decipherable code $\gamma(x^n)$ whose codeword lengths $l_\gamma(u)$, $u \in X^n$, satisfy the relation

$$-\log_m \sigma(u) \le l_\gamma(u) < -\log_m \sigma(u) + 1 \tag{3}$$

(for instance $\gamma(X^n)$ may be obtained by using the Shannon algorithm). This leads to the following theorem:

Theorem 1.1

The Shannon encoding scheme applied on σ yields an uniquely decipherable code $\gamma(X^n) \subseteq W(\Gamma)$ such that

$$H(\pi,\sigma) \le L_n(\pi,\gamma) < H(\pi,\sigma) + \frac{1}{n}.$$

Clearly, the limiting value of $L_n(\pi,\gamma)$ for $n \longrightarrow \infty$ is given by $H(\pi,\sigma)$. Hence, the mismatch of using a code built for one source on another source can be described by the directed divergence

$$D(\pi,\sigma) = H(\pi,\sigma) - H(\pi) = \sum_{x\in X} \pi(x) \log_m \frac{\pi(x)}{\sigma(x)}.$$

The proof of theorem 1.1 was given by Gilbert (1971), however he does not used the concept of inaccuracy for describing this property.

For the further investigations we make the additional assumption that the encoder knows that the true (but unknown) p. d. π is taken from some certain convex and compact class S of probability distributions on X. Let $\gamma(\sigma)$ denote an encoding function satisfying (3). Then define

$$L(\pi,\gamma(\sigma)) =_{df} \lim_{n\to\infty} L_n(\pi,\gamma(\sigma)),$$

$$L(S,\gamma(\sigma)) =_{df} \sup_{\pi\in S} L(\pi,\gamma(\sigma)),$$

$$L(S) =_{df} \inf_{\sigma} \sup_{\pi\in S} L(\pi,\gamma(\sigma)).$$

<u>Theorem 1.2</u>

For any convex and copact class S of probability distributions on X holds

$$L(S,\gamma(\sigma)) = \sup_{\pi\in S} H(\pi,\sigma)$$

and

$$L(S) = H(S) =_{df} \sup_{\pi\in S} H(\pi).$$

Proof:

The first assertion follows directly from theorem 1.1. Since S and the set of all probability distributions on X are convex and compact, and since $H(\pi,\sigma)$ is concave in σ and linear in π we get

$$L(S) = \inf_{\sigma} \sup_{\pi\in S} H(\pi,\sigma) = \sup_{\pi\in S} \inf_{\sigma} H(\pi,\sigma) = \sup_{\pi\in S} H(\pi).$$

Hence, there is an encoding scheme such that for any source $[X,\pi]$ with $\pi\in S$ an average codeword length less than $L(S) = \sup_{\pi\in S} H(\pi)$ is attainable (in the limit).

Therefore the question arises, how a p. d. σ such that an encoding function $\gamma(\sigma)$ is universal for S can be found. For giving an answer take $\pi_0 \in S$ such that $H(\pi_0) = \sup_{\pi \in S} H(\pi)$. It has been shown that for any p. d. $\pi \in S$ the inequality

$$(4) \qquad H(\pi, \pi_0) \leq H(\pi_0)$$

is satisfied (Fischer, 1977b). Hence, for an encoding function $\gamma(\pi_0)$ we get from theorem 1.1 with regard to (4)

$$L(\pi, \gamma(\pi_0)) = H(\pi, \pi_0) \leq H(\pi_0) = \sup_{\pi \in S} H(\pi)$$

for any source $[X, \pi]$ with $\pi \in S$. That is $\gamma(\pi_0)$ is universal with respect to S.

For Campbell's exponential average codeword length

$$L_{t,n}(\pi, \gamma) =_{df} \frac{1}{n} \log_m \sum_{u \in X^n} \pi(u) \cdot m^{t \cdot l_\gamma(u)}, \quad t > 0$$

similar results hold.

<u>Theorem 1.3</u>

For any positve integer n there is an uniquely decipherable code $\gamma(X^n) \subseteq W(\Gamma)$, which is based only on the hypothetical p. d. σ, satisfying

$$H_\alpha(\pi, \sigma) \leq L_{t,n}(\pi, \gamma) < H_\alpha(\pi, \sigma) + \frac{1}{n},$$

where $\alpha =_{df} 1/(1+t)$ and

$$H_\alpha(\pi, \sigma) =_{df} \frac{\alpha}{1-\alpha} \cdot \log_m \sum_{x \in X} \pi(x) \cdot \sigma(x)^{\alpha - 1} + \log_m \sum_{x \in X} \sigma(x)^\alpha .$$

The proof can be found by applying the well-known Shannon algorithm to the auxiliary p. d. $\sigma_\alpha(u) =_{df} \sigma(u)^\alpha / \sum_{u'} \sigma(u')^\alpha$, $u \in X^n$, see Fischer (1977a).

The quantity $H_\alpha(\pi, \sigma)$ is a generalization of inaccuracy and of Rényi's entropy of order α

$$H_\alpha(\pi) = \frac{1}{1-\alpha} \log_m \sum_{x \in X} \pi(x)^\alpha ,$$

since $\lim_{\alpha \to 1} H_\alpha(\pi,\sigma) = H(\pi,\sigma)$ and $H_\alpha(\pi,\pi) = H_\alpha(\pi)$. Defining $\gamma(\sigma_\alpha)$ as an encoding function satisfying (3) when σ is replaced by σ_α one obtaines for the quantity

$$L_t(S) =_{df} \inf_{\sigma} \sup_{\pi \in S} \lim_{n \to \infty} \frac{1}{n} L_{t,n}(\pi, \gamma(\sigma_\alpha))$$

the following result:

$$L_t(S) = H_\alpha(S) =_{df} \sup_{\pi \in S} H_\alpha(\pi)$$

where $\alpha = 1/(1+t)$, $t > 0$ (Fischer, 1977b).

2. SOURCE CODING FOR NOISY CHANNELS

Let $[X,\pi]$ be a discrete memoryless source and σ another p. d. on X as before. For any positive integer n a set $M \subseteq X^n$ will be called a σ-code for $[X,\pi]$, if there is a real number $\varepsilon > 0$ such that

$$M = M_\varepsilon(\sigma) = \{ u \,/\, u \in X^n \wedge \sigma(u) > \varepsilon \} .$$

Let $P(\pi,\sigma) =_{df} \pi(\overline{M_\varepsilon(\sigma)})$ denotes the probability of error that arises when using a σ-code for some unknown source $[X,\pi]$. Fischer (1977a) has shown the following relation concerning $P(\pi,\sigma)$ and the encoding rate $R =_{df} (\log|M|)/n$:

<u>Theorem 2.1</u>

There is a function $E(R,\pi,\sigma)$ such that for any positive integer n

$$P(\pi,\sigma) \leq 2^{-n \cdot E(R,\pi,\sigma)},$$

where the error exponent $E(R,\pi,\sigma)$ is positive, if $R > H(\pi,\sigma)$.+)

Theorem 2.1 can be interpreted as a statement on the error probability which occurs when using some hypothesis σ instead of the true (but unknown) p.d. π for constructing the source code $M = M_\varepsilon(\sigma)$. The error exponent can be represented as

$$E(R,\pi,\sigma) = \sup_{\alpha \in (0,1]} \frac{1-\alpha}{\alpha} \left[R - H_\alpha(\pi,\sigma) \right].$$

Hence, using a π-code, theorem 2.1 is Jelinek's well-known source coding theorem, see Jelinek (1968), theorem 5.2.

+) For all what follows the logarithms are taken to basis 2.

Now suppose again that the unknown π is a member of a certain convex and compact class S of probability distributions on X. Then the quantities

$$P(S,\sigma) =_{df} \sup_{\pi \in S} P(\pi,\sigma)$$

and

$$P(S) =_{df} \inf_{\sigma} \sup_{\pi \in S} P(\pi,\sigma)$$

are important.

Theorem 2.2

1) There exists a function $E_S(R,\sigma)$ such that for any $n > 0$

$$P(S,\sigma) \leq 2^{-n \cdot E_S(R,\sigma)},$$

$$E_S(R,\sigma) > 0, \text{ if } R > \sup_{\pi \in S} H(\pi,\sigma).$$

2) There exists a function $E_S(R)$ such that for any $n > 0$

$$P(S) \leq 2^{-n \cdot E_S(R)},$$

$$E_S(R) > 0, \text{ if } R > H(S) = \sup_{\pi \in S} H(\pi).$$

Proof:

Let $E_S(R,\sigma) =_{df} \inf_{\pi \in S} E(R,\pi,\sigma)$ and choose π^* such that $E(R,\pi^*,\sigma) = E_S(R,\sigma)$. From theorem 2.1 follows $E_S(R,\sigma) > 0$, if $R > H(\pi^*,\sigma)$. Hence, $E_S(R,\sigma)$ is also positive, if $R > \sup_{\pi \in S} H(\pi,\sigma)$.

For proving the second assertion define

$$E_S(R) =_{df} \sup_{\sigma} \inf_{\pi \in S} E(R,\pi,\sigma)$$

and choose σ^* such that $H(S,\sigma^*) = \sup_{\pi \in S} H(\pi,\sigma^*) = \inf_{\sigma} \sup_{\pi \in S} H(\pi,\sigma)$.

From the assertion 1) of theorem 2.2 we get $E_S(R,\sigma^*) > 0$, if $R > H(S,\sigma^*)$. Therefore, $E_S(R) = \sup_{\sigma} E_S(R,\sigma) \geq E_S(R,\sigma^*) > 0$.

It remains to show that

$$H(S,\sigma^*) = H(S).$$

However, as in the proof of theorem 1.2, we can conclude

$$H(S,\sigma^*) = \inf_{\sigma} H(S,\sigma) = \inf_{\sigma} \sup_{\pi \in S} H(\pi,\sigma)$$

$$= \sup_{\pi \in S} H(\pi) = H(S).$$

Theorem 2.2 can be interpreted as an universal coding theorem again. For the p.d. π_0 defined in the last chapter we obtain from theorem 2.1 for any source $[X,\pi]$ with $\pi \in S$ by using a π_0-code

$$P(\pi,\pi_0) \leq 2^{-n \cdot E(R,\pi,\pi_0)},$$

where $E(R,\pi,\pi_0)$ is positive, if $R > H(\pi,\pi_0)$. Since $H(\pi,\pi_0) \leq H(\pi_0) = H(S)$ (4), the last condition is satisfied, if $R > H(S)$. Consequently, a π_0-code is universal relative to S, if we relate the universality of some code to achieving the best possible convergence region $R > H(S)$.

Finally, it should be pointed out that the classical results of Shannon's information theory ($\sigma = \pi$) are not only some special results in our generalized theory. The classical theory is included in our considerations; since $H(\pi) \leq H(\pi,\sigma)$ and $H_\alpha(\pi) \leq H_\alpha(\pi,\sigma)$, it represents the optimum case. For further details see Fischer(1977a).

3. A GENERALIZED MUTUAL INFORMATION FUNCTION

Now let $[X,\pi]$ be a discrete memoryless source and σ a hypothesis as above. Let further a discrete memoryless channel $[X,\varkappa,Y]$ and another channel probability matrix $\omega = (\omega(y/x))_{x \in X,\, y \in Y}$ be given. The matrix ω can be interpreted as the decoder's hypothesis on the unknown or not exactly known true channel probability matrix $\varkappa = (\varkappa(y/x))_{x \in X,\, y \in Y}$. Then Pötschke (1970) has defined the following generalized mutual information function:

$$(5) \qquad I(\pi,\sigma;\varkappa,\omega) =_{df} \sum_{x \in X} \sum_{y \in Y} \pi(x)\varkappa(y/x) \log \frac{\omega(y/x)}{\sum_{x' \in X} \sigma(x')\omega(y/x')}$$

(suppose that $\omega(y/x) > 0$, if $\varkappa(y/x) > 0$). Three special cases of the expression (5) are of interest:

Clearly, for $\sigma = \pi$ and $\omega = \varkappa$ we get Shannon's mutual information

$$I(\pi;\varkappa) = I(\pi,\pi;\varkappa,\varkappa).$$

The function

$$(6) \qquad I(\pi,\sigma;\varkappa) =_{df} I(\pi,\sigma;\varkappa,\varkappa)$$

was already investigated by Kerridge (1961). Mathur (1974) considered some connections between this function and a generalized rate distortion function. For our purposes the last case plays a crucial role:

$$I(\pi;\varkappa,\omega) =_{df} I(\pi,\pi;\varkappa,\omega).$$

Now for any $x \in X$, $y \in Y$ we define the following probability measures:

$$\mu(x,y) =_{df} \pi(x)\,\varkappa(y/x), \qquad \nu(x,y) =_{df} \sigma(x)\,\omega(y/x),$$

$$\pi'(y) =_{df} \sum_x \mu(x,y), \qquad \sigma'(y) =_{df} \sum_x \nu(x,y),$$

$$\varkappa'(x/y) =_{df} \frac{\mu(x,y)}{\pi'(y)}, \qquad \omega'(x/y) =_{df} \frac{\nu(x,y)}{\sigma'(y)}.$$

Further define the average conditional entropies

$$H(\varkappa,\omega/\pi) =_{df} -\sum_x \pi(x) \sum_y \varkappa(y/x) \log \omega(y/x),$$

$$H(\varkappa/\pi) =_{df} H(\varkappa,\varkappa/\pi),$$

$$H(\varkappa',\omega'/\pi') =_{df} -\sum_y \pi'(y) \sum_x \varkappa'(x/y) \log \omega'(x/y),$$

$$H(\varkappa'/\pi') =_{df} H(\varkappa',\varkappa'/\pi').$$

The formulae

$$\begin{aligned} I(\pi;\varkappa) &= H(\pi) - H(\varkappa'/\pi') \\ &= H(\pi') - H(\varkappa/\pi) \\ &= H(\pi) + H(\pi') - H(\mu) \end{aligned}$$

are well-known from Shannon's information theory. The corresponding formulae for the generalized mutual information (5) are

(7) $$I(\pi,\sigma;\varkappa,\omega) = H(\pi,\sigma) - H(\varkappa',\omega'/\pi')$$

(8) $$= H(\pi',\sigma') - H(\varkappa,\omega/\pi)$$

$$= H(\pi,\sigma) + H(\pi',\sigma') - H(\mu,\nu).$$

The relations between the three special cases introduced above can be expressed by the inequalities

$$I(\pi;\varkappa,\omega) \leq I(\pi;\varkappa) \leq I(\pi,\sigma;\varkappa).$$

Moreover, while $0 \leq I(\pi;\varkappa) \leq \log|X|$, the function $I(\pi;\varkappa,\omega)$ has no lower bound and $I(\pi,\sigma;\varkappa)$ has no upper one.

Finally, define the generalized capacity function

$$C(\varkappa,\omega) =_{df} \sup_{\pi} I(\pi;\varkappa,\omega).$$

It is easy to see that

$$0 \leq C(\varkappa,\omega) \leq C(\varkappa),$$

where $C(\varkappa) =_{df} C(\varkappa,\varkappa)$ is the usual channel capacity for discrete memoryless channels.

4. CHANNEL CODING AND DECODING

The problem of universal channel coding and decoding can now be attacked in a similar way as it was done for the corresponding source coding problem. For decoding the outputs of a discrete memoryless channel $[X,\varkappa,Y]$ with unknown channel probability matrix $\varkappa$ we use a decoding algorithm that is based on some hypothetic decoding matrix $\omega = (\omega(y/x))_{x\in X,\, y\in Y}$. With respect to the memoryless character of the given channel we assume, that the channel $[X,\omega,Y]$ is also memoryless.

Let N,n be some arbitrary positive integers. For any given N-code $\mathcal{L} = \{u_1,\dots,u_N\}$, $u_i \in X^n$, $i = 1,\dots,N$, define a ω-decoding system

$$\vartheta = \{(u_i,V_i) \,/\, i = 1,\dots,N\}$$

as follows: Let V_i' be the set of all sequences $v\in Y^n$ such that

$$\omega(v/u_i) = \max_{j\in\{1,\dots,N\}} \omega(v/u_j).$$

Then

$$V_i =_{df} \begin{cases} V_1' & i = 1 \\ V_i' \setminus \bigcup_{j=1}^{i-1} V_j' & i = 2,\dots,N. \end{cases}$$

The following theorem states a result on the asymptotic behavior of the average probability of error

$$P(\mathcal{L},\varkappa,\omega) =_{df} \frac{1}{N}\sum_{i=1}^{N} \varkappa(\overline{V}_i/u_i)$$

that occurs when using a ω-decoding system.

Theorem 4.1

For any discrete memoryless channel $[X,\varkappa,Y]$ and any decoding matrix ω there exist a code $\mathcal{L}$ of $N = 2^{nR}$ messages into sequences of length n of channel-input symbols and an ω-decoding system for $\mathcal{L}$ such that

$$P(\mathcal{L},\varkappa,\omega) \leq 2^{-n\cdot E(R,\varkappa,\omega)}$$

where for $R < C(\varkappa,\omega)$

$$E(R,\varkappa,\omega) > 0.$$

Proof:

Our approach to the bounding of the probability of decoding error will be very similar to the method of Jelinek (1968). First define the error indicator function

$$\varphi_i(v/\mathcal{L}) =_{df} \begin{cases} 1 & \text{if } \omega(v/u_j) \geq \omega(v/u_i) \text{ for some } j \neq i, \\ & u_i,\ u_j \in \mathcal{L} \\ 0 & \text{otherwise.} \end{cases}$$

We then get for any $\delta \in [0,1]$

$$\varphi_i(v/\mathcal{L}) \leq \left[\sum_{j\neq i} \left(\frac{\omega(v/u_j)}{\omega(v/u_i)}\right)^{\frac{1}{1+\delta}}\right]^{\delta}.$$

Since

$$P(\mathcal{L},\varkappa,\omega) \leq \frac{1}{N}\sum_{i=1}^{N} \sum_{v\in Y^n} \varkappa(v/u_i)\cdot\varphi_i(v/\mathcal{L}),$$

we get

$$\text{(9)} \quad E_\eta\left[P(\cdot,\varkappa,\omega)\right] \leq \sum_v \sum_u \eta(u)\,\varkappa(v/u)\,\omega(v/u)^{-\frac{\delta}{1+\delta}} \left[(N-1)\sum_{u'} \eta(u')\,\omega(v/u')^{\frac{1}{1+\delta}}\right]^{\delta}$$

where E_η denotes the expectation with respect to the code selection process governed by some probability distribution η on X, see e. g. Jelinek (1968) pp. 140 - 141. Using the memoryless character of the given channel we get from (9) by further simplification

$$\text{(10)} \quad \inf_\eta E_\eta\left[P(\cdot,\varkappa,\omega)\right] \leq 2^{-n\cdot E(R,\varkappa,\omega)}$$

where

$$E(R,\varkappa,\omega) =_{df} \sup_\eta \sup_{\delta\in[0,1]} \left[E(\delta,\eta,\varkappa,\omega) - \delta R\right],$$

$$E(\delta,\eta,\varkappa,\omega) =_{df} -\log\sum_{x,y}\eta(x)\varkappa(y/x)\omega(y/x)^{-\frac{\delta}{1+\delta}}\left[\sum_{x'}\eta(x')\omega(y/x)^{\frac{1}{1+\delta}}\right]^{\delta}.$$

Since $\inf_{\mathcal{L}} P(\mathcal{L},\varkappa,\omega) \leq E_\eta\left[P(\cdot,\varkappa,\omega)\right]$, there is at least one code $\mathcal{L}$ such that $P(\mathcal{L},\varkappa,\omega)$ is upper bounded by the right hand side of inequality (10). The second assertion of theorem 4.1 follows from the lemma below:

Lemma 4.2

$$\frac{\partial}{\partial \delta} E(\delta, \eta, \varkappa, \omega)\Big|_{\delta=0} = I(\eta; \varkappa, \omega)$$

This equation expressing the connection between the present and the preceding chapter is easy to verify.

A similar approach was first done by Stiglitz (1966), however he has not considered the relations between his approach and generalizations of information measures. Clearly, theorem 4.1 is a generalization of Gallager's familiar random coding theorem, see Gallager (1965). However, Gallager's theorem is not only some special case of theorem 4.1. Since $C(\varkappa) = \sup_{\omega} C(\varkappa, \omega)$ and since $E(R, \pi) =_{df} E(R, \tilde{\pi}, \pi)$ is the best possible exponent (see Fischer (1976)), it describes the optimum case within our considerations.

Now let us assume that the true (but unknown) channel probability matrix $\varkappa$ belongs to some known class C of channel probability matrices on X,Y. Suppose C is convex and compact. Then define

$$P(\varkappa, \omega) =_{df} \inf_{\eta} E_{\eta}\left[P(\cdot, \varkappa, \omega)\right],$$

$$P(C, \omega) =_{df} \sup_{\varkappa \in C} P(\varkappa, \omega),$$

$$P(C) =_{df} \inf_{\omega} \sup_{\varkappa \in C} P(\varkappa, \omega).$$

Theorem 4.3

1) There is a function $E_C(R, \omega)$ such that

$$P(C, \omega) \leq 2^{-n \cdot E_C(R, \omega)}$$

where $E_C(R, \omega)$ is positive, if $R < \inf_{\varkappa \in C} C(\varkappa, \omega)$.

2) There is a function $E_C(R)$ such that

$$P(C) \leq 2^{-n \cdot E_C(R)}$$

where $E_C(R)$ is positive, if $R < \sup_{\omega} \inf_{\varkappa \in C} C(\varkappa, \omega)$.

Proof:

From (10) we get $P(C, \omega) = \sup_{\varkappa \in C} P(\varkappa, \omega) \leq 2^{-n \cdot \inf_{\varkappa \in C} E(R, \varkappa, \omega)}$.

Now for each R, ω let $\varkappa^*$ be such chosen that $E(R, \varkappa^*, \omega) = \inf_{\varkappa \in C} E(R, \varkappa, \omega)$.

Then we can conclude from theorem 4.1, that $E_C(R,\omega) =_{df} \inf_{\varkappa\in C} E(R,\varkappa,\omega)$ $= E(R,\varkappa^*,\omega) > 0$, if $R < C(\varkappa^*,\omega)$. Clearly, the latter is still true, if $R < \inf_{\varkappa\in C} C(\varkappa,\omega)$. Now let ω^* be such, that $\inf_{\varkappa\in C} C(\varkappa,\omega^*) = \sup_{\omega} \inf_{\varkappa\in C} C(\varkappa,\omega)$. Then $E_C(R,\omega^*)$ is positive, if $R < \inf_{\varkappa\in C} C(\varkappa,\omega^*)$. Hence, $E_C(R) =_{df} \sup_{\omega} E_C(R,\omega)$ is positive too.

A similar theorem was proven in the paper of Stiglitz (1966) by using somewhat different error exponent

$$(11)\quad E_C^!(R) =_{df} \sup_{\delta\in[0,1]} \sup_{\eta} \inf_{\varkappa\in C} \left[E(\delta,\eta,\varkappa,\varkappa) - \delta R\right] \leq E_C(R).$$

If there is even equality in (11) is open.

Finally, it should be remarked, that

$$\sup_{\omega} \inf_{\varkappa\in C} C(\varkappa,\omega) = \inf_{\varkappa\in C} \sup_{\omega} C(\varkappa,\omega) = \inf_{\varkappa\in C} C(\varkappa).$$

5. CONCLUDING REMARKS

In the rate distortion theory some similar considerations are possible. Let $[X,\pi]$ be a given discrete memoryless source, Y a reproducing alphabet and d a bounded single-letter distortion measure defined on $X \times Y$, that is $d(x,y) \leq d_0$ for all $x \in X$, $y \in Y$. For any positive integer $D \geq \sum_x \pi(x) \cdot \min_y d(x,y)$ let C_D denote the set of all D-admissible probability matrices $(\varkappa(y/x))_{x\in X,\ y\in Y}$.

Usually, for encoding a source $[X,\pi]$ with respect to some given distortion measure d a random coding argument is used. If the code selection process is governed by some arbitrary p. d. σ' on Y, it is possible to show (following an approach of Omura (1973)) that there exists a code $\mathcal{L}$ of block length n and rate R with average distortion $d(\mathcal{L})$ achieved with code $\mathcal{L}$ satisfying

$$d(\mathcal{L}) \leq D + d_0 \cdot 2^{-n\cdot E(R,D,\pi,\sigma')}$$

where $E(R,D,\pi,\sigma')$ is positive for rates greater than

$$(12)\quad R(D,\pi,\sigma') =_{df} \inf_{\varkappa\in C_D} \left[H(\pi',\sigma') - H(\varkappa/\pi)\right].$$

Clearly, (12) is minimized by choosing $\sigma' = \pi'$, that is

$$\sigma'(y) =_{df} \pi'(y) = \sum_{x\in X} \pi(x)\,\varkappa(y/x), \quad y \in Y.$$

This corresponds to the usual approach in rate distortion theory. However, defining σ' by $\sigma'(y) =_{df} \sum_{x\in X} \sigma(x)\,\varkappa(y/x)$, $y \in Y$, where

σ is some hypothesis on π, we get

$$R(D,\pi,\sigma') = \inf_{\varkappa \in C_D} I(\pi,\sigma;\varkappa).$$

The latter function is a generalization of the usual rate distortion function investigated by Mathur (1974), and $I(\pi,\sigma;\varkappa)$ is Kerridge's mutual information, see formulae (6) and (8).

Finally, it should be pointed out, that further applications of our approach concerning coding theorems on weighted informations are also possible, see Fischer (1975).

ACKNOWLEDGMENT

The results presented here were obtained while I was with the Department of Mathematics, Humboldt University, Berlin. Therefore, I am grateful to all former colleagues for pleasing cooperating. Especially, I am indebted to Prof. Dr. H. Thiele for kindling my interest in information theory, and to Dipl. Math. D. Pötschke for many useful discussions on the subjects treated herein.

REFERENCES

Aczél, J., Ostrowski, A. M.: On the characterization of Shannon's entropy by Shannon's inequality. J.australian math. soc. 16 (1973), part 3, 368 - 374.

Blackwell, D., Breiman, L., Thomasian, A.J. (1959): The capacity of a class of channels. Ann. math. statistics 30 (1959), 1229 - 1241.

Bongard, M. M. (1963): O ponjatii "polesnaja informatsija" (On the concept of "useful information"). Probl. Kibernetiki 9, Moskva 1963

Davisson, L.D. (1973): Universal noiseless coding. IEEE Trans. Inf. Th. IT-19 (1973), 783 - 795.

Dobrushin, R.L. (1959): Optimalnaja peredacha informatsii po kanalu s neiswestnymi parametrami (Optimal information transmission over a channel with unknown parameters). Radiotechnika 4 (1959), 1951 - 1956.

Fischer, P. (1972): On the inequality $\sum p_i \cdot f(p_i) \geq \sum p_i \cdot F(q_i)$. Metrika 18 (1972), 199 - 208.

Fischer, Th. R. M. (1975): Über eine Verallgemeinerung der Bongard-Entropie und Kodierungssätze für bewertete Informationen (On a generalization of the Bongard-entropy and coding theorems for weighted informations). Wiss. Z. Humboldt Univ., Math.-Nat. Reihe 24 (1975), 750 - 755.

——— (1976): Kodierungssätze für Kanäle mit unbekanntem statistischen Verhalten (Coding theorems for unknown channels). Berlin 1976, submitted to EIK.

——— (1977a): Über Verallgemeinerungen der Bongard-Entropie und Kodierungssätze für Quellen mit unbekanntem statistischen Verhalten (On generalisations of the Bongard-entropy and coding theorems for unknown sources). EIK 13 (1977), 125-135.

——— (1977b): Zu einigen verallgemeinerten Entropiebegriffen und ihrer Bedeutung in der Shannonschen Theorie der Informationsübertragung(On some generalized entropies and their role in Shannon's theory of information transmission). Rough copy for a dissertation, Berlin 1977.

Fitingov, B.M. (1966): Optimalnoje kodirowanie pri neiswestnoj i menjajushesja statistike soobshenie (Optimal encoding with unknown and variable message statistics). Probl. Peredachi Informatsii 2 (1966), 3 - 11.

Gallager, R. G. (1965): A simple derivation of the coding theorem and some applications. IEEE Trans. Inf. Th. IT-11(1965), 3 - 18

Gilbert, E. N. (1971): Codes based on inaccurate source probabilities. IEEE Trans. Inf. Th. IT-17 (1971), 304 - 314.

Goppa, W.D. (1975): Uniwersalnoje dekodirowanie dlja simmetrichnych kanalow (Universal decoding for symmetric channels). Probl. Peredachi Informatsii 11 (1975), 15 - 22.

Gray, R. M., Neuhoff, D.L., Shields, P. C. (1975): A generalization of Ornstein's $\bar{d}$ distance with applications to information theory. Ann. Prob. 3 (1975), 315 - 328.

Jelinek, F. (1968): Probabilistic information theory. McGraw-Hill, New York 1968.

Kannappan, Pl. (1972): On directed divergence and inaccuracy. Z. f. Wahrscheinlichkeitstheorie u. verw. Geb. 25 (1972), 49 - 55.

Kerridge, D. F. (1961): Inaccuracy and inference. J. Roy. Statist. Soc. Ser. B 23 (1961), 184 - 194.

Mathur, Y. D. (1974): Information improvement and computation of the rate-distortion function under true and inaccurate input distributions. EIK 10 (1974), 553 - 564.

Nath, P. (1968): Inaccuracy and coding theory. Metrika 13 (1968), 123 - 135.

Neuhoff, D. L., Gray, R.M., Davisson, L.D. (1975): Fixed rate universal block source coding with fidelity criterion. IEEE Trans. Inf. Th. IT-21 (1975), 511 - 523.

Ng, C. T. (1974): Representation for measures of information with the branching property. Information Control 24 (1974), 45-56.

Omura, J.K. (1973): A coding theorem for discrete-time sources. IEEE Trans. Inf. Th. IT-19 (1973), 490 - 498.

Pötschke, D. (1970): Zum pragmatischen Entropiebegriff (On the concept of pragmatic entropy). Diplomarbeit, Humboldt-Universität, Berlin 1970.

——— (1974): A statistical interpretation of the B-rate of information theory. Seventh Prague Conference, Prague 1974.

Sakrison, D. J. (1969): The rate distortion function of a class of sources. Inform. Control 15 (1969), 165 - 195.

Stiglitz, I. G. (1966): Coding for a class of unknown channels. IEEE Trans. Inf. Th. IT-12 (1966), 189 - 195.

Wolfowitz, J. (1961): Coding theorems of information theory. Springer, Berlin, Göttingen, Heidelberg 1961, 2nd ed. 1964.

Ziv, J. (1972): Coding of sources with unknown statistics. IEEE Trans. Inf. Th. IT-18 (1972), 384 - 394.

Technical University of Dresden,
Department of Mathematics

DDR - 8027 Dresden
Mommsenstraße 13

German Democratic Republic

ON REALIZABILITY OF STOCHASTIC PROCESSES

Peter Gaenssler, Winfried Stute

Bochum

ABSTRACT

Let $\xi = (\xi_t)_{0\leq t\leq 1}$ be a stochastic process on some probability space $(\Omega, \mathscr{F}, \mathbb{P})$ and let X be a fixed class of real functions on the unit interval. Then, by definition, ξ is said to have a realization in X, if there exists a process $\eta = (\eta_t)_{0\leq t\leq 1}$ with sample paths in X and such that ξ and η have the same finite dimensional distributions. It is the purpose of this paper to point out the strong interdependence between realizability in X and the behaviour of a corresponding modulus function.

SOME CRITERIA FOR REALIZABILITY OF STOCHASTIC PROCESSES

Let $\xi = (\xi_t)_{0\leq t\leq 1}$ be a stochastic process defined on some probability space $(\Omega, \mathscr{F}, \mathbb{P})$, and let X be a fixed class of real-valued functions on the unit interval I. The process ξ is said to have a realization in X if there exists a process $\eta = (\eta_t)_{0\leq t\leq 1}$ (not necessarily defined on the same probability space) with sample paths in X and such that ξ and η have the same finite dimensional distributions. Hence both processes are stochastically equivalent.

It is well-known that ξ has a realization in X if and only if $\mathscr{L}\{\xi\}^*(X) = 1$, where $\mathscr{L}\{\xi\}$ is the distribution of ξ and * denotes outer measure. The process η is usually defined to be the coordinate process on X.

Since in general such a condition is not at all easy to verify, more accessible criteria are needed. For the class X = C of all continuous functions on I the following result is due to Mann (1951).

<u>Theorem 1</u>. For ξ to have a realization in X it is necessary and sufficient that

$$(*)\begin{cases}\text{for all } \varepsilon,\eta>0 \text{ there exists some } \delta=\delta(\varepsilon,\eta)>0 \text{ such that for} \\ \text{each finite subset } S \text{ of the unit interval} \\ \mathbb{P}(\{\omega\in\Omega: \omega^X(\xi(\omega),\delta,S)>\varepsilon\})\le\eta.\end{cases}$$

Here $\xi(\omega)$ denotes the sample path of ξ at $\omega\in\Omega$ and, for each $f\in\mathbb{R}^I$,

$$\omega^X(f,\delta,S) = \omega^C(f,\delta,S) := \sup_{\substack{|t-s|<\delta \\ t,s\in S}} |f(t)-f(s)|$$

is the modulus of continuity of f on $S\subset I$. A condition like (*) is useful since it is only based on the finite dimensional distributions of ξ. Furthermore, it is worthwhile noting that the space C enters into the proof only via its modulus function ω^C. It is therefore natural to expect the validity of Theorem 1 also for spaces X having a corresponding modulus function ω^X, i.e. for which

$$f\in X \Leftrightarrow \lim_{\delta\to 0} \omega^X(f,\delta,I) = 0.$$

In this generality Theorem 1 is due to Gaenssler (1974). For the definition of a general modulus function,and the complete statement and proof of this result, we refer the reader to Gaenssler and Stute (1977). In order to verify (*) for a particular X one needs suitable criteria in terms of the lower dimensional marginal distributions, rather than the class of all finite dimensional distributions. It is the aim of this paper to investigate such conditions for various cases of interest. Most of the results may be found elsewhere. However, the usefulness of the present setup stems from the fact that it points out the strong interdependence between realizability in X and the stochastic behaviour of the associated modulus function.

As a second example we consider the space $\hat{D}$ of all functions f on I with right and left limits, and with either $f(t) = f(t-0)$ or $f(t) = f(t+0)$, at each $0\le t\le 1$. The corresponding modulus function is then defined by

$$\omega^{\hat{D}}(f,\delta,S) = \sup_{\substack{t-\delta\le t_1<t<t_2\le t+\delta \\ t_1,t,t_2\in S}} \min\{|f(t)-f(t_1)|, |f(t_2)-f(t)|\}.$$

Finally, let D' denote the space of all functions f on I which have a continuous derivative $f'(t)$ for all $0\le t\le 1$. The associated modulus function is now given by

$$\omega^{D'}(f,\delta,S) = \sup_{\substack{t-\delta\le t_1<t<t_2\le t+\delta \\ t_1,t,t_2\in S}} \left| \frac{f(t)-f(t_1)}{t-t_1} - \frac{f(t_2)-f(t)}{t_2-t} \right|.$$

In order to get manageable criteria for realizability in the cases considered above one has to study the behaviour of ω^X on two- and three-point parameter sets, respectively. For this we define

$$W_C(\delta,\varepsilon) := \sup_{|t-s|\leq\delta} \mathbb{P}(\{|\xi_t-\xi_s| > \varepsilon\}),$$

$$W_{\hat{D}}(\delta,\varepsilon) := \sup_{t-\delta\leq t_1<t<t_2\leq t+\delta} \mathbb{P}(\{\min(|\xi_t-\xi_{t_1}|, |\xi_{t_2}-\xi_t|) > \varepsilon\})$$

$$W_{D'}(\delta,\varepsilon) := \sup_{t-\delta\leq t_1<t<t_2\leq t+\delta} \mathbb{P}(\{|\frac{\xi_t-\xi_{t_1}}{t-t_1} - \frac{\xi_{t_2}-\xi_t}{t_2-t}| > \varepsilon\}).$$

Theorem 2. Let X be either C, $\hat{D}$ or D', and suppose that for some non-decreasing function q the following two integrals exist:

$$(**)\begin{cases} (q) & \int_0^1 q(\delta)\delta^{-1}d\delta < \infty \\ (W_Xq) & \int_0^1 W_X(\delta,q(\delta))\delta^{-2}d\delta < \infty. \end{cases}$$

Then ξ has a realization in X.

Condition (q) is satisfied for all q which are sufficiently small near zero. On the other hand, since $W_X(\delta,\varepsilon)$ is nonincreasing in ε, condition (W_Xq) prevents q from being too small there.

For X = C, an equivalent form of Theorem 2 may be found in Loève's (1963) book, the integral conditions being replaced by corresponding series convergence conditions. The function $W_{\hat{D}}$ was first considered by Chentsov (1956), and later by Gihman and Skorohod (1974). The result for X = D' is contained in the monograph of Cramér and Leadbetter (1967).

Though these results have been known for a long time, the above approach is more informative in that it gives a unified interpretation of the second integral convergence condition. The proof proceeds by showing that for given positive ε and η there exists a set $\Omega_0 \in \mathscr{F}$ with $\mathbb{P}(\Omega_0) > 1-\eta$ and a positive δ such that for all $\omega\in\Omega_0$

(1) $\omega^X(\xi(\omega),\delta,T) \leq \varepsilon$.

Here T is a countable dense subset of the unit interval containing 0 and 1. For example, T is the set of all dyadic rationals in I if X = C or D'. For verifying (*) it suffices to observe that, if (**) is satisfied, then for each $0<t<1$ there exists a sequence $\{t^n\}$ in T such that $t^n \to t$ and $\xi_{t^n} \to \xi_t$ $\mathbb{P}$-almost surely. If $S = \{t_1,\dots,t_k\}$ is a fixed finite subset of I, the corresponding approximation holds for each point in S. We therefore obtain sets $S_n = \{t_1^n,\dots,t_k^n\}$ such that $t_i^n \to t_i$ and $\xi_{t_i^n} \to \xi_{t_i}$ $\mathbb{P}$-almost surely for all i=1,...,k. This implies that on a set of probability larger than $1-\eta$ one has

$$\omega^X(\xi(\cdot),\ \delta/2,S) \le \limsup_{n\to\infty} \omega^X(\xi(\cdot),\delta,S_n) \le \omega^X(\xi(\cdot),\delta,T) \le \varepsilon.$$

Hence it suffices to show (1). For X = C or $\hat{D}$ a proof may be found in Gaenssler and Stute (1977). Here we shall prove (1) for the case X = D'. Let

$$A_k^n := \{|2\xi(k2^{-(n+1)}) - \xi((k-1)2^{-(n+1)}) - \xi((k+1)2^{-(n+1)})| \le q(2^{-n})2^{-(n+1)}\},$$

$k=1,\dots,2^{n+1}-1$, where $\xi(t) := \xi_t$. Put

$$D^n := \bigcap_{m\ge n} \bigcap_{k=1}^{2^{m+1}-1} A_k^m.$$

We have

$$\mathbb{P}(\complement D^n) \le \sum_{m\ge n} 2^{m+1} W_{D'}(2^{-(m+1)}, q(2^{-m})).$$

By condition $(W_{D'}, q)$ the right-hand side can be made arbitrarily small for all large enough $n \in \mathbb{N}$. Furthermore, $\sum q(2^{-r}) < \infty$ by condition (q). One may therefore find, for each choice of positive ε and η, a finite n such that

$$\mathbb{P}(\complement D_n) < \eta \text{ and } \sum_{r\ge n} q(2^{-r}) < \varepsilon/2.$$

Let $\delta = 2^{-(n+2)}$. If f is a function on I then for all $0 \le a < b \le 1$

$$\frac{f(c)-f(a)}{c-a} = \frac{f(b)-f(a)}{b-a} + \frac{1}{2}\left[\frac{f(c)-f(a)}{c-a} - \frac{f(b)-f(c)}{b-c}\right],$$

where c = (a+b)/2. Repeated application of this formula shows that on D^n

$$\left|\frac{\xi(k2^{-(n+1)})-\xi((k-1)2^{-(n+1)})}{2^{-(n+1)}} - \frac{\xi((s+1)2^{-(n+m)})-\xi(s2^{-(n+m)})}{2^{-(n+m)}}\right| \le \sum_{r=n}^{n+m-1} q(2^{-r})$$

for all $k=1,\dots,2^{n+1}-1$ and $(k-1)2^{-(n+1)} \le s2^{-(n+m)} \le (s+1)2^{-(n+m)} \le (k+1)2^{-(n+1)}$. By choice of n this establishes that (1) is satisfied for all $\omega \in D^n$, with T being the set of all dyadic rationals. This completes the proof of Theorem 2 for X = D'.

The following applications of Theorem 2 are well-known and may be simply derived by means of the Markov inequality by choosing appropriate q's. For example, ξ has a realization in

$X = C$, if

$$\mathbb{E}(|\xi_t-\xi_s|^p) \le K|t-s|^{1+\varepsilon} \qquad \text{(Kolmogorov)}$$

or

$$\mathbb{E}(|\xi_t-\xi_s|^p) \le K|t-s|/(\log|t-s|)^{1+r},\ r > p \qquad \text{(Loève)},$$

where p,K and ε are positive constants. More generally, one may consider processes ξ fulfilling a Lipschitz condition

$$\mathbb{E}(|\xi_t-\xi_s|^p) \le f(|t-s|),$$

where f is nondecreasing on I. Then ξ has a realization in C if, e.g.,

$$\int_0^1 q(\delta)^{-p}\delta^{-2}f(\delta)d\delta < \infty$$

for some q satisfying (q). For related criteria see also Hahn and Klass (1977).

For $X = \hat{D}$ and $X = D'$ the moment conditions (without the logarithmic refinements) are as follows.

$X = \hat{D}$:

$$\mathbb{E}(|\xi_t-\xi_{t_1}|^{p_1}|\xi_{t_2}-\xi_t|^{p_2}) \le K|t_2-t_1|^{1+\varepsilon},\ 0 \le t_1 < t < t_2 \le 1 \qquad \text{(Chentsov)}$$

$X = D'$:

$$\mathbb{E}(|2\xi_t-\xi_{t+h}-\xi_{t-h}|^p) \le K|h|^{1+p+\varepsilon} \qquad \text{(Cramér and Leadbetter)}.$$

The last result follows from the observation that Theorem 2 is still valid if $W_{D'}$ is replaced by $\tilde{W}_{D'}$, where $\tilde{W}_{D'}$ is defined in the same way as $W_{D'}$ but with the supremum extended over the set of all equidistant points. Indeed, only equidistant points were needed in the above proof; thus it carries over word for word to the $\tilde{W}_{D'}$-case.

We now consider the space C_o of absolutely continuous functions on I. The associated modulus function is given by

$$\omega^{C_o}(f,\delta,S) = \sup_{\substack{0\le t_1<t_1'\le\ldots\le t_s<t_s'\le 1\\ s\in\mathbb{N},\ldots\in S\\ \sum\limits_{\ldots}(t_i-t_i')<\delta}} \sum_{i=1}^{s}|f(t_i)-f(t_i')|.$$

Let q be a nondecreasing function on I. Put

$$H_q(\delta) := \sup_{y>0} \frac{q(\delta y)}{y},\ 0 < \delta < 1.$$

<u>Theorem 3.</u> Suppose that for some q

(2) $\lim\limits_{\delta\to 0} H_q(\delta) = 0$

and $(W_C q)$ of Theorem 2 are satisfied. Then ξ has a realization in C_o.

Clearly, (2) is strictly stronger than (q) of Theorem 2. The proof of Theorem 3 is similar to that of Theorem 2 for X = C. The sets A_k^n and D^n are now defined as

$A_k^n := \{|\xi(k2^{-(n+1)}) - \xi((k+1)2^{-(n+1)})| \le q(2^{-n})\}$, $k=0,\dots,2^{n+1}-1$,

$D^n := \bigcap_{m \ge n} \bigcap_{k=0}^{2^{m+1}-1} A_k^m$, $n \in \mathbb{N}$. By $(W_C q)$, $\mathbb{P}(D^n)$ can be made arbitrarily close to 1 for all large enough n. Let $\delta = 2^{-(n+1)}$. For each set S: $0 \le t_1 < t_1' \le t_2 < t_2' \le \dots \le t_s < t_s' \le 1$ of finitely many dyadic rationals in I with $\sum(t_i'-t_i) < \delta$, there exists an $r \in \mathbb{N}$ such that every point of S is a member of the $2^{-(n+r)}$-grid in I. On D^n we have

$$|\xi(t_i')-\xi(t_i)| \le c_i q(2^{-(n+r-1)}),$$

where c_i is the number of intervals of the form $[\frac{j}{2^{n+r}}, \frac{j+1}{2^{n+r}}]$ contained in $[t_i,t_i']$. Since $\sum(t_i'-t_i)<\delta$, it follows that $\sum c_i < 2^{r-1}$ and therefore

$$\sum|\xi(t_i')-\xi(t_i)| \le 2^{r-1}q(2^{-(n+r-1)}) \le H_q(2^{-n}).$$

By (2) the last term can be made arbitrarily small for all large enough n. This completes the proof of Theorem 3.

Condition (2) is in particular satisfied for $q(\delta) = \delta$. Hence ξ has a realization in C_o if, for example,

$$\mathbb{E}(|\xi_t-\xi_s|^p) \le K|t-s|^{1+p+\varepsilon} \text{ for some } p \ge 0 \text{ and } \varepsilon > 0.$$

There are many important spaces X for which a modulus function does not exist. To obtain sufficient conditions for realizability in these cases it is often necessary to consider a second space X_1, say, with $X \subset X_1$ and such that ξ is already known to have a realization in X_1. The problem is then to find extra conditions on the distribution of ξ which guarantee that on a set of probability one, the sample paths of ξ belong to X.

For example, if D is the subspace of $\hat{D}$ consisting of all right-continuous functions on I, then ξ has a realization in D iff

ξ has a realization in $\hat{D}$

and ξ is right-continuous in probability, i.e.

$$\lim_{h \downarrow 0} \mathbb{P}(\{|\xi_{t+h}-\xi_t| > \varepsilon\}) = 0 \text{ for all } 0 \le t < 1 \text{ and } \varepsilon > 0.$$

Next, we consider the space C_g of functions fulfilling a Lipschitz condition

$$|f(t+h)-f(t)| = \mathcal{O}(g(h)) \text{ as } h \to 0,$$

where g is a preassigned nonnegative nondecreasing function on I.

Theorem 4. Suppose that $(W_C q)$ holds for some nondecreasing function q such that

(3) $\sum_{r\geq n} q(2^{-r}) \leq K'g(2^{-n})$ for all $n\in\mathbb{N}$.

Then ξ has a realization in C_g.
For the proof, we may assume without loss of generality that ξ is a process with continuous sample paths. Let D^n be defined as in the proof of Theorem 3. It is easy to see that for $\delta := 2^{-(n+1)}$ one has

$$\omega^C(\xi(\cdot),\delta,T) \leq 3 \sum_{r\geq n} q(2^{-r}) \leq 3K'g(2^{-n}) \text{ on } D^n$$

and therefore, by continuity,

$$\omega^C(\xi(\cdot),\delta,I) \leq 3K'g(2^{-n}) \text{ on } D^n.$$

Since $\mathbb{P}(\liminf_{n\to\infty} D^n) = 1$ by the Borel-Cantelli Lemma, this shows that $\xi(\cdot)\in C_g$ with probability one.

Condition (3) is in particular satisfied for $g(\delta) = q(\delta) = \delta^\alpha$, $0<\alpha\leq 1$. For this choice of q, $(W_C q)$ holds for each process fulfilling

$$\mathbb{E}(|\xi_t-\xi_s|^p) \leq K|\xi_t-\xi_s|^{1+r} \text{ for some } r/p>\alpha.$$

CONCLUDING REMARKS

Let us remark that Theorem 1 remains true if I is replaced by any separable metrizable parameter set S. However, in order to derive sufficient conditions for realizability in terms of the lower dimensional marginal distributions, some geometrical properties of the set S are needed. For example, if S is the unit cube in the k-dimensional Euclidean space, certain extensions of Theorems 2-4 are available. In more general situations appropriate entropy conditions may instead serve as an adequate tool.

The authors are especially grateful to David Pollard for his careful reading of the manuscript.

REFERENCES

Chentsov, N.N. (1956): Weak convergence of stochastic processes whose trajectories have no discontinuities of the second kind and the heuristic approach to the

Kolmogorov-Smirnov tests. Theor. Probability Appl. 1 (1956), 140-144.

Cramér, H. and Leadbetter, M.R. (1967): Stationary and related stochastic processes. Wiley, New York (1967).

Gaenssler, P. (1974): On the realization of stochastic processes by probability distributions in function spaces. To appear in: Trans. of the Seventh Prague Conference, Prague 1974.

Gaenssler, P. and Stute, W. (1977): Wahrscheinlichkeitstheorie. Springer, Berlin-Heidelberg-New York (1977).

Gihman, I.I. and Skorohod, A.V. (1974): The theory of stochastic processes I. Die Grundlehren der mathematischen Wissenschaften in Einzeldarstellungen, Band 210, Springer, Berlin-Heidelberg-New York (1974).

Hahn, M.G. and Klass, M.J. (1977): Sample continuity of square-integrable processes. Ann. Probability 5 (1977), No. 3, 361-370.

Loève, M. (1963): Probability theory. 3rd edition, van Nostrand, Princeton (1963).

Mann, H.B. (1951): On the realization of stochastic processes by probability distributions in function spaces. Sankhyā Ser. A 11 (1951), 3-8.

Mathematisches Institut
Ruhr-Universität Bochum
Universitätsstraße 150
D-4630 Bochum
Bundesrepublik Deutschland

TESTS FOR HOMOGENEITY OF SCALE
AGAINST ORDERED ALTERNATIVES

Z. Govindarajulu, Lexington and G.D. Gupta, Carbondale

ABSTRACT

A locally most powerful rank test (LMPRT) for the ordered scale alternatives is derived, assuming that the location parameters of the populations are all equal but unknown. A parametric test based on the likelihood derivative method is also obtained for the ordered scale alternatives. Asymptotic distributions of both the statistics are derived and the statistics are compared via the Pitman efficiency criterion. It is surmised that the asymptotic efficiency of the LMPRT relative to the likelihood derivative test procedure is less than one and tends to unity as the number of samples becomes large. A heuristic class of rank tests is also proposed for the above hypothesis-testing problem, where certain constants are chosen so as to maximize the efficacy. An asymptotically distribution-free test is also proposed for the case when the locations are unequal and unknown.

1. INTRODUCTION

While comparing two or more populations, we may be interested in detecting differences in scale (dispersion) parameters of the populations. The F-test and the Bartlett's H-test are widely used for comparing the variances of normal populations. Also, there are some c-sample nonparametric tests for homogeneity of scale; however these tests, except when c=2, are not optimal if the alternatives are ordered. In certain biological experiments, the distribtuions of the treatment effects do not differ in location, however, there may be differences in scale. Puri (1965) has proposed a class of tests for the homogeneity of scale against ordered alternatives. Govindarajulu and Haller (1976) have considered optimal tests against ordered Lehmann alternatives and also propose a class of test statistics that are weighted sums of Chernoff-Savage (1958) type of rank order statistics. It is of further interest to derive LMP rank tests for other alternatives and compare them with the existing ones.

2. LOCALLY MOST POWERFUL RANK TESTS

Let X_{ij}, $j=1,\ldots,n_i$ be independently distributed random variables with common d.f. $F_i(x)$, where

(2.1) $$F_i(x) = F((x-\mu_i)\Theta_i),$$

where μ_i is location parameter and $\Theta_i > 0$ is the scale parameter $(i=1,2,\ldots,k)$. We assume that $\mu_i = \mu$, $i=1,2,\ldots,k$. The hypotheses of interest is the null hypothesis

$$H_o:\ \Theta_1 = \Theta_2 = \ldots = \Theta_k,$$

against the alternative

(2.2) $$H_1:\ \Theta_1 < \Theta_2 < \ldots < \Theta_k,$$

with strict inequality for at least one pair of Θ's. Let R_{ij} denote the rank of X_{ij} among X_{ij}, $j=1,2,\ldots,n_i$; $i=1,2,\ldots,k$. When using ranks, without loss of generality, we can assume $\mu = 0$. Let $N = \sum_{i=1}^{k} n_i$. Then we are led to the following theorem.

<u>Theorem 2.1</u> Under the regularity conditions, a locally most powerful rank test of H_o against the simple alternative H_Δ: $\Theta_i = (1 + c_i\Delta)\Theta_o$ with normal alternatives is given by: reject H_o when

(2.3) $$S_{1N} = \sum_{i=1}^{k} \sum_{j=1}^{n_i} c_i\, E[X^2_{R_{ij},N}] > s_{1\alpha},$$

where $X_{u,N}$ is the u-th order statistic in a sample of size N drawn from the standard normal distribution,

$$P(S_{1N} > s_{1\alpha}|H_o) \leq \alpha,$$

and α is the level of significance.

<u>Proof</u>: the result follows from a direct application of an extension of Theorem II.4.8 of Hájek and Sidák (1967). For the normal alternatives

$$\log f(\underline{x};\Delta) = K - (1/2) \sum_{i=1}^{k} \sum_{j=1}^{n_i} x_{ij}^2 [(1+c_i\Delta)\Theta_o]^{-1}.$$

Differentiating with respect to Δ and taking limits as $\Delta \to 0$, we get the desired statistic.

The choice of the constants c_i depends on the nature of the alternative which must be guarded against. For example, one could choose $c_i=i$, $(i=1,2,\ldots,k)$, in order to test for the increasing trend in the scale parameter.

2.1 <u>Asymptotic Distribution of S_{1N}</u>

Let us redefine the normal score statistic S_{1N} as

(2.4) $$S_{1N} = N^{-1/2} \sum_{i=1}^{k} c_i \sum_{j=1}^{n_i} E_o[X^2_{R_{ij},N}]$$
$$= N^{-1/2} \sum_{i=1}^{k} c_i\, n_i\, T_{i,N},$$

where

(2.5) $$T_{i,N} = n_i^{-1} \sum_{j=1}^{N} E_o[X^2_{j,N}]\, Z^{(i)}_{N,j},$$

and as defined by Govindarajulu, et al. (1967), $Z^{(i)}_{N,j} = 1$ if the j-th smallest observation from the combined sample of size N is from the i-th sample and $Z^{(i)}_{N,j}$ is equal to zero otherwise.

$T_{i,N}$, $i=1,2,\ldots,k$ are Chernoff-Savage (1958) type statistics. Govindarajulu, et al. (1967) have obtained generalizations of theorems of Chernoff and Savage (1958) on asymptotic normality of the statistics.

Consider the sequence of alternatives $\Theta_i = 1 + c_i N^{-1/2}$, $i=1,2,\ldots,k$. Then from Theorem 7.4 and corollary 5.2.1 of Govindarajulu, et al. (1967), it follows that the variables $N^{1/2}\,(T_{i,N} - \mu_{Ni})$, $i=1,2,\ldots,k$, have a joint asymptotic normal distribution with means zero and $[(1-\lambda_i)/\lambda_i]A^2$ for variances and $-A^2$ for co-variances, where

(2.6) $$A^2 = \int_0^1 J^2(u)du - [\int_0^1 J(u)du]^2,$$

(2.7) $$\mu_{Ni} = \int_{-\infty}^{\infty} J(\Sigma\, \lambda_i\, F_i(x))\; d\, F_i(x),$$

where

(2.8) $$J(u) = [\Phi^{-1}(u)]^2,$$

$\Phi(u)$ is the standard normal distribution function, and $\lambda_i = n_i/N$, $i=1,2,\ldots,k$, where for all N, the inequalities $0 < \lambda_0 \le \cdots \le \lambda_k \le 1 - \lambda_0$ hold for some fixed $\lambda_0 \le 1/k$.

Since, S_{1N} is a linear combination of variables that are asymptotically normal, it is asymptotically normally distributed with the normalizing constants $\mu(S_{1N})$ and $\sigma_o^2(S_{1N})$, where

(2.9) $$\mu(S_{1N}) = N^{-1/2} \sum_{i=1}^{k} n_i c_i \int_{-\infty}^{\infty} [\Phi^{-1}(\sum_{j=1}^{k} \lambda_j F_j(x))]^2\; d\, F_i(x),$$

and $$\sigma_o^2(S_{1N}) = [\sum_i \lambda_i(1-\lambda_i)c_i^2 - \sum_{i\neq i'} \Sigma\, c_i c_{i'} \lambda_i \lambda_{i'}]A^2$$
$$= [\Sigma\, \lambda_i\, c_i^2 - (\Sigma\, \lambda_i\, c_i)^2]A^2.$$

Since
$$A^2 = \int_0^1 [\Phi^{-1}(u)]^4\, du - [\int_0^1 (\Phi^{-1}(u))^2]du]^2 = 2,$$

we have

(2.10) $$\sigma_o^2(S_{1N}) = 2[\Sigma\, \lambda_i\, c_i^2 - (\Sigma\, \lambda_i\, c_i)^2].$$

2.2 Efficacy of S_{1N}

Consider the sequence of alternatives $\Theta_i = 1 + c_i N^{-1/2}$ $i=1,2,\ldots,k$. Then the efficacy of a test statistic T_N is defined as (for the definition of efficacy see Chernoff and Savage (1958) p. 980)

(2.11) $$e(T_N) = [\mu'(T_N)\big|_{\xi=0}]^2/\sigma_o^2(T_N),$$

where $\xi = \Delta N^{-1/2}$.

Differentiating (2.9) with respect to $\xi = \Delta/N^{1/2}$, we have

$$\mu'(S_{1N}) = N^{-1/2}\sum_{i=1}^{k} n_i c_i \int_{-\infty}^{\infty} \frac{d}{d\xi}[\Phi^{-1}(\sum_{j=1}^{k} \lambda_j F(x\Theta_j))]^2 dF(x\Theta_i)$$

$$= N^{-1/2}\sum_{i=1}^{k} n_i c_i \int_{-\infty}^{\infty} \frac{d}{d\xi}[\Phi^{-1}(\sum_{j=1}^{k} \lambda_j F(\Theta_j Y/\Theta_i))]^2 dF(y).$$

Hence,

$$\mu'(S_{1N})\big|_{\xi=0} = 2N^{-1}\sum_{i=1}^{k} n_i c_i[\Sigma \lambda_j(c_i-c_j)I$$

$$= 2[(\Sigma \lambda_i c_i)^2 - \Sigma \lambda_i c_i^2]I,$$

where

(2.12) $$I = \int_{-\infty}^{\infty} y f^2(y)\, \Phi^{-1}(F(y))[\phi(\Phi^{-1}(F(y))]^{-1}dy,$$

where ϕ denotes the standard normal density. Therefore, using (2.11) the efficacy of S_{1N} is given by

(2.13) $$e(S_{1N} = 2[\Sigma \lambda_i c_i^2 - (\Sigma \lambda_i c_i)^2]I^2.$$

If the sample sizes are equal, that is, $\lambda_i = \lambda$, $(i=1,2,\ldots,k)$, then the expression for the efficacy becomes

(2.14) $$e(S_{1N}) = 2k^{-1}\sum_i (c_i - \bar{c})^2 I^2,$$

where $\bar{c} = k^{-1}\sum_i c_i$, and I is defined by (2.12).

3. THE LIKELIHOOD DERIVATIVE TEST

In this section we derive a likelihood derivative test for the homogeneity of variances against the ordered alternatives. It is obtained by differentiating the log likelihood function with respect to the parameter of interest, and evaluating it at the null value of the parameter. Nuisance parameters are replaced by their maximum likelihood estimators under H_o. The procedure was first proposed by Rao (1948). Neyman (1959) has shown that, under certain conditions, the likelihood derivative test (also called c(α) test) is locally asymptotically most powerful.

Let X_{ij}, $j=1,2,\ldots,n_i$ be independently normally distributed with common unknown mean μ and variance Θ_i $(i=1,2,\ldots,k)$. Then the log likelihood function of $\underline{X} = (X_{11}, X_{12},\ldots,X_{1n_1},\ldots,X_{k1},\ldots,X_{k,n_k})$ under the alternative $\Theta_i = (1 + \Delta c_i)\Theta_o$, where Θ_o is specified, is given by

(3.1) $$\log f(x;\mu,\Theta_i) = \sum_{i=1}^{k} \log (2\pi) - \sum_{i=1}^{k} (n_i/2) \log \theta_i.$$
$$- (1/2) \sum_{i=1}^{k} \sum_{j=1}^{n_i} (x_{ij} - \mu)^2 \Theta_i^{-1}.$$

Differentiating (3.1) with respect to Δ and evaluating at $\Delta = 0$, we get

$$\frac{\partial}{\partial\Delta} \log f(x;\mu,\Theta_o)\Big|_{\Delta=0} = - (1/2) \sum n_i c_i + (2\Theta_o)^{-1} \sum_i c_i \sum_j (x_{ij} - \mu)^2.$$

Let $\hat{\mu}$ be the m.l.e. of μ under H_o. Then the likelihood derivative test is to reject H_o if

(3.2) $$\tilde{S}_{2N} = N^{-1/2} \sum_i c_i \sum_j (X_{ij} - \hat{\mu})^2 > s_{2\alpha}$$

where $s_{2\alpha}$ is chosen such that under H_o

$$P(\tilde{S}_{2N} > s_{2\alpha}) \leq \alpha$$

where α is the level of significance. Under H_o, $\hat{\mu} = \bar{X} = N^{-1} \sum_i \sum_j X_{ij}$. Hence the statistic is given by

(3.3) $$\tilde{S}_{2N} = N^{-1/2} \sum_i c_i \sum_j (X_{ij} - \bar{X})^2.$$

We write $\tilde{S}_{2N}$ as

(3.4) $$\tilde{S}_{2N} = N^{-1/2} \sum_{i=1}^{k} c_i[\sum_{j=1}^{n_i} (X_{ij}-\mu)^2 - n_i(\bar{X}_i-\mu)^2 + n_i(\bar{X}_i-\bar{X})^2]$$
$$= N^{-1/2} \sum_{i=1}^{k} c_i \sum_{j=1}^{n_i} (X_{ij}-\mu)^2 - N^{-1/2} \sum_{i=1}^{k} n_i c_i(\bar{X}_i-\mu)^2$$
$$+ N^{-1/2} \sum_{i=1}^{k} n_i c_i(\bar{X}_i-\bar{X})^2.$$

Each of the last two terms of (3.4) is positive. Further,

$$E[N^{-1/2} \sum c_i n_i(\bar{X}_i-\mu)^2] = N^{-1/2} \sum c_i \Theta_i \to 0 \ (N\to\infty),$$

and

$$E[N^{-1/2} \sum n_i c_i (\bar{X}_i-\bar{X})^2] = N^{-1/2} \sum n_i c_i E(\bar{X}_i-\bar{X})^2$$
$$= N^{-1/2} \sum n_i c_i [(\Theta_i/n_i)$$
$$+ (\sum \lambda_i \Theta_i)/N - 2\Theta_i/N]$$
$$= N^{-1/2} [\sum c_i \Theta_i + (\sum \lambda_i \Theta_i)(\sum \lambda_i c_i)$$
$$- 2 \sum \lambda_i c_i \Theta_i]$$

which tends to zero as N tends to infinity. Therefore, both of the last two terms of (3.4) tend to zero in probability. Hence the test statsitic is asymptotically equivalent to

(3.5) $$S_{2N} = N^{-1/2} \sum_{i=1}^{k} c_i \sum_{j=1}^{n_i} (X_{ij}-\mu)^2$$

Let us rewrite S_{2N} as

(3.6) $$S_{2N} = \sum_{i=1}^{k} \lambda_i^{1/2} c_i Z_i,$$

where $Z_i = n_i^{-1/2} \sum_{j=1}^{n_i} (X_{ij}-\mu)^2$. If X_{ij} has a finite fourth moment, by the central limit theorem, Z_i is asymptotically normally distributed. Since S_{2N} is a linear combination of asymptotically normally distributed random variables, it is asymptotically normally distributed with mean $\mu(S_{2N})$ and variance $\sigma^2(S_{2N})$, where

(3.7) $$\mu(S_{2N}) = \sum_{i=1}^{k} \lambda_i^{1/2} c_i n_i^{1/2} \Theta_i,$$

and

(3.8) $$\sigma^2(S_{2N}) = \sum_{i=1}^{k} \lambda_i c_i^2\{E(X_{ij}-\mu)^4 - \Theta_i^2\}.$$

Under H_o, the variance of S_{2N} is given by

(3.9) $$\sigma_o^2(S_{2N}) = \sum_{i=1}^{k} \lambda_i c_i^2 [E(Y^4) - 1] \Theta_o,$$

where Y is a parameter-free random variable.

3.1 Efficacy of S_{2N}

Consider the sequence of alternatives $\Theta_i = 1 + \Delta c_i N^{-1/2}$, i=1,2,...,k, where without loss of generality we have assumed $\Theta_o = 1$. Differentiating (2.8) with respect to $\xi = \Delta N^{-1/2}$ and evaluating at $\xi = 0$, we obtain

(3.10) $$e(S_{2N}) = (\sum \lambda_i c_i^2)/[E(Y^4) - 1].$$

For equal sample-size case, the efficacy reduces to

(3.11) $$e(S_{2N}) = k^{-1}(\sum c_i^2)/[E(Y^4) - 1],$$

and for equally spaced alternatives i.e., c_i=i, i=1,2,...,k), we have

(3.12) $$e(S_{2N}) = (k+1)(2k+1)/6[E(Y^4) - 1].$$

4. OTHER TESTS FOR ORDERED SCALE ALTERNATIVES

Govindarajulu and Haller (1976) have proposed a class of statistics for equal sample sizes which is a weighted sum of Chernoff-Savage type of statistics. We consider unequal sample sizes and define the statistic as

(4.1) $$S_{3N} = N^{-1/2} \sum_{i=1}^{k} b_i n_i \psi_i$$

where

$$n_i^{-1} \psi_i = \int_{-\infty}^{\infty} J_N(\sum_j \lambda_j F_j(x)) \, d F_i(x), \quad i=1,2,\ldots,k.$$

For $0 < u < 1$, the function $J_N(u)$ converges in Lebesque measure to an absolutely continuous function $J(u)$ with $|J'(u)| \leq K[u(1-u)]^{-3/2+\delta}$ for some $0 < \delta < 1/2$, where K is a fixed constant, and the b_i's are not all equal. Thus S_{3N} is a heuristic test statistic and the constants b_i $(i=1,\ldots,k)$ can be chosen so as to maximize the efficacy of the statistic. This differs from the locally most powerful rank test S_{1N} in the sense that in S_{1N} the constants c_i $(i=1,\ldots,k)$ are optimal for the given ordered normal alternative. When $J(u) = [\Phi^{-1}(u)]^2$, S_{3N} becomes S_{1N} with a different set of constants. Furthermore, $J(u) = (u-1/2)^2$, $|u-1/2|$, and $-\log(1-u)$ leads to Mood, Ansari-Bradley, and Savage type of statistics respectively.

From Theorem 7.4 and corollary 5.2.1 of Govindarajulu, et al. (1967) and following the approach of section (2.1), it follows that S_{3N} is asymptotically normally distributed with normalizing constants $\mu(S_{3N})$ and $\sigma_o^2(S_{3N})$, where

(4.2) $$\mu(S_{3N}) = N^{-1/2} \sum_{i=1}^{k} n_i\, b_i \int_{-\infty}^{\infty} J\Big(\sum_{j=1}^{k} \lambda_j F_j(x)\Big)\, d F_i(x),$$

and

(4.3) $$\sigma_o^2(S_{3N}) = [\Sigma \lambda_i b_i^2 - (\Sigma \lambda_i b_i)^2]A^2,$$

where A^2 is defined by (2.6).

Let us consider the same alternative $\theta_i = 1 + \Delta c_i/N^{1/2}$, $i=1,2,\ldots,k$. Then, differentiating (4.2) with respect to $\xi = \Delta N^{-1/2}$ we obtain

(4.4) $$u'(S_{3N})\Big|_{\xi=0} = \sum_i \lambda_i b_i \{\sum_j \lambda_j (c_i - c_j)\} I_1,$$

where

(4.5) $$I_1 = \int_{-\infty}^{\infty} y\, f^2(y)\, J'(F(y))dy.$$

Hence the efficacy of S_{3N} is given by

(4.6) $$e(S_{3N}) = \frac{[\Sigma \lambda_i b_i \sum_j \lambda_j (c_i - c_j)]^2 I_1^2}{[\Sigma \lambda_i b_i^2 - (\Sigma \lambda_i b_i)^2]A^2}.$$

Let $c_i^* = \sum_j \lambda_j (c_i - c_j)$, $i=1,2,\ldots,k$. Then from the Cauchy-Schwarz inequality we have

$$(\Sigma \lambda_i b_i c_i^*)^2 \leq (\Sigma \lambda_i b_i^2)(\Sigma \lambda_i c_i^{*2})$$

with equality for $b_i = \beta c_i^*$, $i=1,2,\ldots,k$, where β is an arbitrary constant. Hence substituting $b_i = \beta c_i^*$ in (4.6), and noting that

$$\Sigma \lambda_i b_i = \beta \Sigma \lambda_i c_i^* = \beta (\sum_i \lambda_i c_i - \sum_j \lambda_j c_j) = 0,$$

the maximum efficacy of S_{3N} is given by

(4.7) $$e(S_{3N}) = (\Sigma \lambda_i c_i^{*2}) I_1^2/A^2.$$

As a special case, if $J(u) = [\Phi^{-1}(u)]^2$, then $I_1^2 = 4I^2$, $A^2 = 2$, and

$$e(S_{3N}) = 2(\Sigma \lambda_i c_i^{*2})I^2 = 2[\Sigma \lambda_i c_i^2 - (\Sigma \lambda_i c_i)^2]I^2,$$

where I is defined by (2.12).

We notice that the maximum efficacy of S_{3N} occurs for the choice of constants $b_i = \Sigma \lambda_j(c_i - c_j)$, i=1,2,...,k. In the special case when $J(u) = [\Phi^{-1}(u)]^2$, the maximum efficacy of S_{3N} is equal to the efficacy of S_{1N}.

Substituting $b_i = \sum_j \lambda_j(c_i - c_j) = c_i - \Sigma \lambda_j c_j$, in the expression for S_{3N}, we get

$$S_{3N} = N^{-1/2} \sum_i^u c_i n_i \psi_i - \text{const.}$$

Therefore, if the function J is the square of the inverse of the normal distribution function, S_{3N} and S_{1N} are equivalent provided the constants are $b_i = c_i - (\Sigma \lambda_i c_i)$; i=1,2,...,k.

For equal sample-sizes and for equally-spaced constants ($c_i = i$, i=1,2,...,k), the efficacies of S_{3N} specialized to Savage, Ansari-Bradley and Mood types of scores are given by Govindarajulu and Haller (1976), and are presented in Table 1 along with the efficacies of S_{1N} and S_{2N}. Efficacies of these statistics for the normal and exponential distributions are given in Table 2.

TABLE 1

Efficacies of Various Tests Under Equally Spaced Ordered Scale Alternatives

Test	Efficacy
S_{2N}	$(k+1)(2k+1)/6[E(y^4) - 1]$
S_{1N}	$(k^2-1)\{\int_{-\infty}^{\infty} y\, f(y)\, \Phi^{-1}(F(y))[\phi(\Phi^{-1}(F(y))]^{-1}dy\}^2/6$
S_{3N} : Scores	
Savage	$(k^2-1)\{\int_{-\infty}^{\infty}[y\, f^2(y)/\{1 - F(y)\}]dy\}^2$
Mood	$15(k^2-1)\{\int_{-\infty}^{\infty} y\, f^2(y)[2F(y) - 1]dy\}^2$
Ansari-Bradley	$4(k^2-1)\{\int_{-\infty}^{\xi_{.5}} y\, f^2(y)dy - \int_{\xi_{.5}}^{\infty} y\, f^2(y)dy\}^2$

Here Φ denotes the standard normal c.d.f., ϕ the standard normal density, $E(\cdot)$ the expected value of $(\cdot)$, $\xi_{.5}$ the median of F(y).

TABLE 2

Efficacies of Test Statistics Under
Normal and Exponential Ordered Scale Alternatives

Test	Distributions	
	Normal	Exponential
S_{1N}	$(k^2 - 1)/6$	$(k^2-1)/12(5.65)$
S_{2N}	$(k+1)(2k+1)/12$	$(k+1)(2k+1)/6(23)$
S_{3N} : Scores		
Savage	$(k^2-1)(.35)/12$	$(k^2-1)/12$
Mood	$5(k^2-1)/4\ \pi^2$	$(k^2-1)/12(7.174)$
Ansari-Bradley	$(k^2-1)/\pi^2$	$(k^2-1)/12(8.928)$

5. ASYMPTOTIC RELATIVE EFFICIENCIES

Therefore, from (2.14) and (3.11), the A.R.E. of the normal score test S_{1N} with respect to the likelihood derivative test S_{2N} is given by

(5.1)
$$\mathrm{ARE}(S_{1N}, S_{2N}) = \frac{2[\Sigma\ \lambda_i\ c_i^2 - (\Sigma\ \lambda_i\ c_i)^2]I^2}{(\Sigma\ \lambda_i\ c_i^2)/(EY^4 - 1)}$$

where I is defined by (2.12).

Considering the case of equal sample-sizes we obtain

(5.2)
$$\mathrm{ARE}(S_{1N}, S_{2N}) = \frac{2(EY^4 - 1)\ \sum_i (c_i - \bar{c})^2\ I^2}{\Sigma\ c_i^2},$$

$$= \frac{(EY^4 - 1)(k - 1)\ I^2}{(2k + 1)}, \text{ when } c_i = ic\ (i=1,\ldots,k)$$

$$= \frac{(2(k - 1)}{(2k + 1)} \text{ (for normal case)}$$

where $\bar{c} = (1/k)\ \Sigma\ c_i$. Clearly $\mathrm{ARE}(S_{1N}, S_{2N})$ tends to unity as k tends to infinity.

For small values of k the exact efficiency results are given in Table 3.

TABLE 3

Asymptotic Efficiency of S_{1N} Relative to S_{2N}

k	3	4	5	6	8	10	∞
Normal	.37	.67	.73	.77	.82	.86	1.00
Exponential	.68	.68	.74	.78	.83	.87	1.02
Double Exponential	3.17	3.70	4.03	4.27	4.57	4.75	5.55

6. NORMAL SCORE TEST WHEN LOCATIONS ARE UNKNOWN AND UNEQUAL

Let X_{ij}, $j=1,2,\ldots,n_i$ be independently distributed random variables with common continuous d.f. $F_i(x)$, where

(6.1) $$F_i(x) = F[(x-\mu_i)\Theta_i]$$

where μ_i is the location parameter and $\Theta_i > 0$ is the scale parameter and $\mu_i \neq \mu_j$ for at least one pair (i,j). Consider the combined sample $(X_{ij} - \mu_i)$. $(j=1,2,\ldots,n_i;\ i=1,2,\ldots,k)$, and let S_N be a test based on the ranks of $(X_{ij} - \mu_i)$ where μ_i is known. Let $\hat{\mu}_i$ be a consistent estimate of μ_i such that $N^{1/2}(\hat{\mu}_i - \mu_i)$ is bounded in probability $(i=1,\ldots,k)$. Let S_N^* denote the test based on the combined sample $(X_{ij} - \hat{\mu}_i)$, $(j=1,2,\ldots,n_i;\ i=1,2,\ldots,k)$. Then for $k = 2$, Raghavachari (1965) has shown that the limiting distribution of the modified test statistic S_N^* is the same as that of S_N under fairly general conditions on the underlying distributions $F_i(x)$, namely, when

(i) $F_i(x)$, $i=1,2,\ldots,k$, are symmetric about their respective location parameters, and

(ii) $f_i(x)/\phi[\Phi^{-1}(F_i))]$, $i=1,2,\ldots,k$, are bounded for all x.

His results can trivially be extended to the case $k > 2$. The modified test is asymptotically distribution-free for a fairly general class of alternatives. Therefore, the test statistic S_{1N} given by (2.4) can be used to test the homogeneity of scale parameters against ordered alternatives with some modifications, although the modified test will be no longer distribution-free. Efficiency results for underlying normal distribution given in Table 3 remain unchanged.

REFERENCES

Chernoff, H. and Savage, I.R. (1958): Asymptotic normality and efficiency of certain nonparametric test statistics. Ann. Math. Statist. 29, 972-994.

Govindarajulu, Z. and Haller, H.S. (1977): C-sample tests of homogeneity against ordered alternatives. Proceedings of the Symposium to honour Jerzy Neyman, (Ed. R. Bartoszynski, et al.) Polish Scientific Publishers, Warszawa, 91-102.

Govindarajulu, Z., Lecam, L. and Raghavachari, M. (1967): Generalizations of theorems of Chernoff and Savage on asymptotic normality of nonparametric test statistics. Proceedings Fifth Berkeley Symposium on Mathematical Statistics and Probability. University of California Press, Vol. 1, 609-638.

Hájek, J. and Sidák, Z. (1967): Theory of Rank Tests. Academic Press, New York.

Neyman, J. (1959): Optimal asymptotic tests of composite statistical hypotheses. Probability and Statistics: The Herald Cramér Volume (Ed. W. Grenander). Almquist and Wiksell, Stockholm, 213-234.

Puri, M.L. (1965): Some distribution-free k-sample rank tests of homogeneity against ordered alternatives. Comm. Pure Appl. Math. 18, 51-63.

Ragharachari, M. (1965): The two-sample scale problem--when the locations are unknown. Ann. Math. Statist. 36, 1236-1242.

Rao, C.R. (1948): Large sample tests of statistical hypotheses concerning several parameters with application to problems of estimation. Proc. Camb. Philos. Soc. 44, 50-57.

Department of Statistics
University of Kentucky
Lexington, Kentucky 40506 U.S.A.

Department of Mathematics
Southern Illinois University
Carbondale, Illinois 62801 U.S.A.

A MATHEMATICAL MODEL FOR THE RESIDENCE TIME OF AEROSOL PARTICLES REMOVED BY PRECIPITATION SCAVENGING

Jan Grandell and Henning Rodhe

Stockholm

ABSTRACT

Precipitation scavenging is an important process for removing aerosol particles from the atmosphere. In an earlier paper we presented a simple model for this process, where the scavenging intensity, which is almost proportional to the precipitation intensity, was regarded as a random process. Here we shall generalize this model, hopefully making it more realistic. We shall also consider properties of estimates based on real precipitation data.

1. INTRODUCTION

1.1 The physical background

A knowledge about the times spent in the atmosphere by particles or gases emitted by human activities is of fundamental importance in connection with the study of many air pollution problems. Particularly when considering travel times that are comparable with or larger than the average residence time of the pollutant in the atmosphere, the influence of the removal processes on the air concentration and on the fallout pattern is profound. As an example one may mention the problem of sulfur dispersion over Europe where the travel distances of interest are up to a few thousand kilometers (corresponding to a few days' travel time).

For some gases and also for particles with a diameter larger than a few μm there are also other removal processes than precipitation scavenging that contribute to limiting the atmosphereic residence time, e.g. direct absorption at the surface and sedimentation (only for large particles). However, for particles in the size range 0.1 - 1 μm, which are those of main concern in most air pollution problems, precipitation scavenging is likely to be the most important process (Garland, 1978). The actual uptake of particles into the droplet can take place either inside the cloud, for example during the condensation process, or when the raindrop falls from the cloud to the ground. For a discussion about the microphysical processes involved reference is made to the paper by Garland (1978).

Earlier attempts to model the precipitation removal process were essentially based on the amounts of precipitation (Junge, 1963). No consideration was given to the frequency of occurence of precipitation periods. In an earlier paper (Rodhe and Grandell, 1972) we formulated a model where the frequency of occurence of precipitation and dry periods as well as the scavenging intensity during such periods were explicitly considered. This model has been used for example by Fisher (1975, 1978) in connection with his estimates of long range transport of sulfur pollutants over Europe.

In this paper we present a generalization of our previous model. A basic assumption is a direct proportionality between precipitation intensity and scavenging intensity. In this way precipitation data can be used to estimate the parameters of the model. Experimental data so far available are in general accordance with such an assumption (Engelmann, 1968 ; Precipitation Scavenging, 1970). In the sequel we shall always discuss in terms of scavenging intensity.

1.2 Randomly fluctuating scavenging intensity

Let $\{\lambda(t):t\in R\}$ be a stochastic process describing the variation of the scavenging intensity. Let the random variable T be the residence time of a particle, i.e. the time to the removal of a particle from the atmosphere since it entered the atmosphere. We shall always let t be zero at the epoch when a particle enters the atmosphere.

In this very general setting it is, of course, difficult to say very much about T. We shall, however, show that the residence time is systematically underestimated if the random variation of the scavenging intensity is left out of considerations. To make this statement precise we shall introduce some notions.

Let X_1 and X_2 be non-negative random variables with <u>survival</u> functions $G_i(x) = Pr\{X_i > x\}$. We say that X_1 is <u>larger in distribution</u> than X_2 if $G_1(x) \geq G_2(x)$ for all $x \geq 0$ and use the notation $X_1 \overset{d}{\geq} X_2$. Since $E(X_i^k) = \int_0^\infty x^k d(1-G_i(x)) = k\int_0^\infty x^{k-1}G_i(x)dx$ it follows from $X_1 \overset{d}{\geq} X_2$ that $E(X_1^k) \geq E(X_2^k)$.

<u>Theorem 1</u>

Let T be the residence time when the scavenging intensity varies according to some stochastic process $\lambda(t)$ and let S be the residence time when the scavenging intensity is non-random and equal to $E\lambda(t)$. Then $T \overset{d}{\geq} S$.

Proof.

We have $\Pr\{T > x\} = E(\exp\{-\int_0^x \lambda(y)dy\}$. Since the function e^{-x} is convex it follows from Jensen's inequality that $\Pr\{T > x\} \geq \exp(-E\{\int_0^x \lambda(y)dy\}) = \exp(-\int_0^x E(\lambda(y))dy) = \Pr\{S>x\}$, i.e. $T \overset{d}{\geq} S$.

1.3 The mathematical model

From now on we shall make more specific assumptions about $\lambda(t)$. We shall consider alternating precipitation and dry periods. The length of these periods are independent random variables. Let the length of a precipitation (dry) period be a random variable $\tilde{T}_p$ $(\tilde{T}_d)$. During a precipitation period we have a certain scavenging intensity $\tilde{\lambda}$ which also is a random variable. That scavenging intensity is independent of all other random variables except of the length of the period in which it acts. Because of all independence assumptions the process $\lambda(t)$ is defined by the random variable $\tilde{T}_d$ and the random vector $(\tilde{T}_p, \tilde{\lambda})$ if suitable initial conditions are added.

Let us define the following five random variables.

T_{bp} (T_{bd}) = the residence time for a particle entering the atmosphere at the beginning of a precipitation (dry) period.

T_p (T_d) = the residence time for a particle entering the atmosphere arbitrarily but during a precipitation (dry) period.

T = the residence time for a particle entering the atmosphere arbitrarily.

Let further G_{bp}, G_{bd}, G_p, G_d and G be the corresponding survival functions.

From the point of view of application we believe that T is the most interesting variable. We may think of a stream of particles entering the atmosphere at a constant rate, for example through a chimney. If we consider all particles entering during a very long time interval and pick out one of these by random, then the residence time of that particle is described by T. If we further observe that the particle happened to enter during a precipitation (dry) period, then its residence time is described by T_p (T_d). One may think of applications where T_p and T_d are of interest themselves. For example, the situation where one wants to avoid adverse ecological effects due to local fallout by stopping the emissions during periods of precipitation. The interpretation of T_{bp} and T_{bd} is obvious. We shall in section 2.2.1 discuss how the different initial conditions give rise to different models for the intensity process $\lambda(t)$.

In our earlier model (Rodhe and Grandell, 1972) $\tilde{T}_p$ and $\tilde{T}_d$ were exponentially distributed and $\tilde{\lambda}$ was non-random and equal to λ_p during all precipitation periods. Put $\tau_p = E(\tilde{T}_p)$ and $\tau_d = E(\tilde{T}_d)$. We showed that

(1) $$E(T) = \frac{\tau_d^2}{\tau_d + \tau_p} + \frac{\tau_d + \tau_p}{\tau_p \lambda_p}$$

and

$$G(t) = Ke^{-\varkappa_1 t} + (1-K)e^{-\varkappa_2 t}$$

where

$$\varkappa_1 = \frac{1}{2}\left(\frac{1}{\tau_d} + \frac{1}{\tau_p} + \lambda_p\right) - \sqrt{\frac{1}{4}\left(\frac{1}{\tau_d} + \frac{1}{\tau_p} + \lambda_p\right)^2 - \frac{\lambda_p}{\tau_d}}$$

$$\varkappa_2 = \frac{1}{2}\left(\frac{1}{\tau_d} + \frac{1}{\tau_p} + \lambda_p\right) + \sqrt{\frac{1}{4}\left(\frac{1}{\tau_d} + \frac{1}{\tau_p} + \lambda_p\right)^2 - \frac{\lambda_p}{\tau_d}} .$$

$$K = \frac{\frac{1}{\tau_d} + \frac{1}{\tau_p} + \frac{\tau_d \lambda_p}{\tau_d + \tau_p} - \varkappa_1}{\varkappa_2 - \varkappa_1} .$$

In our paper we used precipitation data - continuous records for one year from one station in Sweden - to estimate the parameters of that model. For comparison, we intend to use, in the first hand, the same precipitation data to estimate the parameters of the present more general model. The result will be presented in a separate paper.

In our derivations we also allowed the particeles to be removed from the atmosphere during dry periods. We then assumed λ_d to be the scavenging intensity during dry periods. In the present model it could then be natural to assume that we always have a certain intensity λ_d and that $\lambda(t)$ is the increase of the scavenging intensity due to precipitation so that $\lambda_d + \lambda(t)$ is the model for the scavenging intensity. Then

$$\Pr(T > t) = e^{-\lambda_d} E(\exp\{-\int_0^t \lambda(s)ds\})$$

so our results about the survival functions are easily modified to this situation. Unfortunately it does not seem possible to modify our results about $E(T)$ in a simple way.

At least with the modification mentioned above the present model ought to have applications to reliability problems. In fact, Gaver (1963) studied a model related to ours with such applications in mind. For a discussion of his model and other related ones we refer to Grandell (1976, 40-52) and to the references given there.

2. MEAN VALUES AND SURVIVAL FUNCTIONS

2.1 Results

Let $\tilde{T}_p$ and $\tilde{T}_d$ have distribution functions F_p and F_d and assume for simplicity that they are absolutely continuous with density functions f_p and f_d A suitable way to describe the possible dependence between $\tilde{T}_p$ and $\tilde{\lambda}$ is to consider the conditional distribution function U_t defined by

$$U_t(x) = \Pr\{\tilde{\lambda} \leq x \mid \tilde{T}_p = t\} .$$

<u>Theorem 2</u>

Put

$$\tau_p = E(\tilde{T}_p)$$

$$\tau_d = E(\tilde{T}_d)$$

$$\sigma_d^2 = Var(\tilde{T}_d)$$

$$A = E(e^{-\tilde{\lambda}\tilde{T}_p})$$

$$B = E\left(\frac{1 - e^{-\tilde{\lambda}\tilde{T}_p}}{\tilde{\lambda}}\right)$$

$$C = E\left(\frac{e^{-\tilde{\lambda}\tilde{T}_p} - 1 + \tilde{\lambda}\tilde{T}_p}{\tilde{\lambda}^2}\right)$$

and assume all these quantities to be finite.

Then

(2) $$E(T_{bp}) = \frac{B + \tau_d A}{1 - A}$$

(3) $$E(T_{bd}) = \frac{B + \tau_d}{1 - A}$$

(4) $$E(T_p) = \frac{1}{\tau_p}\left(C + B\frac{B + \tau_d}{1 - A}\right)$$

(5) $$E(T_d) = \frac{\sigma_d^2 - \tau_d^2}{2\tau_d} + \frac{B + \tau_d}{1 - A}$$

(6) $$E(T) = \frac{1}{\tau_d + \tau_p}\left(\frac{\sigma_d^2 - \tau_d^2}{2} + C + \frac{(B + \tau_d)^2}{1 - A}\right) .$$

Although these formulas are rather complicated, at least in comparison with (1), we believe that they might be useful, since the quantities entering in them are quite easy to estimate from real data. Properties of such estimates shall be discussed in section 3. It is also possible to calculate the variances of the residence times. The resulting formulas are, however, so horrible that, although they may be defended in the same way, we do not have confidence in their usefulness and we shall not give them.

Now we shall consider the survival functions. We have not managed to calculate them exactly, but we shall give approximations for large values of t.

Theorem 3

Suppose there exists a solution $\varkappa$ to the equation

$$E(e^{\varkappa\tilde{T}_d})\,E(e^{(\varkappa-\tilde{\lambda})\tilde{T}_p}) = 1 \tag{7}$$

Put

$$\begin{aligned} \alpha &= E(e^{\varkappa\tilde{T}_d}) \\ \beta &= E(\tilde{T}_d\, e^{\varkappa\tilde{T}_d}) \\ \gamma &= E(\tilde{T}_p\, e^{(\varkappa-\tilde{\lambda})\tilde{T}_p}) \\ \delta &= E\Big(\frac{e^{(\varkappa-\tilde{\lambda})\tilde{T}_p}-1}{\varkappa-\tilde{\lambda}}\Big) \end{aligned} \tag{8}$$

and assume all these quantities to be finite. Assume further that

$$E(e^{\varkappa\tilde{T}_p}) < \infty \ . \tag{9}$$

Then

$$\lim_{t\to\infty} e^{\varkappa t} G_a(t) = K_a \ , \quad a = bp, bd, p, d, -,$$

where

$$K_{bp} = \frac{\delta + \frac{\alpha-1}{\alpha\varkappa}}{\alpha\gamma + \beta/\alpha} \tag{10}$$

$$K_{bd} = \alpha K_{bp} \tag{11}$$

$$K_p = \frac{\alpha\delta}{\tau_p} K_{bp} \tag{12}$$

$$K_d = \frac{\alpha-1}{\tau_d\varkappa} K_{bp} \tag{13}$$

$$K = \frac{\alpha\delta + \frac{\alpha-1}{\varkappa}}{\tau_d + \tau_p} K_{bp} \tag{14}$$

The condition (9) is not necessary, but it seems rather harmless in comparison with (8), at least in the Swedish climatic conditions where in general dry periods are longer than precipitation periods.

2.2 Proofs of theorems 2 and 3

2.2.1 Dependence on initial conditions

Now we shall consider how the different initial conditions influence the scavenging intensity process. Let $\mathcal{T} = \{\tilde{T}_d^{(k)}, (\tilde{T}_p^{(k)}, \tilde{\lambda}^{(k)});\ k = \pm 1, \pm 2, \ldots\}$ be a set of independent random variables such that $\tilde{T}_d^{(k)} \overset{d}{=} \tilde{T}_d$ and $(\tilde{T}_p^{(k)}, \tilde{\lambda}^{(k)}) \overset{d}{=} (\tilde{T}_p, \tilde{\lambda})$ where $\overset{d}{=}$ means equal in distribution. We shall use $\lambda_{bp}(t)$, $\lambda_{bd}(t)$, $\lambda_p(t)$, $\lambda_d(t)$ and $\lambda(t)$ as notations for the resulting intensity processes.

Let $\tilde{T}_d$ and $(\tilde{T}_p, \tilde{\lambda})$ be independent of $\mathcal{T}$, i.e. of all variables in $\mathcal{T}$. Then the statistical properties of $\lambda_{bp}(t)$ and $\lambda_{bd}(t)$ follow of Figures 1 and 2.

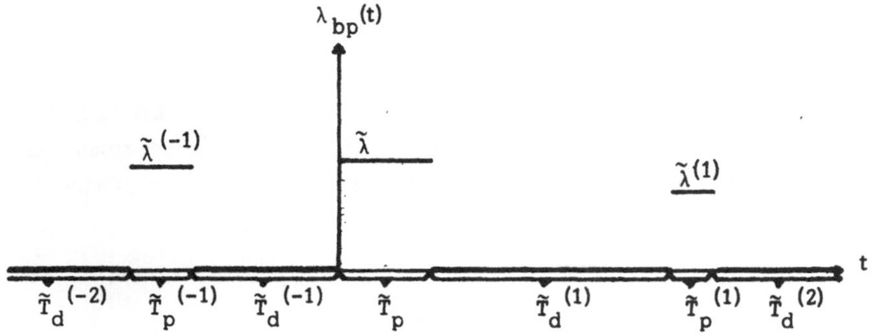

FIG. 1

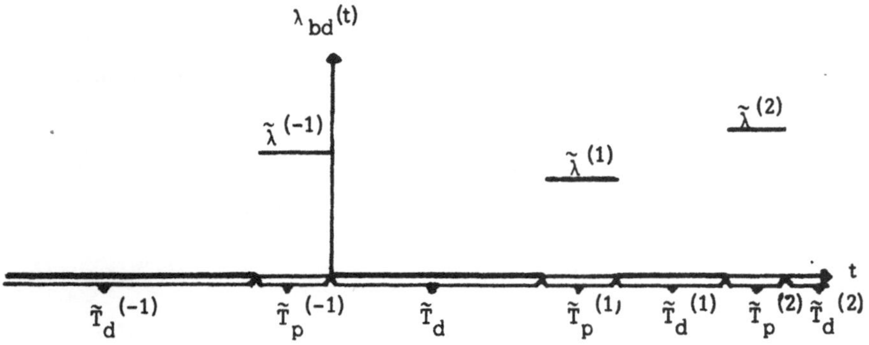

FIG. 2

Now we shall consider $\lambda_d(t)$ and $\lambda_p(t)$. The problem is that the period in which a particle happens to enter does not have the same statistical properties as the other ones. The reason is that 'a long period has a better chance to cover the entrance epoch than a short one'.

Consider $\lambda_p(t)$ and let the random variables $\tilde{L}_p$, $\tilde{W}_p$ and $\tilde{\ell}$ be defined by Figure 3.

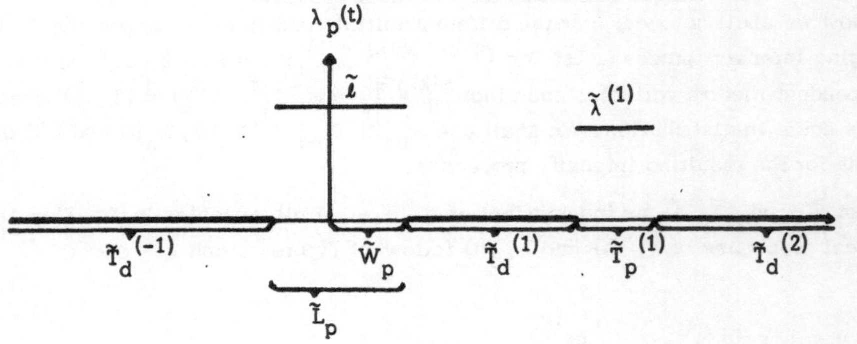

FIG. 3

Thus $\tilde{L}_p$ is the length of the precipitation period at which the particle happens to enter and $\tilde{W}_p$ is the remaining length of that period. From our assumptions and from e.g. Feller (1971, pp. 369-371) it follows that $(\tilde{L}_p, \tilde{W}_p, \tilde{\ell})$ is independent of $\mathfrak{T}$ and that

(15) $$K_p(t) = \Pr\{\tilde{L}_p \leq t\} = \frac{\int_0^t s f_p(s)ds}{\tau_p}$$

(16) $$H_p(t) = \Pr\{\tilde{W}_p \leq t\} = \frac{\int_0^t (1-F_p(s))ds}{\tau_p}$$

(17) $$\Pr\{\tilde{L}_p \leq t \mid \tilde{W}_p = s\} = \begin{cases} 0 & \text{if } t < s \\ \dfrac{F_p(t) - F_p(s)}{1-F_p(s)} & \text{if } t \geq s \end{cases}$$

In general $\tilde{\ell}$ and $\tilde{\lambda}$ have different distributions since $\tilde{T}_p$ and $\tilde{L}_p$ have different distributions but we have

$$\Pr\{\tilde{\ell} \leq x \mid \tilde{L}_p = t\} = U_t(x)$$

Consider $\lambda_d(t)$ and define $\tilde{W}_d$ similar as $\tilde{W}_p$, see Figure 4.

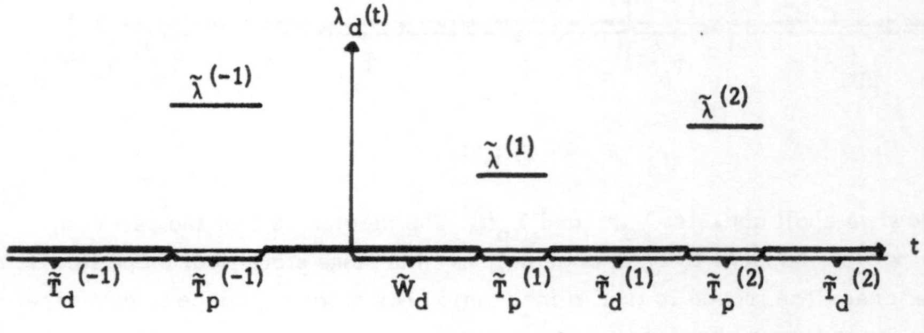

FIG. 4

In this case $\tilde{W}_d$ and $\mathcal{T}$ are independent. Further

$$H_d(t) = \Pr\{\tilde{W}_d \leq t\} = \frac{\int_0^t (1-F_d(s))ds}{\tau_p}$$

quite as before.

Finally we consider $\lambda(t)$. From Feller (1971, p. 380) it follows that the probability for a particle, which enters the atmosphere at arbitrary time, to enter the atmosphere at a precipitation (dry) period is $\frac{\tau_p}{\tau_d + \tau_p}$ $\left(\frac{\tau_d}{\tau_d + \tau_p}\right)$. Thus we obtain $\lambda(t)$ by randomization between $\lambda_p(t)$ and $\lambda_d(t)$ according to these probabilities. This means that

(18) $$G(t) = (\tau_d G_d(t) + \tau_p G_p(t)/(\tau_d + \tau_p).$$

2.2.2 Fundamental equations for the survival functions

Let us start with some notations. Let $a_1, \ldots, a_n$ be non-negative functions on R_+ such that $\int_0^\infty a_i(t)dt < \infty$. The functions $a_1 a_2$, $a_1 + a_2$ and $a_1 * a_2$ are defined by

$$a_1 a_2(t) = a_1(t)a_2(t), \quad (a_1+a_2)(t) = a_1(t) + a_2(t) \quad \text{and} \quad (a_1 * a_2)(t) = \int_0^t a_1(t-s)a_2(s)ds.$$

Note that

(19) $$\int_0^\infty (a_1 * a_2 * \ldots * a_n)(t)dt = \prod_{k=1}^n \int_0^\infty a_k(t)dt$$

and

(20) $$\int_0^\infty t(a_1 * a_2)(t)dt = \int_0^\infty t a_1(t)dt \int_0^\infty a_2(t)dt + \int_0^\infty a_1(t)dt \int_0^\infty t a_2(t)dt$$

and that, if $b(t) = e^{\varkappa t}$,

(21 $$b(a_1 * \ldots * a_n) = ba_1 * ba_2 * \ldots * ba_n .$$

Let us now consider the variables T_{bd} and T_{bp}. From the definition of $\lambda_{bd}(t)$ it follows that

(22) $$T_{bd} \stackrel{d}{=} \tilde{T}_d + T_{bp}$$

where $\tilde{T}_d$ and T_{bp} are independent.

Thus we have

$$1 - G_{bd}(t) = \int_0^t (1-G_{bp}(t-s))\, f_d(s)ds$$

or

(23) $$G_{bd} = 1-F_d(t) + G_{bp} * f_d(t).$$

Now consider G_{bp}. From Figure 1 one realizes that

(24) $$\Pr\{T_{bp} > t \mid (\tilde{T}_p, \tilde{\lambda})\} = \begin{cases} \exp(-\tilde{\lambda}\, t) & \text{if } t < \tilde{T}_p \\ \exp(-\tilde{\lambda}\tilde{T}_p)\, G_{bd}(t-\tilde{T}_p) & \text{if } t > \tilde{T}_p \end{cases}$$

Define the conditional Laplace-transform $\hat{u}_s(t)$ for $\tilde{\lambda}$ given $\tilde{T}_p = s$ by

$$\hat{u}_s(t) = E(e^{-\tilde{\lambda}t} \mid \tilde{T}_p = s) = \int_0^\infty e^{-xt} U_s\{dx\}.$$

Integration of (24) with respect to $\tilde{\lambda}$ given $\tilde{T}_p = s$ yields

$$\Pr\{T_{bp} > t \mid \tilde{T}_p = s\} = \begin{cases} \hat{u}_s(t) & \text{if } t \le s \\ \hat{u}_s(s)\, G_{bd}(t-s) & \text{if } t > s \end{cases}$$

and thus

(25) $$G_{bp}(t) = \int_t^\infty \hat{u}_s(t) f_p(s)ds + \int_0^t \hat{u}_s(s) G_{bd}(t-s) f_p(s)ds$$

For convenience we define the function

$$v(s,t) = \int_s^\infty \hat{u}_y(t) f_p(y)dy.$$

If we put $\hat{u}(s) = \hat{u}_s(s)$ and $v(t) = v(t,t)$ we can write (25) on the form

(26) $$G_{bp}(t) = v(t) + G_{bd} * \hat{u} f_p(t) .$$

Putting (23) and (26) together we get

(27) $$G_{bp} = v(t) + (1-F_d) * \hat{u} f_p(t) + G_{bp} * f_d * \hat{u} f_p(t)$$

which is a (defective) renewal equation for G_{bp}.

Now we consider the variables T_d and T_p. Like in (22) we have

(28) $$T_d \overset{d}{=} \tilde{W}_d + T_{bp}$$

where $\tilde{W}_d$ and T_{bp} are independent and thus

$$G_d(t) = 1 - H_d(t) + G_{bp} * h_d(t)$$

where $h_d(t) = H_d'(t)$.

From the definition of H_d we get

(29) $$G_d(t) = \frac{1}{\tau_d}\left(\int_t^\infty (1-F_d(s))ds + G_{bp} * (1-F_d)(t)\right) .$$

Put

$$\check{u}_s(t) = E(e^{-\tilde{\lambda}t} \mid \tilde{W}_p = s)$$

$$\check{u}(t) = \check{u}_t(t)$$

$$h_p(t) = H_p'(t) .$$

A similar reasoning as in the derivation of (25) yields

(30) $$G_p(t) = \int_t^\infty \check{u}_s(t) h_p(s)ds + G_{bd} * \check{u}h_p(t)$$

From (17) we get

$$\check{u}_s(t) = \int_s^\infty \hat{u}_y(t) \frac{f_p(y)}{1-F_p(s)} dy = \frac{v(s,t)}{1-F_p(s)} .$$

Putting this into (30) yields

(31) $$G_p(t) = \frac{1}{\tau_p} \left(\int_t^\infty v(s,t)ds + G_{bd} * v(t) \right) .$$

2.2.3 Proof of theorem 2

Integration of (27) yields

$$ET_{bp} = \int_0^\infty v(t)dt + \tau_d \int_0^\infty \hat{u}f_p(t)dt + ET_{bp} \cdot 1 \cdot \int_0^\infty \hat{u}f_p(t)dt .$$

From the definition of $\hat{u}$ it follows that $\int_0^\infty \hat{u}f_p(t)dt = A$ and further

(32) $$\int_0^\infty v(t)dt = \int_0^\infty \int_t^\infty \hat{u}_y(t) f_p(y)dy\,dt = \int_0^\infty f_p(y)dy \int_0^y E(e^{-\tilde{\lambda}t} \mid \tilde{T}_p=y)dt = \int_0^\infty E\left(\frac{1-e^{-\tilde{\lambda}y}}{\tilde{\lambda}} \mid \tilde{T}_p=y\right)dy = B$$

Thus $E(T_{bp}) = B + \tau_d A + AE(T_{bp})$ and (2) follows.
From (22) we get $E(T_{bd}) = \tau_d + E(T_{bp})$ and thus (3) follows.

Since

$$E(\tilde{W}_d) = \frac{1}{\tau_d} \int_0^\infty x(1-F_d(x))dx = \frac{E(\tilde{T}_d^2)}{2\tau_d} = \frac{\sigma_d^2 + \tau_d^2}{2\tau_d}$$

it follows from (28) that (5) is true.

With similar calculations as in the derivation of (32) we get $\int_0^\infty \int_t^\infty v(s,t)ds = C$ and (4) follows by integration of (31). Finally (6) follows from (18).

2.2.4 Proof of theorem 3

We shall first consider G_{bp} and use the technique described by Feller (1971, pp. 375-376). Consider (27) and assume that there exists a $\varkappa$ such that

$$\int_0^\infty e^{\varkappa t} \cdot (f_d * \hat{u}f_p)(t)dt = 1.$$

From (19) and (21) it follows that this is the same as to assume (7). Then we have

$$\lim_{t\to\infty} e^{\varkappa t} G_{bp}(t) = \frac{\int_0^\infty e^{\varkappa t}(v(t) + (1-F_d) * \hat{u}f_p(t))dt}{\int_0^\infty t e^{\varkappa t}(f_d * \hat{u}f_p(t))dt}$$

provided $e^{\varkappa t}(v(t) + (1-F_d) * \hat{u}f_p(t))$ is directly Riemann integrable and the quantities are finite.

From (19) and (21) it follows that

$$\int_0^\infty e^{\varkappa t}(v(t) + (1-F_d) * \hat{u}f_p(t))\,dt = \delta + \frac{\alpha-1}{\varkappa} \cdot E\, e^{(\varkappa-\tilde{\lambda})\tilde{T}_p} = \delta + \frac{\alpha-1}{\varkappa} \cdot \frac{1}{\alpha}.$$

From the assumptions it follows that this is finite. It is not difficult to realize that $v(t) + (1-F_d) * \hat{u}f_p(t)$ is decreasing and the direct Riemann integrability follows from Grandell (1976, p. 46).

From (20) we get $\int_0^\infty t e^{\varkappa t}(f_d * \hat{u}f_p(t))dt = \beta \cdot \frac{1}{\alpha} + \alpha\gamma$ and (10) follows.

From (8) and (9) it follows (Grandell 1976, p. 47) that (11), (12) and (13) are true. Finally (14) follows from (18).

3. ESTIMATION OF PARAMETERS

In reality the distributions of $\tilde{T}_d$ and $(\tilde{T}_p, \tilde{\lambda})$ are unknown. Thus we must furnish ourselves with observations on these variables from precipitation data. We shall assume that we have n observations on each of $\tilde{T}_d$ and $(\tilde{T}_p, \tilde{\lambda})$ and that n is reasonably large. This assumption is not quite unquestionable since it is natural to observe the precipitation during a fixed period. The n ought to be regarded as a random variable. For practical applications this matters very little, since in the derivations we may replace the ordinary central limit theorem by its generalization to sums of a random number of random variables. Therefore we shall keep the assumption of non-random n.

In order to estimate these parameters properly one would need observations on precipitation and dry periods experienced by a parcel of air as it is carried along by the wind. Such observations are generally very difficult to obtain and one is left with data from fixed locations. For precipitation from convective clouds or from moving frontal systems the difference may not be too important. However, for orographic precipitation, the occurence of which is dependent upon the location of mountains, there may be a very important difference. For a further discussion of this point we refer to Rodhe and Grandell (1972, pp. 444, 446).

3.1 The mean removal time

Let $X = (X_1, \ldots, X_6)$ be the random vector $(\tilde{T}_d, \tilde{T}_d^2, \tilde{T}_p, e^{-\tilde{\lambda}\tilde{T}_p}, (1-e^{-\tilde{\lambda}\tilde{T}_p})/\tilde{\lambda}, (e^{-\tilde{\lambda}\tilde{T}_p} - 1 + \tilde{\lambda}\tilde{T}_p)/\tilde{\lambda}^2)$.

Let $X^{(1)}, \ldots, X^{(n)}$ be independent random vectors distributed as X and let $x^{(1)}, \ldots, x^{(n)}$ be observations of $X^{(1)}, \ldots, X^{(n)}$. Put $\bar{x} = \left(\frac{1}{n}\sum_1^n x_1^{(i)}, \ldots, \frac{1}{n}\sum_1^n x^{(i)}\right)$.

Now we restrict ourselves to ET. Define the function $e: R \to R$ by

$$e(x) = \frac{1}{x_1 + x_3}\left(\frac{1}{2}x_2 - x_1^2 + x_6 + \frac{(x_5 + x_1)^2}{1 - x_3}\right).$$

Then it is seen that $E(T) = e(EX)$. A natural estimate of ET is formed by $e(\bar{x})$. Further we have (Cramér 1945, p. 366) $\sqrt{n}(e(\bar{X}) - E(T)) \xrightarrow{d} W$ as $n \to \infty$ where $\xrightarrow{d}$ means 'tends in distribution to' and where W is a normally distributed random variable with $EW = 0$ and $Var(W) = \sigma^2$. Here $\sigma^2 = \sum_{i=1}^{6}\sum_{j=1}^{6}\mu_{i,j}e_i e_j$, where $\mu_{i,j} = Cov(X_i, X_j)$ and $e_i = \frac{\partial e(x)}{\partial x_i}(E(X))$.

Replacing $\mu_{i,j}$ and e_i with their natural estimates we can form an estimate σ^* of σ. Note that $\mu_{i,j} = 0$ for $i = 1,2$ and $j = 3,4,5,6$. Thus an approximative 95% confidence interval for E(T) is given by $e(\bar{x}) \pm 2\sigma^*/\sqrt{n}$.

3.2 The survival function

Let $Z = (Z_1, Z_2, Z_3)$ be the random vector $(\tilde{T}_d, \tilde{T}_p, \tilde{\lambda})$.

Let $Z^{(1)}, \ldots, Z^{(n)}$ be independent random vectors distributed as Z and let $z^{(1)}, \ldots, z^{(n)}$ be observations of $Z^{(1)}, \ldots, Z^{(n)}$. Define the stochastic process $Y_n(x)$ by

$$Y_n(x) = \left(\frac{1}{n}\sum_{k=1}^{n} e^{xZ_1^{(k)}}\right)\cdot\left(\frac{1}{n}\sum_{k=1}^{n} e^{(x-Z_3^{(k)})Z_2^{(k)}}\right).$$

Let $y_n(x)$ be the observation of $Y_n(x)$ and put $y(x) = E(e^{xZ_1}\cdot e^{(x-Z_3)Z_2})$. Note that $E(Y_n(x)) = y(x)$.

From (7) it follows that $\varkappa$ is the solution of $y(x) = 1$. Thus the solution $\varkappa^*$ of $y_n(x) = 1$ is a natural estimate of $\varkappa$.

Let us now consider G(t) for large values of t. A natural estimate of G(t) is, according to theorem 3, given by $K^*\exp(-\varkappa^* t)$ where K^* is formed from K by replacing τ_d, τ_p, α, β, γ, δ and $\varkappa$ with their natural estimates. With natural estimates we mean $\tau_d^* = \bar{z}_1$, $\alpha^* = \frac{1}{n}\sum_{k=1}^{n}\exp(\varkappa^* z_1^{(k)})$ and so on.

Define the random variables $\varkappa_n$, corresponding to $\varkappa^*$, as the solution of $Y_n(x) = 1$.

Theorem 4

Assume that $y(2\varkappa) < \infty$. Then $\sqrt{n}(\varkappa_n - \varkappa) \xrightarrow{d} W$ as $n \to \infty$ where W is a normally distributed random variable with $E(W) = 0$ and $Var(W) =$

$$\sigma^2 = \frac{Var(e^{\varkappa\tilde{T}_d}) + \alpha^4 Var(e^{(\varkappa-\tilde{\lambda})\tilde{T}_p})}{(\beta + \alpha^2\gamma)^2}.$$

For large values of t the uncertainty of $\varkappa^*$ is more important than the uncertainty of K^*. An approximative 95% confidence interval for G(t) is thus given by $K^*\exp((-\varkappa^* \pm 2\sigma^*/\sqrt{n})t)$ where σ^* is the natural estimate of σ. Because of the construction of the interval we may consider all t larger than some t_o simultaneously without changing the level.

Before proving theorem 4 we shall give a lemma.

Lemma

Let $Y_n(x)$, $y(x)$, $\varkappa_n$ and $\varkappa$ be defined as above. Then $\sqrt{n}(\varkappa_n - \varkappa)$ tends in distribution to the same limit as $-\frac{\sqrt{n}(Y_n(\varkappa)-1)}{y'(\varkappa)}$ when $n \to \infty$.

The reason for extracting this lemma is that it essentially is a very special case of a general result due to Högfeldt and Rosén, see Högfeldt (1977). They use a more general notion of convergence. Using the special structure of our processes we shall give a direct proof, which is similar to the proof of asymptotic normality of maximum likelihood estimates given by Cramér (1945, pp. 500-503).

Proof of lemma.

In the proof all statements about random quantities are meant to hold for almost all realizations of $\{Z^{(k)};\ k = 1,2,\ldots\}$.

Since $y(2\varkappa) < \infty$ it follows that $Y_n(x) \to y(x)$ and $Y_n'(x) \to y'(x)$ for $x < 2\varkappa$. Note that $Y_n'(x)$ is increasing in x and that $Y_n'(0) > 0$.

Since $y'(\varkappa) \in (0,\infty)$ the lemma follows if we prove that $(1-Y_n(\varkappa))/(\varkappa_n - \varkappa) \to y'(\varkappa)$.

Now $1-Y_n(\varkappa) = Y_n(\varkappa_n) - Y_n(\varkappa) = (\varkappa_n - \varkappa)Y_n'(\varkappa + \theta_n(\varkappa_n - \varkappa))$ for some $\theta_n \in [0,1]$ and thus $|\varkappa_n - \varkappa| \leq |Y_n(\varkappa)-1|/Y_n'(0)$. Since $Y_n'(0) \to y'(0) > 0$ and $Y_n(\varkappa) \to 1$ we have $\varkappa_n \to \varkappa$ as $n \to \infty$. Further $|[(1-Y_n(\varkappa))/(\varkappa_n - \varkappa)] - y'(\varkappa)| = |Y_n'(\varkappa + \theta_n(\varkappa_n - \varkappa)) - y'(\varkappa)| \leq |Y_n'(\varkappa) - y'(\varkappa)| + |Y_n'(\varkappa_n) - y'(\varkappa)|$ and thus it only remains to prove that $|Y_n'(\varkappa_n) - y'(\varkappa)| \to 0$. This follows since $\varkappa_n \to \varkappa$ and since $y'(x)$ is continuous for $x < 2\varkappa$.

Proof of theorem 4.

From the lemma and from Cramér (1945, p. 366) the theorem follows after simple calculations. The assumption $y(2\varkappa) < \infty$ guarantees σ^2 to be finite.

REFERENCES

Cramér H. (1945): Mathematical methods of statistics. Almqvist & Wiksell, Uppsala 1945 and Princeton University Press, Princeton 1946.

Engelmann R.J. (1968): The calculation of precipitation scavenging. In: Meteorology and atomic energy, USAEC, Springfield 1968, 208-221.

Feller W. (1971) : An introduction to probability theory and its applications. Vol. II. 2nd. ed., John Wiley and Sons, New York 1971.

Fisher B.E.A. (1975) : The long range transport óf sulphur dioxide. Atmospheric Environment 9 (1975), 1063-1070.

——— (1978) : The calculation of long term sulphur deposition in Europe. To appear in Atmospheric Environment 12 (1978).

Garland J.A. (1978) : Dry and wet removal of sulphur from the atmosphere. To appear in Atmospheric Environment 12 (1978).

Gaver D.P. (1963) : Random hazard in reliability problems. Technometrics 5 (1963), 211-226.

Grandell J. (1976) : Doubly stochastic Poisson processes. Lecture Notes in Math. 529, Springer-Verlag, Berlin 1976.

Högfeldt P. (1977) : On the asymptotic behaviour of first passage time processes and certain stopped stochastic processes. In: Abstracts of communications T.3. Second Vilnius conf. on prob. theory and math. stat., Vilnius 1977, 86-87.

Junge C.E. (1963) : Air chemistry and radioactivity. Academic Press, New York 1963.

Precipitation scavenging (1970) : Proc. Symp. Richland, Washington, USAEC, Springfield 1970.

Rodhe H. and Grandell J. (1972) : On the removal time of aerosol particles from the atmosphere by precipitation scavenging. Tellus 24 (1972), 442-454.

Jan Grandell
Department of Mathematics
The Royal Institute of Thechnology
S-100 44 Stockholm 70, Sweden

Henning Rodhe
Department of Meteorology
University of Stockholm
S-106 91 Stockholm, Sweden

OPTIMAL CONTROL OF ONE DIMENSIONAL NON-CONSERVATIVE QUASI-DIFFUSION PROCESSES

Jürgen Groh

Jena

ABSTRACT

An extension of the work of P.Mandl concerning the optimal control of time-homogeneous diffusion processes in one dimension is given. Instead of a classical second order differential operator, Feller's generalized differential operator $D_m D_p^+$ with a nondecreasing weight function m is used as infinitesimal generator. In this manner an optimal control of a wider class of one dimensional Markov processes - including diffusion processes as well as birth-death processes - is realized.

1. INTRODUCTION

In his monograph P.Mandl (1968) gives a complete investigation of the optimal control of classical one dimensional diffusion processes on a compact interval. He describes a diffusion process X in terms of its infinitesimal generator, which is an ordinary second order differential operator

$$A = a(d^2/dx^2) + b(d/dx)$$

appropriately restricted by Feller's boundary conditions (see Feller (1952)). In order to control the process X the diffusion and drift coefficients $a = a(x,z) > 0$, $b = b(x,z)$ are supposed to depend on a parameter z from some compact parameter space J. Now a control is a "nice" function s on the state space I with values in J, and the controlled process X^s is defined by its infinitesimal generator

$$A_s = a(x,s(x))(d^2/dx^2) + b(x,s(x))(d/dx),$$

together with some (non-controlled) boundary conditions. After introducing the notion of a "cost" for the process X^s an optimality criterion is given. In Mandl's book the problem to compute the minimal expected cost is solved completely. Also a necessary and sufficient condition for the existence of an optimal control, and a method of its calculation is given.

It is well known from Feller's work that the classical differential operator A is not the most general diffusion process generator, but this is a generalized differential operator $D_m D_p^+$. It was shown in Groh (1972), (1975) that $D_m D_p^+$ with a nondecreasing function m, possibly restricted by certain boundary conditions, is an infinitesimal generator, related to a so-called quasi-diffusion process. These processes are the natural generalizations of boths diffusion as well as birth-and-death processes. Roughly spoken, all trajectories of these processes have the nearest neighbour property (see section 2).

In the present note we generalize Mandl's approach to optimal control to the more general case of non-conservative quasi-diffusion processes. Here a direct control of the functions m and p in $D_m D_p^+$ turned out to be very complicated. Therefore we consider infinitesimal generators of the form

$$A = a\, D_m D_p^+ + b\, D_p^-$$

with fixed scales m and p and controlled coefficients $a > 0$ and b. As in earlier papers (see Groh (1978a)) we treat our control problem for processes on a compact state space with regular boundary points, which are killed at this points with a positive probability. But now we are free from all unnecessary conditions.

For a more complicated control space in case of diffusion processes compare with Puterman (1974), for pure jump processes (with "great jumps") see Pliska (1975) and for piecewise increasing jump processes with a negative drift deMorais (1976).

2. QUASI-DIFFUSION PROCESSES

Let M be a nondecreasing right continuous function on the bounded interval $I := [r_0, r_1] \subset \mathbb{R}$, continuous in the boundary points r_0 and r_1. This function defines a nonnegative measure on I. We add - if necessary - to the support of this measure the points r_0, r_1 and denote this compact by E. Further, we set $\mathring{E} := E \setminus \{r_0, r_1\}$, and E_+^β (E_-^β) denotes the set of all right(left)-side limit points in E; E^β stands for the union $E_+^\beta \cup E_-^\beta$, and E_Δ denotes the set $\{x \in E : \Delta^- M(x) := M(x) - M(x-0) > 0\}$ of all discontinuity points of M. Let $\vec{C}$ be the set of all real valued bounded functions f on E such that f(x+0) (f(x-0)) exists for all $x \in E_+^\beta$ (E_-^β), $f(x) = f(x-0)$ for all $x \in \mathring{E} \setminus E_\Delta$, and $f(x) = f(x+0)$ for all $x \in \mathring{E} \setminus E_-^{\beta}$ with $\Delta^- M(x) = 0$. Moreover, we assume that all $f \in \overleftarrow{C}$ have outside of the set E_Δ finitely many discontinuities only. Let C be the set of all continuous functions on E, $\overleftarrow{C}_E$ (C_E) denotes the set of all functions $g: I \to \mathbb{R}$, linear on each component of the open set $I \setminus E$ such that the restriction $g|E$ is contained in $\overleftarrow{C}$ (C). With $\overline{f}$ or $(f)^-$ we denote the (unique) extension of a function $f \in \overleftarrow{C}$ to a function in $\overleftarrow{C}_E$. Clearly, for $f \in C$, we have $\overline{f} \in C_E$. Finally, we choose a strictly isotone function $P \in C_E$, and we assume $P(r_0) = M(r_0) = 0$.

In order to define the generalized differential operator $D_M D_P^+ : C \to \overleftarrow{C}$ we choose some reals $\varkappa_i, \vartheta_i, \pi_i, \sigma_i \geq 0$ $(i = 0,1)$ with

$$\varkappa_0 + \varkappa_1 > 0, \qquad \varkappa_i + \vartheta_i + \pi_i > 0,$$

and $\sigma_i > 0$ in the case of an isolated boundary point $r_i \notin E^{\beta}$ $(i = 0, 1)$. Further, μ_i $(i = 0,1)$ stands for a probability measure on E with $\mu_i\{r_i\} = 0$. We shall say that the function $f \in C$ is contained in the domaine $D(D_M D_P^+)$ of the operator $D_M D_P^+$ if there exist a function $g \in \overleftarrow{C}$ and constants $\alpha, \beta \in \mathbb{R}$ such that

$$\overline{f}(x) = \int_{r_0}^{x} \int_{r_0}^{y} \overline{g}(u)\,dM(u)\,dP(y) + \alpha + \beta P(x) \qquad (x \in I);$$

in this case we define

$$(D_M D_P^+ f)(x) := \begin{cases} g(x) & (x \in \mathring{E}) \\ \lim\limits_{y \to r_i} g(y) & (x = r_i \in E^{\beta};\ i = 0,1) \\ -\sigma_i^{-1} \Big\{ \varkappa_i f(r_i) + \vartheta_i \int_E [f(r_i) - f(y)]\,d\mu_i(y) + \\ \qquad + (-1)^{i+1} \pi_i (D_P^+ f)(r_i) \Big\} & (x = r_i \notin E^{\beta};\ i = 0,1), \end{cases}$$

$$(D_P^+ f)(x) := \beta + \int_{r_0}^{x} \overline{g}(u)\,dM(u) \qquad (x \in I),$$

$$(D_P^- f)(x) := (D_P^+ f)(x-0) \qquad (x \in I).$$

(All integrals are Lebesgue-Stieltjes integrals, and $\int_{r_0}^{x}$ stands for an integration over the interval $[r_0, x]$; for every function h we understand by $h(r_0 - 0)$ the value $h(r_0)$.) Because of the requirement $g \in \overleftarrow{C}$ the correspondence $f \to D_M D_P^+ f$ is unique; $D_M D_P^+$ is a linear operator. According to W.Feller the functions M and P are called the canonical measure and the canonical scale, respectively. In his terminology, both boundaries are <u>regular</u>.

For every function $f \in D(D_M D_P^+)$ the following rules are valid

$$\int_{r_0}^{x} (D_M D_P^+ f)^-(u)\,dM(u) = (D_P^+ f)(x) - (D_P^+ f)(r_0) \qquad (x \in I)$$

$$\int_{r_0}^{x} (D_P^+ f)(y)\,dP(y) = f(x) - f(r_0) \qquad (x \in I).$$

We define on $D(D_M D_P^+)$ two linear functionals Φ_i $(i = 0,1)$ by

$$\Phi_i(f) := \varkappa_i f(r_i) + \vartheta_i \int_E [f(r_i) - f(y)]\, d\mu_i(y) + \\ + (-1)^{i+1}\pi_i (D_P^+ f)(r_i) + \sigma_i (D_M D_P^+ f)(r_i).$$

From our assumption $\varkappa_0 + \varkappa_1 > 0$, at most one of the functionals Φ_i is non-conservative (see Mandl (1968), Remark 5, p.47). In Groh (1972) we have proved that the restriction of the operator $D_M D_P^+$ to the set

$$\{f \in D(D_M D_P^+) : D_M D_P^+ f \in C,\ \Phi_0(f) = \Phi_1(f) = 0\}$$

is the infinitesimal generator of a Fellerian, strong Markov process $X = (x_t, \zeta, M_t, P_x)$ on the state space E (in the sense of Dynkin (1963)). Also many assertions, made in Mandl's book - especially most results of chapters II, III are valid for the generalized operator $D_M D_P^+$ with a not necessarily strongly isotone canonical measure M. Many properties of the trajectories are valid in the general case: they are right continuous, without discontinuities of second kind and have the nearest neighbour property for $\mathring{E}$ in the sense, that for every time t with $0 < t < \zeta$ there does not exist a point between $x_{t-0} \in \mathring{E}$ and x_t belonging to the state space, see Groh (1975). Also, the behaviour of the trajectories in the boundaries is closely related to classical diffusion processes. We call the process X a quasi-diffusion process.

3. THE COST OF A QUASI-DIFFUSION PROCESS

Following Mandl (1968), the cost associated with the trajectory x_t of the process X has three components: the cost c(x) arising from the quasi-continuous (nearest neighbour) movement through a point x, the cost $\nu_i(x)$ caused by the jumps from r_i into x and the cost $\lambda_i = \lambda(r_i)$ connected with the termination of the process in r_i. Suppose the function $c: E \to \mathbb{R}$ is contained in $\vec{C}$ and ν_i $(i = 0,1)$ is integrable with respect to the measure μ_i. If we denote by $\varphi_i(x)$ $(x \in E;\ i = 0,1)$ the random number of jumps made by the trajectory x_t from the boundary r_i into the set $[r_0, x] \cap E$, the total cost is

$$V = \int_0^\zeta c(x_t)\,dt + \sum_{i=0}^1 \int_E \nu_i(y)\varphi_i(dy) + \lambda(x_{\zeta-0}).$$

We remark that $E_x\zeta := \int \zeta\, dP_x < \infty$ $(x \in E)$ and that the expected time between two successive jumps from the boundaries is positive. (With E_x we denote the mathematical expectation corresponding to the measure P_x.) In the following assertion a characterization of the expected total cost

$$v(x) := E_x V \qquad (x \in E)$$

is given.

THEOREM 1. The expected total cost v is the unique solution of the equation

$$(D_M D_P^+ v)(x) + c(x) = 0 \qquad (x \in \mathring{E})$$

<u>which satisfies the boundary conditions</u>

$$\varkappa_i v(r_i) - \vartheta_i \int_E [v(r_i) - v(y)] d\mu_i(y) + (-1)^{i+1} \pi_i (D_P^+ v)(r_i) -$$
$$- \sigma_i c(r_i) - \vartheta_i \int_E \nu_i(y) d\mu_i(y) - \varkappa_i \lambda_i = 0 \qquad (i = 0,1).$$

The proof is given in an analogous manner like that of Theorem 1 in Mandl (1968), p.149. Although Mandl's theorem concerns with the expected discounted cost with discount rate $\lambda > 0$, an extension to the case $\lambda = 0$ is possible here, because of $\varkappa_0 + \varkappa_1 > 0$ the process X is non-conservative (see Mandl (1968), Remark 1, p.152).

4. THE CONTROLLED PROCESS

Let m be a fixed canonical measure and p be a related fixed canonical scale on I (all notations and definitions of section 2 are preserved analogously also in case of the functions m and p). Further, let $J \subset \mathbb{R}$ a compact set and $a(\cdot,\cdot)$, $b(\cdot,\cdot)$, $c(\cdot,\cdot)$ real valued continuous functions on the compact $I \times J$. Additionally, the function a is supposed to be strictly positive, and it holds the relation

$$\text{(D)} \qquad 1 - a(x,z)^{-1} b(x,z) \Delta^- m(x) > 0 \qquad (x \in E, z \in J).$$

We remark that the values of the functions a, b, c are relevant to our control problem on the compact $E \times J \subset I \times J$ only, but for technical reasons we define they on the whole set $I \times J$.

A function $s: E \to J$ which is an element of $\overleftarrow{C}$ is called a <u>control</u>; with S we denote the set of all controls.

To define for any given $s \in S$ the operator A_s and the functionals Φ_i^s (i = 0,1) we choose as in section 2 fixed reals $\varkappa_i$, ϑ_i, π_i, σ_i and a probability measure μ_i, and define the functions a_s and b_s by

$$a_s(x) := a(x, \bar{s}(x)), \qquad b_s(x) := b(x, \bar{s}(x)) \qquad (x \in I).$$

For any $f \in D(D_m D_p^+)$ we set

$$(A_s f)(x) := a_s(x)(D_m D_p^+ f)(x) + b_s(x)(D_p^- f)(x) \qquad (x \in \mathring{E})$$

$$(A_s f)(r_i) := \lim_{x \to r_i} (A_s f)(x) \qquad (r_i \in E^\beta;\ i = 0,1)$$

$$(A_s f)(r_i) := -\sigma_i^{-1} \Big\{ \varkappa_i f(r_i) + \vartheta_i \int_E [f(r_i) - f(y)] d\mu_i(y) +$$
$$+ (-1)^{i+1} \pi_i (D_p^+ f)(r_i) \Big\} \qquad (r_i \notin E^\beta;\ i = 0,1)$$

$$\Phi_i^s(f) := \varkappa_i f(r_i) + \vartheta_i \int_E [f(r_i) - f(y)] d\mu_i(y) +$$
$$+ (-1)^{i+1} \pi_i (D_p^+ f)(r_i) + \sigma_i (A_s f)(r_i) \qquad (i = 0,1).$$

In the sequel we will show that the operator A_s permits a representation as generalized differential operator and that a suitable restriction determines some quasi-diffusion process.

THEOREM 2. The operator A_s, restricted to the set

$$D(A_s) := \{f \in D(D_m D_p^+) : A_s f \in C, \Phi_0^s(f) = \Phi_1^s(f) = 0\}$$

is the infinitesimal generator of a quasi-diffusion process $X^s = (x_t, \zeta, M_t, P_x^s)$ on the state space E.

REMARKS. 1. In fact, it is possible to choose the process (x_t, ζ, M_t, P_x^s) such that the first three components in this quadruple are independent of the control s.

2. If we consider $A_s f$ in a point $x \in E_\Delta$,

$$(A_s f)(x) = \Delta m(x)^{-1} a(x,s(x))(D_p^+ f)(x) - (\Delta m(x)^{-1} a(x,s(x)) - b(x,s(x)))(D_p^- f)(x),$$

we find the necessity of condition (D) (= dispersiveness), because the operator A_s must satisfy the maximum principle.

To p r o v e Theorem 2 we formulate at first an auxiliary assertion, for the proof see Hildebrandt (1959) and Groh (1978b).

LEMMA 1. The function B_s, defined by

$$B_s(x) = \exp\{\int_{r_0}^x a_s(u)^{-1} b_s(u) dm(u)\} \prod_{u<x} [1 - a_s(u)^{-1} b_s(u) \Delta^- m(u)]^{-1} \cdot \exp(-a_s(u)^{-1} b_s(u) \Delta^- m(u)) \qquad (x \quad I),$$

is positive, bounded, and the unique solution of the equation

$$(4.1) \qquad B_s(x) = 1 + \int_{r_0}^x B_s(u) a_s(u)^{-1} b_s(u) dm(u) \qquad (x \in I).$$

Moreover, for the function $1/B_s$, it holds the equation

$$(4.2) \qquad B_s(x)^{-1} = 1 - \int_{r_0}^x B_s(u-0)^{-1} a_s(u)^{-1} b_s(u) dm(u) \qquad (x \in I).$$

The convergence and positivity of the infinite products depends essentially on the above condition (D).

Now we define a canonical measure M_s and a canonical scale P_s by the relations

$$M_s(x) := \int_{r_0}^x a_s(u)^{-1} B_s(u) dm(u) \qquad (x \in I),$$

$$P_s(x) := \int_{r_0}^x B_s(y)^{-1} dp(y) \qquad (x \in I).$$

By the aid of integration by parts (see Hinderer (1975), p.132) we conclude for every $f \in D(D_m D_p^+)$ the relation

$$\int_{r_0}^x \int_{r_0}^y (A_s f)^-(u) dM_s(u) dP_s(y) =$$

$$= \int_{r_0}^x [\int_{r_0}^y B_s(u) d\{(D_p^+ f)(u)\} + \int_{r_0}^y (D_p^+ f)(u-0) dB_s(u)] dP_s(y) =$$

$$= f(x) - f(r_0) - (D_p^+ f)(r_0) P_s(x).$$

That means $f \in D(D_{M_s} D_{P_s}^+)$ and

$$(D_{M_s} D_{P_s}^+ f)(x) = (A_s f)(x) \qquad (x \in \mathring{E}).$$

By the aid of formula (4.2) we show through integration by parts for such functions f

$$\int_{r_0}^{x}\int_{r_0}^{y} a(u)^{-1}\left[(D_{M_s}D_{P_s}^{+}f)(u) - b(u)B_s(u-0)^{-1}(D_{P_s}^{+}f)(u-0)\right]dm(u)dp(y) =$$
$$=\int_{r_0}^{x}\left[\int_{r_0}^{y} B_s(u)^{-1}d\{(D_{P_s}^{+}f)(u)\} + \int_{r_0}^{y}(D_{P_s}^{+}f)(u-0)d\{B_s(u)^{-1}\}\right]dp(y) =$$
$$= f(x) - f(r_0) - (D_{P_s}^{+}f)(r_0)p(x).$$

Thus we have $D(D_m D_p^+) = D(D_{M_s} D_{P_s}^+)$. From this we conclude

$$D(A_s) = \{f \in D(D_{M_s} D_{P_s}^+) : D_{M_s} D_{P_s}^+ f \in C,\ \Phi_0^s(f) = \Phi_1^s(f) = 0\},$$

and, according to section 2 the operator $(A_s, D(A_s))$ is the infinitesimal operator of a quasi-diffusion process. We remark that because of the requirement $A_s f \in C$ the domaine $D(A_s)$ depends really on the control s.

For fixed control s and cost components

$$c_s(x) := c(x,s(x)), \quad \nu_i(x)\ (x \in E), \quad \lambda(r_i) \quad (i = 0,1)$$

the random cost caused by the process X^s is

$$V_s = \int_0^{\zeta} c_s(x_t)dt + \sum_{i=0}^{1}\int_E \nu_i(y)\varphi_i(dy) + \lambda(x_{\zeta-0}).$$

An application of the above two theorems permits us to characterize the expected cost from the process $X^s = (x_t, \zeta, M_t, P_x^s)$

$$v_s(x) := E_x^s V_s = \int V_s dP_x^s \qquad (x \in E)$$

in the following manner.

THEOREM 3. <u>The function v_s is the unique solution of the equation</u>

$$(4.3) \qquad a_s(x)(D_m D_p^+ v_s)(x) + b_s(x)(D_p^- v_s)(x) + c_s(x) = 0 \qquad (x \in \mathring{E}),$$

<u>satisfying the boundary conditions</u>

$$(4.4) \qquad \begin{aligned} &\varkappa_i v_s(r_i) + \vartheta_i \int_E [v_s(r_i) - v_s(y)]d\mu_i(y) + (-1)^{i+1}\pi_i(D_p^+ v_s)(r_i) - \\ &- \sigma_i c_s(r_i) - \vartheta_i \int_E \nu_i(y)d\mu_i(y) - \varkappa_i\lambda_i = 0 \qquad (i = 0,1). \end{aligned}$$

5. OPTIMALITY

Now we formulate our optimization problem. Let us define for this the <u>minimal expected cost</u> by

$$v^*(x) := \inf_{s \in S} v_s(x) \qquad (x \in E).$$

A control s^* is called optimal if $v^*(x) = v_{s^*}(x)$ $(x \in E)$. In the following a characterization of the minimal expected cost as a solution of a nonlinear boundary value problem is given. Also we derive a necessary and sufficient condition for the optimality of a control. For abbrevation we set $\gamma_i := \min(c(r_i,z)\ ;\ z \in J)$ $(i = 0,1)$.

THEOREM 4. I. <u>The function v^* is the unique solution of the equation</u>

$$(5.1) \qquad (D_m D_p^+ v)(x) + \min_{z \in J}\{a(x,z)^{-1}[b(x,z)(D_p^- v)(x) + c(x,z)]\} = 0 \qquad (x \in \mathring{E}),$$

<u>satisfying the boundary conditions</u>

$$(5.2) \qquad \begin{aligned} &\varkappa_i v(r_i) + \vartheta_i \int_E [v(r_i) - v(y)]d\mu_i(y) + (-1)^{i+1}\pi_i(D_p^+ v)(r_i) - \\ &- \sigma_i\gamma_i - \varkappa_i\lambda_i - \vartheta_i \int_E \nu_i(y)d\mu_i(y) = 0 \qquad (i = 0,1). \end{aligned}$$

II. A control s is optimal if and only if

$$a_s(x)^{-1}[b_s(x)(D_p^-v^*)(x) + c_s(x)] =$$
$$= \min_{z\in J}\{a(x,z)^{-1}[b(x,z)(D_p^-v^*)(x) + c(x,z)]\} \qquad (x \in \mathring{E})$$

and $\quad \sigma_i(c_s(r_i) - \gamma_i) = 0 \qquad (i = 0,1).$

Before proving Theorem 4, we prepare some definitions and lemmas. At first we set for $x \in I$, $z \in J$ and $w \in \mathbb{R}$

$$\beta(x,z) := b(x,z)/a(x,z), \qquad \gamma(x,z) := c(x,z)/a(x,z),$$
$$\Psi(x,w) := -\min_{z\in J}[\beta(x,z)w + \gamma(x,z)].$$

As in Mandl (1968), p.161 we can show that the function Ψ satisfies a Lipschitz condition with the Lipschitz constant

$$L := \max(|\beta(x,z)| : x \in I, z \in J).$$

LEMMA 2. For each $w_0 \in \mathbb{R}$ the equation

$$w(x) = w_0 + \int_{r_0}^{x}\Psi(u,w(u-0))dm(u) \qquad (x \in I)$$

has a unique solution $x \to w(x,w_0)$ $(x \in I)$.

This assertion is proved in Groh (1978b) via Banach's fixed point principle. The crucial point is that we use a weighted norm $\|f\|_L := \sup(h_L(x)|f(x)| : x \in I)$ where the weight function h_L is the unique solution of the equation

$$h_L(x) = 1 + 2L\int_{r_0}^{x} h_0(u-0)dm(u),$$

see Lemma 1. By the aid of the same lemma we can prove also the next assertion.

LEMMA 3. The function $w_0 \to w(x,w_0)$ $(w_0 \in \mathbb{R})$ is strongly isotone and continuous for every fixed $x \in I$. If we denote for $x \in I$

$$\beta_-(x) := \min(\beta(x,z) : z \in J), \qquad \beta_+(x) := \max(\beta(x,z) : z \in J)$$

then it holds $\quad \lim_{w_0\to\pm\infty} w_0^{-1}w(x,w_0) = Q_{\mp}(x)^{-1}$

uniformly in $x \in I$, where the functions $Q_{\pm}$ are the unique solutions of the equations

$$Q_{\pm}(x) = 1 + \int_{r_0}^{x} Q_{\pm}(u)\beta_{\pm}(u)dm(u) \qquad (x \in I).$$

The unicity and positivity of this solutions follows from condition (D). The following assertion may be proved by analyzing of all components of the function v_s (see Mandl (1968), Theorem 1, p.149 and chap. II, §5), beginning with an estimation of the functions B_s. This functions are uniformly bounded with respect to all $s \in S$, $x \in E$.

LEMMA 4. The cost functions v_s, defined in Theorem 3, have uniformly bounded derivatives from the right, i.e. it holds

$$|(D_p^+v_s)(x)| < k \qquad (s \in S, x \in E)$$

for some constant k.

Now we are able to p r o v e Theorem 4. Let for the sake of definiteness $\varkappa_1 > 0$, i.e. the processes X^s will be killed at the right boundary point r_1 with positive probability.

1°. We show that there exists a unique function v_0 satisfying (5.1) and (5.2). By Lemma 5 each solution of (5.1) has the form

$$v_0(x) = K + \int_{r_0}^{x} w(y,w_0)dp(y) \qquad (x \in E). \tag{5.3}$$

We will se that the boundary conditions (5.2) are valid for a unique choice of the constants K and w_0. Denote

$$N_i := \vartheta_i \int_E \nu_i(y)d\mu_i(y) + \sigma_i\gamma_i + \varkappa_i\lambda_i \qquad (i = 0,1).$$

Inserting (5.3) into (5.2) we obtain

$$\varkappa_0 K - \vartheta_0 \int_E\int_{r_0}^{x} w(y,w_0)dp(y)d\mu_0(x) - \pi_0 w_0 = N_0 \tag{5.4}$$

$$\varkappa_1 K + (\vartheta_1 + \varkappa_1)\int_{r_0}^{r_1} w(y,w_0)dp(y) - \vartheta_1 \int_E\int_{r_0}^{x} w(y,w_0)dp(y)d\mu_1(x) + \\ + \pi_1 w(r_1,w_0) = N_1. \tag{5.5}$$

We define the linear continuous functional Θ by

$$\Theta(f) := \varkappa_0(\vartheta_1 + \varkappa_1)\int_{r_0}^{r_1} f(y)dp(y) + \varkappa_1\vartheta_0\int_E\int_{r_0}^{x} f(y)dp(y)d\mu_0(x) + \\ + \varkappa_1\pi_0 f(r_0) + \varkappa_0\pi_1 f(r_1) - \varkappa_0\vartheta_1\int_E\int_{r_0}^{x} f(y)dp(y)d\mu_1(x).$$

For a positive function f the value $\Theta(f)$ is also positive. Therefore the strict isotony of the function $w_0 \to w(\cdot,w_0)$ implies the strict isotony of $w_0 \to \Theta(w(\cdot,w_0))$ $(w_0 \in \mathbb{R})$. Eliminating K from (5.4), (5.5) we obtain

$$\Theta(w(\cdot,w_0)) = \varkappa_0 N_1 - \varkappa_1 N_0. \tag{5.6}$$

Also the continuity of $w_0 \to w(\cdot,w_0)$ implies the continuity of $w_0 \to \Theta(w(\cdot,w_0))$. From Lemma 3 we obtain

$$\lim_{w_0\to\pm\infty} w_0^{-1}\,\Theta(w(\cdot,w_0)) = \Theta(Q_\mp(\cdot)^{-1}).$$

The function $w_0 \to \Theta(w(\cdot,w_0))$ is strictly increasing, tending to $\pm\infty$ for $w_0 \to \pm\infty$. Consequently, $\Theta(w(\cdot,w_0)) = \varkappa_0 N_1 - \varkappa_1 N_0$ holds for exactly one w_0. The constant K is determined by (5.5).

2°. For a fixed control $s \in S$ we denote $\tilde{v}(x) := v_s(x) - v_0(x)$ and

$$q(x) := (D_m D_p^+ \tilde{v})(x) + \beta(x,s(x))(D_p^- \tilde{v})(x) \qquad (x \in E). \tag{5.7}$$

By Theorem 3, (4.3) and equation (5.1) we have

$$q(x) \leq (D_m D_p^+ v_s)(x) + \beta(x,s(x))(D_p^- v_s)(x) + \gamma(x,s(x)) - \\ - \left\{(D_m D_p^+ v_0)(x) + \min_{z\in J}[\beta(x,z)(D_p^- v_0)(x) + \gamma(x,z)]\right\} = 0.$$

Subtracting boundary conditions (5.2) from (4.4) we obtain

$$\varkappa_i\tilde{v}(r_i) + \vartheta_i\int_E[\tilde{v}(r_i) - \tilde{v}(y)]d\mu_i(y) + (-1)^{i+1}\pi_i(D_p^+ v)(r_i) - \\ - \sigma_i(c_s(r_i) - \gamma_i) = 0 \qquad (i = 0,1). \tag{5.8}$$

If we set $\tilde{c}_s(x) := - a_s(x)q(x) \quad (x \in \mathring{E})$, $\quad \tilde{c}_s(r_i) := c_s(r_i) - \gamma_i$

(i = 0,1), $\tilde{\nu}_i(x) := 0$ $(x \in E;\ i = 0,1)$, and $\tilde{\lambda}_i := 0$ (i = 0,1), the equations (5.7) and (5.8) are of the same form as (4.3) and (4.4). Therefore, Theorem 3 implies

$$\tilde{v}(x) = E_0^s \int_0^{\zeta} \tilde{c}_s(x_t)dt.$$

The function $\tilde{c}_s$ is nonnegative, from this it follows $\tilde{v}(x) \geq 0$ $(x \in E)$ and, finally,

(5.9) $$v_0(x) \leq v_s(x) \qquad (s \in S, x \in E).$$

3^0. Our next aim is the construction of a sequence of controls $s_n \in S$ such that $v_{s_n}(x) \to v_0(x)$ $(x \in E)$. Let for $x \in I$

(5.10) $$z_-(x) := \min\{z : B(x,z)w(x-0,w_0) + \gamma(x,z) = -\Psi(x,w(x-0,w_0))\}$$

Following an analogous argument as in Mandl (1968), p.163 we conclude that the function z_- is lower semi-continuous in all points $x \in E \setminus E_\Delta$ and lower semi-continuous from the left side for all $x \in E_\Delta$. Therefore we can find a sequence of functions $z_n : E \to J$ (n = 1,2,...) belonging to $\bar{C}$ and pointwise converging to z_-. If we define

$$s_n(x) := \begin{cases} z_n(x) & (x \in \mathring{E}) \\ \min(z : c(r_i,z) = \gamma_i) & (x = r_i;\ i = 0,1), \end{cases}$$

then it holds for the controls s_n the relation

$$\lim_{n \to \infty} s_n(x) = z_-(x) \qquad (x \in \mathring{E})$$

and also

$$c(r_i, s_n(x)) = \gamma_i \qquad (i = 0,1).$$

Now we set

(5.11) $$w_n(x) := (D_p^+ v_{s_n})(x) \qquad (x \in I).$$

From Lemma 4 the $|w_n(x)|$ $(x \in E;\ n = 1,2,\ldots)$ are uniformly bounded. Thus we can choose $(s_n)_n$ such that $(w_n(r_0))_n$ is convergent to some value $\tilde{w}(r_0)$. By Theorem 3 it holds for $x \in I$

$$w_n(x) = w_n(r_0) - \int_{r_0}^x [B(u,\bar{s}_n(u))w_n(u-0) + \gamma(u,s_n(u))]dm(u).$$

According to Groh (1978b) and the condition (D), this linear equation has the unique solution

$$w_n(x) = D_n(x)\Big[w_n(r_0) - \int_{r_0}^x D_n(u-0)^{-1}\gamma(u,\bar{s}_n(u))dm(u) -$$

$$- \sum_{u \leq x} D_n(u)^{-1}B(u,\bar{s}_n(u))\Delta^- m(u)\gamma(u,\bar{s}_n(u))\Delta^- m(u)\Big] \qquad (x \in I),$$

in which $D_n > 0$ is the unique solution of the equation (see Lemma 1)

$$D_n(x) = 1 - \int_{r_0}^x D_n(u-0)B(u,\bar{s}_n(u))dm(u) \qquad (x \in I).$$

From this it follows the existence of the limits

$$\lim_{n \to \infty} w_n(x) =: \tilde{w}(x) \qquad (x \in I)$$

and the relation

$$\tilde{w}(x) = \tilde{w}(r_0) - \int_{r_0}^x [B(u,z_-(u))\tilde{w}(u-0) + \gamma(u,z_-(u))]dm(u) \qquad (x \in I).$$

On the other hand we have from Lemma 2 and (5.10)

$$w(x,w_0) = w_0 - \int_{r_0}^{x}[\beta(u,z_-(u))w(u-0,w_0) + \gamma(u,z_-(u))]dm(u) \quad (x \in I),$$

and consequently,

$$\tilde{w}(x) - w(x,w_0) = [\tilde{w}(r_0) - w_0] - \int_{r_0}^{x}\beta(u,z_-(u))[\tilde{w}(u-0) - w(u,w_0)]dm(u) \quad (x \in I).$$

Once more according to Lemma 1 and condition (D) we see that the (unique) solution of this equation is either nonnegative or nonpositive on I. The relations

$$\Theta(w_n) = \varkappa_0 N_1 - \varkappa_1 N_0 \qquad (n = 1,2,\dots)$$

imply $\Theta(\tilde{w}) = \varkappa_0 N_1 - \varkappa_1 N_0$ and therefore, comparing (5.6) $\Theta(\tilde{w}(\cdot) - w(\cdot,w_0)) = 0$. The positivity of the functional Θ implies $\tilde{w}(x) = w(x,w_0)$ $(x \in I)$. Thus we have shown

(5.12) $$\lim_{n\to\infty} w_n(x) = w(x,w_0) \qquad (x \in I).$$

From Theorem 3, (4.4) and (5.11) it follows

$$\varkappa_1 v_{s_n}(r_0) + (\vartheta_1 + \varkappa_1)\int_{r_0}^{r_1} w_n(y)dp(y) - \vartheta_1\int_E\int_{r_0}^{x} w_n(y)dp(y)d\mu_1(x) + \pi_1 w_n(r_1) = N_1.$$

Letting $n\to\infty$, we conclude from (5.3), (5.12) that

$$\lim_{n\to\infty} v_{s_n}(r_0) = K$$

$$\lim_{n\to\infty} v_{s_n}(x) = \lim_{n\to\infty}(v_{s_n}(r_0) + \int_{r_0}^{x} w_n(y)dp(y)) = v_0(x) \qquad (x \in E).$$

This and (5.9) imply $v_0(x) = \inf_{s\in S} v_s(x) = v^*(x)$ $(x \in E)$.

4°. The argumentation concerning optimal controls is the same as in Mandl (1968), p.166-167.

COROLLARY. If the function z_-, defined by the relation (5.10) is an element of $\vec{C}$, then the function s, defined by

$$s(x) := \begin{cases} z_-(x) & (x \in \mathring{E}) \\ \min(z : c(r_i,z) = \rho_i) & (x = r_i;\ i = 0,1) \end{cases}$$

represents an optimal control.

REFERENCES

Dynkin,E.B. (1963) : Markovskie processy (Markov processes). Nauka, Moskwa 1963; engl. transl.: Springer, Berlin 1965.

Feller,W. (1952) : The parabolic differential equations and the associated semi-groups of transformations. Ann. Math. 55 (1952), 468-519.

Groh, J. (1972) : Eine Klasse eindimensionaler Markov-Prozesse. Dissertation, TU Dresden 1972.

——— (1975) : Über eine Klasse eindimensionaler Markov-Prozesse. Math.Nachrichten 65 (1975), 125-136.

——— (1978a) : On the optimal control of one-dimensional quasi-diffusion processes. to appear in Math.Nachrichten.

——— (1978b) : On a nonlinear Stieltjes integral equation and a generalized Gronwall inequality in one dimension. to appear.

Hildebrandt, T.H. (1959) : On systems of linear differentio-Stieltjes-integral equations. Illinois J.Math. 3 (1959), 352-373.

Hinderer, K. (1975) : Grundbegriffe der Wahrscheinlichkeitstheorie, 2.Aufl., Springer, Berlin-Heidelberg-New York 1975.

Mandl, P. (1968) : Analytical treatment of one-dimensional Markov processes. Academia, Prague, and Springer, Berlin-Heidelberg-New York 1968.

deMorais, P.R. (1976) : Optimal control of a storage system. Ph.D. Dissertation, Northwestern Univ., Evanston (Illinois) 1976.

Pliska, S.R. (1975) : Controlled jump processes. Stoch.Proc.Appl. 3 (1975), 259-282.

Puterman, M.L. (1974) : Sensitive discount optimality in controlled one-dimensional diffusions. Ann.Prob. 2 (1974), 408-419.

Friedrich-Schiller-Universität Jena
Sektion Mathematik

DDR-69 Jena, UHH
German Democratic Republic

INFORMATION IN TRUNCATED EXPONENTIAL SAMPLES

Gisela Härtler

Berlin

ABSTRACT

The amount of information for distinction between two exponential distributions depends on sample size, number of failures, quotient of parameters, truncation time, and location in the parameter space. In the paper the influence of truncation is considered. The result is a very simple expression allowing valuation of the applicability of truncated exponential samples with respect to the true parameter values. The result is of interest to practical reliability engineering.

INTRODUCTION

In reliability engineering the exploration of highly truncated exponential samples is a common practice. Here truncation will be understood as fixed time truncation, also known as type I censoring. If the truncation time is kept fixed, the number of failures in samples of constant size decreases with increasing reliability. Correspondingly, the amount of information is expected to decrease. The problem is of practical interest because reliability of most electronic parts has reached a very high level (if λ denotes the failure rate of exponential populations, the reached values are $\lambda < 10^{-6}h^{-1}$), whereas the observation time will only exceptionally be longer than some thousand hours.

THE AMOUNT OF INFORMATION

The dependence of the amount of information, I, on the parameter λ of the true life distribution and on the truncation time t^* will be calculated according to the approach, given by Kullback (1959). Let be

(1) $$F(x) = 1 - e^{-\lambda x}, \quad x \geq 0, \quad \lambda > 0,$$

the exponential distribution, and let be considered the discrimination problem $H_1: \lambda = \lambda_1$ against $H_2: \lambda = \lambda_2$, $\lambda_1 \neq \lambda_2$. Let be $l_i(x)$ the likelihood-function of the parameters λ_i, $i = 1,2$, and μ_i an appropriate measure on the subspace $E^r: 0 \leq x_{j_n} \leq t^*$, $j = 1,2,\ldots,r$, following from the complete sample space R^n by truncation at time t^*, where n and r denote sample size and number of failures, respectively.

According to Kullback the mean amount of information in favour of H_1 is given by

(2) $$I(1:2/E^r) = \frac{1}{\mu_1(E^r)} \int_{E^r} \log \frac{l_1(x)}{l_2(x)} \, d\mu_1(x)$$

with

(3) $$l_i(x) = C \lambda_i^r e^{-\lambda_i[\sum_1^r x_j + t^*(n-r)]}, \quad C > 0, \; i = 1,2$$

and

(4) $$d\mu_1(x) = l_1(x)dx.$$

The basis of the used information measure is the likelihood-ratio allowing the following interpretation in the Bayesian sense

(5) $$\log \frac{l_1(x)}{l_2(x)} = \log \frac{P\{H_1/x\}}{P\{H_2/x\}} - \log \frac{P\{H_1\}}{P\{H_2\}},$$

where $P\{H_i/x\}$ and $P\{H_i\}$ denote the a-posteriori and a-priori probabilities of the hypotheses H_i ,respectively. From this viewpoint (2) expresses the mean total information by sampling with given truncation t^*. Using ln instead of log from (2) follows for the case of exponential sampling with truncation

$$(6) \quad I(1:2/E^r) = \ln\left(\frac{\lambda_1}{\lambda_2}\right)^r + r\,\frac{\lambda_2-\lambda_1}{\lambda_1} + n(\lambda_2-\lambda_1)t^* - r(\lambda_2-\lambda_1)\frac{t^*}{F(t^*)}.$$

THE RELATIVE AMOUNT OF INFORMATION

The aim of our analysis, however, is to investigate the influence of the truncation time t^* on I for a given reliability level. Therefore the true reliability of the investigated population is assumed as given by $F_1(x)$. Substituting the number of failures, r, by its expectation $E[r] = nF_1(t^*)$, the expected amount of information follows as

$$(7) \qquad E[I] = nF_1(t^*)[\ \ln\frac{\lambda_1}{\lambda_2} + \frac{\lambda_2}{\lambda_1} - 1].$$

Let be given n and $\lambda_2 = a\,\lambda_1$ with fixed $a > 0$, and let be further

$$(8) \qquad K = n(a - \ln a - 1),$$

then (7) becames very simple

$$(9) \qquad E[I] = K(1 - e^{-\lambda_1 t^*}).$$

The maximum of (9) is equal K. It will be reached if t^* tends to infinity. So the relative expected amount of information will be given by

$$(10) \qquad E[I]_{rel.} = 1 - e^{-\lambda_1 t^*}.$$

From (10) the loss of information by truncation can be estimated very easily. Assuming e.g. a population with $\lambda_1 = 10^{-6}h^{-1}$ has to be investigated by an experiment of 1000 h, it becames evident that the relative expected amount of information will be less than 1 per mille of the maximum information.

REFERENCES

Kullback S., Information Theory and Statistics, Wiley, New York - London, 1959

Academy of Sciences of the GDR
Central Institute of Electron Physics

Mohrenstrasse 40/41
108 Berlin
German Democratic Republic

RANK CORRELATION COEFFICIENTS AND ORDERINGS ON THE SPACE OF PERMUTATIONS

Tomáš Havránek, Dan Pokorný

Prague

ABSTRACT

In the present paper various orderings on the space of permutations are studied in order to define some general decision rules used in methods of hypothesis formation; especially the question of monotonicity of rank statistics with respect to orderings defined is investigated as well as various relations between these orderings. Computational aspects are **emphasized**.

INTRODUCTION

In the theory of rank correlations various orderings on the space of permutations are studied since Daniels (1944). If we observe n independent pairs of random variables (X_i, Y_i) having the same two dimensional density, we can suppose that in such samples there are no ties. Denote R_i the rank of X_i among the first coordinates $X_1, \ldots, X_n$ and Q_i the rank of Y_i among the second coordinates $Y_1, \ldots, Y_n$. Put $R = \langle R_1, \ldots, R_n \rangle$ and $Q = \langle Q_1, \ldots, Q_n \rangle$. Rank statistics for testing independence are functions of $\langle R, Q \rangle$. We restrict ourselves, from clear intuitive reasons, to statistics invariant under permutations applied simultaneously to the two vectors R and Q. Such statistics are functions of the rank vector $R^o = Q \circ R^{-1}$. R^o maps the sample space into the set $\underline{P}^n$ of all permutations of the natural numbers $\{1, \ldots, n\}$.

1.1 Denotation. By letters p,q,r we denote permutations from $\underline{P}^n$. Each permutation, say p, is represented by vector $\langle p(1), \ldots, p(n) \rangle$.

1.2 All rank statistics for testing independence and especially rank correlation coefficients can be investigated as function on $\underline{P}^n$, i.e. they can be viewed as diversity measures on a reference set; the reference set is here $\underline{P}^n$. If we forget for a moment the probabilistic point of view, we can formulate some intuitive demands for such a statistic to be reasonable; this is done, for example, in Kendall (1948). Such a statistic:

(i) have to attend a maximum on identical permutation $i = \langle 1,\dots,n \rangle$;

(ii) must not increase with increasing separation of permutations from i.

The second condition is vague. We must specify the notion of increasing separation from i; this was done by various authors in various ways. In fact, it means to define a partial ordering on the space of permutations. Clearly, such an ordering must have the permutation i as the greatest element.

1.3 Denote by NI(p) the number of inversions in the permutation p, i.e. the number of pairs $1 \leq i,j \leq n$ such that $i<j$ and $p(i) > p(j)$. Now we add a further condition to reasonable orderings: if $p \leq q$ then $NI(p) \geq NI(q)$.

1.4 Kendall (1948) studied an ordering denoted here $<^s$ (s - strong) : we write $p_1 \xrightarrow{s} p_2$ if there are two indices $i<j$ such that $p_1(i) = p_2(j) < p_1(j) = p_2(i)$ and $p_1(k) = p_2(k)$ for $k \neq i,j$. Now we write $p <^s q$ if there exists a chain of permutations $q = p_o \xrightarrow{s} p_1 \xrightarrow{s} p_2 \xrightarrow{s} \dots p_m \xrightarrow{s} p$. Kendall proved that under this ordering a class of general correlation coefficients introduced by Daniels (1944) is nondecreasing. This class include Spearman and Kendall rank correlation coefficients. The same ordering was studied by Savage (1957) , c.f. 3.4 here.

1.5 Yanagimoto and Otake (1969) used an ordering defined by interchanging neighbours; thus $p_1 \xrightarrow{w} p_2$ if there is an index i, $1 \leq i < n$, such that $p_1(i) = p_2(i+1) < p_1(i+1) = p_2(i)$ and $p_1(k) = p_2(k)$ for $k \neq i,i+1$. The ordering $<^w$ (w - weak) is then defined by the same way as $<^s$ (transitive closure). The dual ordering to $<^w$ can be defined as follows: $p <^{w'} q$ iff $p^{-1} <^w q^{-1}$.

We have two basic facts concerning just defined orderings:

(i) If $p <^w q$ or $p <^{w'} q$ then $p <^s q$;

(ii) $p <^s q$ iff $p^{-1} <^s q^{-1}$.

Hence $<^s$ has a nice intuitive property of self duality.

1.6 In the context of mechanized hypothesis formation based on rank methods "arithmetical" orderings of permutations have been investigated by Hájek and Havránek (1977a), Havránek (1976). These orderings are used to define classes of generalized quantifiers in observational calculi. For these notions see the works just mentioned as well as Hájek and Havránek (1977b). In fact, one has studied decision functions having the following property:

If d is a decision function with the range $\{0,1\}$ and $\leq$ the used ordering then we demand that $p \leq q$ and $d(p) = 1$ imply $d(q) = 1$.

Since computer methods making a great number of such decisions have been considered, the computer complexity of the questions of the form " $p \leq q$? " is of interest. Further information on these methods can be found in Hájek and Havránek (1977b), Havránek (1978) and Havránek and Vosáhlo (1978).

The first of these orderings is inspired by Spearman correlational coefficient:

(i) $p \leq^{d} q$ if for $k = 1,\ldots,n-1$ card $\{i;\ |p(i) - i| = k\} \geq$ card $\{i;\ |q(i) - i| = k\}$ (d - difference).

The second and third one are weaker forms of $\leq^{d}$:

(ii) $p \leq^{wd} q$ if for $i = 1,\ldots,n$ $|p(i) - i| \geq |q(i) - i|$

and (iii) $p \leq^{swd} q$ if $p \leq^{wd} q$ and $|\mathrm{sign}(i-p(i)) - \mathrm{sign}(i-q(i))| \leq 1$ (s - sign). The strict form $<$ is defined by the clear way.

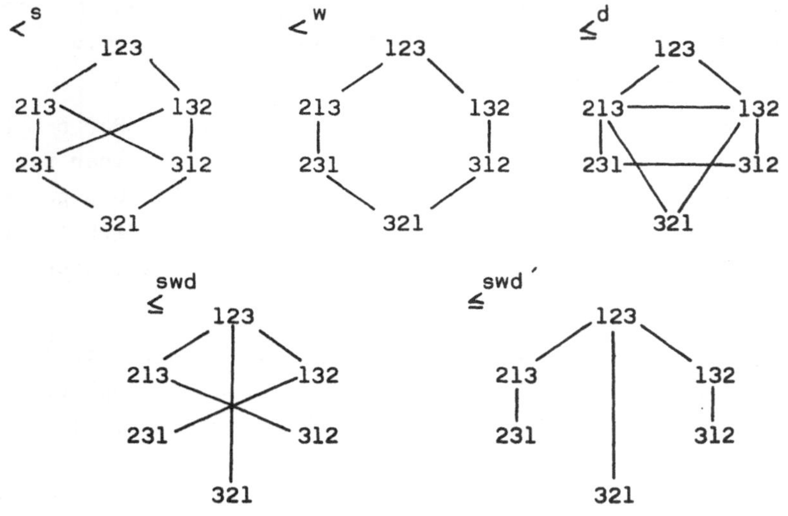

FIG. 1

Immediatelly, from the definition we see that $p \leq^{swd} q$ implies $p \leq^{wd} q$ and $p \leq^{wd} q$ implies $p \leq^{d} q$. Dual orderings $\leq^{swd}$ and $\leq^{wd}$ are defined in the usual way.

1.7 The aim of the present paper is to test the just described orderings with respect to their compatibility with some general desired properties as monotonicity of known rank statistics or number of inversions (cf.1.3) and consider some of their computational properties. Main results describes relations between considered orderings on the space of permutations; as far as the authors know such relations have not yet been in the scope of interest of combinatorics; cf. Flaschmeyer (1972) or Knuth (1968).

On the fig. 1 the considered orderings are presented for n = 3. Some basic relations between these orderings are clear from this figure.

RELATIONS BETWEEN ORDERINGS

2.1 <u>Proposition</u>. $\leq^{d}$ is selfdual.

Proof. $|p^{-1}(j) - j| = k$ iff $|p(i) - i| = k$ for $p(i) = j$.

2.2 <u>Proposition</u>. If $p <^{d} q$ then $S(p) < S(q)$ for any statistic of the form $S(p) = - \sum_{i=1}^{n} |i - p(i)|^{k}$, $k \in R$, $k \geq 1$.

2.3 <u>Fact</u>. $p \leq^{d} q$ does not imply $NI(p) \geq NI(q)$. For n = 2,3,4 the implication holds; for n = 5 we have: if $p <^{d} q$ then $NI(p) \geq NI(q)$. For n = 6 put $p = \langle 3,1,5,2,6,4 \rangle$ and $q = \langle 3,2,1,6,5,4 \rangle$. We have $p <^{d} q$ and $NI(p) = 5 < NI(q) = 6$.

2.4 <u>Fact</u>. $p \leq^{wd} q$ does not imply $NI(p) \geq NI(q)$. For n = 2, ...,5 the implication holds; for n = 6 use the just described p,q; it is easy to see that $p \leq^{wd} q$. From 31 575 wd-comparable pairs for n = 6 only two pairs have the property $p <^{wd} q$ and $NI(p) > NI(q)$.

The results 2.3 and 2.4 were obtained by checking all cases by a computer program.

2.5 <u>Corrollary</u>. Neither $p <^{d} q$ nor $p <^{wd} q$ implies $p <^{s} q$.

Proof. From 2.8 of Kendall (1955) we know that $p <^{s} q$ implies $t(p) < t(q)$, where t is Kendall correlation coefficient, i.e. a decreasing function of NI(p).

2.6 Consider the quadrant statistic

$$S(p) = \frac{1}{4} \sum_{i=1}^{n} (\text{sign}(i-\tfrac{n+1}{2}) + 1)(\text{sign}(p(i)-\tfrac{n+1}{2}) + 1) .$$

<u>Fact</u>. $p \leq^{wd} q$ does not imply $S(p) \leq S(q)$.

Clearly $p = \langle 2,1,4,3\rangle \leq^{wd} q = \langle 1,3,2,4\rangle$ but $S(p) = 2 > S(q) = 1$.

2.7 The class of statistics monotone with respect to $\leq^{wd}$ is slightly greater than the class of statistics monotone with respect to $\leq^{d}$. It contains, for example, statistics of the form $S(p) = \sum_{i=1}^{n} f(i)\, g(i-p(i))$, where $g(x)$ is a decreasing function of $|x|$.

2.8 Fact. No of the following four implications hold: $p \geq^{w} q \Rightarrow p \geq^{swd} q$, $p \geq^{swd} q \Rightarrow p \geq^{w} q$, $p \geq^{w} q \Rightarrow p \geq^{swd} q$, $p \geq^{swd} q \Rightarrow p \geq^{w} q$. Consider, for example, $p = \langle 3,4,1,2\rangle$ and $q = \langle 1,4,3,2\rangle$. Then $p \leq^{swd} q$ and $p \leq^{swd} q$ but neither $p \leq^{w} q$ nor $p \leq^{w} q$.

2.9 Theorem. For each $p,q \in \underline{P}^n$ $p <^{swd} q$ implies $p <^{s} q$.

Proof. We shall use the following fact: $p <^{s} q$ iff p and q are distinct and if for $m = 1,\dots,n$ we rearrange the first m members of permutations in the increasing order, i.e. $p(i_1) < \dots < p(i_m)$ and $q(j_1) < \dots < q(j_m)$, then $p(i_k) \geq q(j_k)$ for $k = 1,\dots,m$ (Yanagimoto and Okamoto, 1969). Now we assume $p <^{swd} q$. For $m = n$ we have clearly the desired property. For $m = n-1$: we have $|n - q(n)| \leq |n - p(n)|$. If $|p(n) - n| = 0$ then $p(n) = q(n) = n$ and the rest for $i = 1,\dots,n-1$ has the desired property. If $|n - p(n)| > 0$ then $p(n) < n$ and hence $n - p(n) > 0$ and $q(n) \geq p(n)$. If we omit $p(n)$ and $q(n)$ we omit from p a greater value. The rest has the desired property.

Suppose that for $m = n-k$ the proof is finished. Consider the case $m = n-k-1$. If $|p(n-k) - (n-k)| = 0$, the proof is trivial. If $n-k - p(n-k) > 0$ then $q(n-k) \geq p(n-k)$ and we can proceed as in the similar case for $m = n-1$.

The case $n-k - p(n-k) < 0$ is more complicated: Let the first $n-k = m+1$ members of each permutation be rearranged:

(1) $q(i_1) < q(i_2) < \dots < q(i_c) < \dots < q(i_{m+1})$

(2) $p(j_1) < p(j_2) < \dots < p(j_d) < \dots < p(j_{m+1})$.

We know: $q(i_a) \leq p(j_a)$ for $a = 1,\dots,m+1$. Let $m+1 = i_c = j_d$. From $\leq^{swd}$ we know that $m+1 \leq q(m+1) \leq p(m+1)$ and for the present case, moreover, $m+1 < p(m+1)$. I.e. $m+1 \leq q(i_c) \leq p(j_d)$ and $m+1 < p(j_d)$. We shall discuss the relation between c and d:

(i) If $c=d$, the proof is trivial.

(ii) If $c > d$ then to preserve inequalities $q(i_a) \leq p(j_a)$ after omiting $q(i_c)$ and $p(j_d)$, in the original system

$$\begin{array}{ccccccccccc} q(i_1) < & \ldots & < q(i_d) < & q(i_{d+1}) < & \ldots & < q(i_{c-1}) < & q(i_c) < & \ldots & q(i_{m+1}) \\ \wedge\!| & & \wedge\!| & \wedge\!| & & \wedge\!| & \wedge\!| & & \wedge\!| \\ p(j_1) < & \ldots & < p(j_d) < & p(j_{d+1}) < & \ldots & < p(j_{c-1}) < & p(j_c) < & \ldots & p(j_{m+1}) \end{array}$$

the following inequalities must hold: for $a = d,\ldots,c-1$ $q(i_a) \leq p(j_{a+1})$. This is true, because $q(i_a) \leq p(j_a) < p(j_{a+1})$.

(iii) If $c < d$, we have the following pattern:

$$\begin{array}{l} m+1 \leq q(i_c) < q(i_{c+1}) < \ldots\ q(i_{x-1}) < q(i_x) < \ldots < q(i_{d-1}) < q(i_d)\ \ldots \\ (3) \\ \quad p(j_c) < p(j_{c+1}) < \ldots\ p(j_{x-1}) < p(j_x) < \ldots < p(j_{d-1}) < p(j_d)\ \ldots \end{array}$$

For $a = c+1,\ldots,d$ we must have $q(i_a) \leq p(j_{a-1})$. Clearly for $a \geq c$ $p(j_a) \geq q(i_a) \geq m+1$ (see (3)). Let for an x be $q(i_x) > p(j_{x-1})$. If we omit $q(i_c)$ and $p(j_d)$ (i.e. $q(m+1)$ and $p(m+1)$) then $q(i_x)$ is the $(m-x+2)$th value in the first row of (3) w.r.t. the increasing ordering; similarly $p(j_{x-1})$ is the $(m-x+2)$th value in the second row of (3). Clearly $\min\{p(j_x), p(j_{x+1}),\ldots,p(j_{m+1})\} \leq p(j_{x-1})$. On the other hand, for an a such that $x \leq a \leq m+1$, we have $q(i_x) \leq q(i_a) \leq p(j_a)$ (from $\leq^{swd}$: $m+1 \leq q(i_a)$) and hence $q(i_x) \leq \min\{p(j_x),\ldots,p(j_{m+1})\} \leq p(j_{x-1})$. The proof is finished .

STATISTICAL PROPERTIES OF THE $<^s$ ORDERING

3.1 It is clear from the previous considerations that the ordering $<^s$ plays the main role in defining monotonocity of rank statistics for testing independence versus positive dependence. As the alternative hypothesis we shall use the following hypothesis:

$$X_i = X_i^* + \Delta Z_i$$
$$Y_i = Y_i^* + \Delta Z_i\ , \quad \Delta \neq 0$$

where X_i^*, Y_i^*, Z_i are mutually independent random variables and their distributions are independent of i. Assume that X_i^* and Y_i^* have densities f and g of a know type and the distribution of Z_i is arbitrary. See Hájek and Šidák (1967), §4.11 of chapter II. (Z_i must have finite and non zero variance.)

We can consider the locally most powerful tests of indepedence against the just defined family of alternative hypotheses. If f' and g' are continuous almost everywhere and $\int |f'(x)|\, dx < \infty$ and $\int |g'(x)|\, dx < \infty$, then such a test has the critical region given by

$$S(R,Q) = \sum_{i=1}^{n} a_n(R_i,f)\, a_n(Q_i,g) \geq k$$

where $a_n(i,f)$ and $a_n(i,g)$ are scores defined by the expression

$$a_n(i,f) = n \binom{n-1}{i-1} \int_0^1 \varphi(u,f)\, u^i (1-u)^{n-i}\, du ,$$

where

$$\varphi(u,f) = \frac{f'(F^{-1}(u))}{f(F^{-1}(u))} , \quad 0 < u < 1 ,$$

with $F^{-1}(u) = \inf\{x;\ F(x) \geq u\}$ and $F(x) = \int_{-\infty}^{x} f(y)dy$.
Similarly for $a_n(i,g)$. Clearly we can use $S^*(R^o) = S(\,i, Q \circ R^{-1})$ instead of $S(R,Q)$.

3.2 <u>Theorem</u>. If f and g are strongly unimodal then S^* is non-decreasing with respect to $<^s$.
Proof. If f is strongly unimodal then $\varphi(u,f)$ is a non-decreasing function of u. $a_n(i,f)$ can be expressed as $a_n(i,f) = E\,\varphi(U_n^{(i)},f)$ where $U_n^{(i)}$ is the i-th member of an ordered sample $U_n^{(1)} < \ldots < U_n^{(n)}$ from the uniform (0,1) distribution. Hence $a_n(i,f)$ is a non decreasing function of i. Similarly for $a_n(i,g)$. Now, if we consider permutations p and q such that $q <^s p$, i.e. there are $i < j$ such that $q(i) = p(j) < g(j) = p(i)$ and $q(k) = p(k)$ for $k \neq i,j$. Then $a_n(i,f)a_n(q(i),g) + a_n(j,f)\, a_n(q(j),g) \geq a_n(i,f)a_n(q(j),g) + a_n(j,f)\, a_n(q(i),g)$.

3.3 <u>Proposition</u>. The same result is obtained if we use aproximate scores $a_n(i,f) = \varphi(i/(n+1),f)$ and $a_n(i,g) = \varphi(i/(n+1),g)$.

3.4 Savage (1957) considers probabilistic properties of $<^s$ under trend alternatives, i.e. he considers regression problems. His results could be directly translated for our positive dependency symmetric alternatives. Consider the joint density $h_\Delta(x,y)$ of X_i, Y_i . Let

$$D(x_1,y_1,x_2,y_2) = \begin{vmatrix} h(x_1,y_1) , & h(x_1,y_2) \\ h(x_2,y_1) , & h(x_2,y_2) \end{vmatrix} \geq 0$$

for each $x_1 < x_2$ and $y_1 < y_2$ and let the inequality be strict on a set with positive measure.

3.5 <u>Theorem</u>. Under the just described conditions $p <^s q$ implies $P(R^o = p) < P(R^o = q)$.
Proof is a direct translation of the proof of Savage's Theorem 1

for two-dimensional densities. Consider that $P(R^o=p) - P(R^o=q)$ reduces to

$$\int_{x_1< x_2,\, y_1< y_2} D(x_1,x_2,y_1,y_2)\ dx_1dx_2dy_1dy_2 \quad .$$

3.6 Remark. Some other conditions could be used. The present condition is rather general; it covers ,e.g., densities of the form $h(\Delta)\ f(x)\ g(y)\ \ \exp(\Delta^2 xy)$.

COMPLEXITY

4.1 From computational point of view it is necessary to consider the complexity of questions of the form " $p \leq q$? " (for various orderings $\leq$). If we use procedures with generalized quantifiers defined on the basis of an ordering $\leq$ on $\underline{P}^n$ in a method of hypotheses formation, we can use the information that $p \leq q$ holds for some deduction steps. But this information must not be too expensive from the pointof view of computer time. As usual, we shall measure here computer complexity by number of steps needed to decide whether p q or not. As measured computer steps serve comparisons.

In the present context also comparison with quick algorithm for computing paired comparison measures on permutations constructed by Leathers (1977). The expected time for two variables (as well as two permutations) is O(n log n).

4.2 Let $p,q \in \underline{P}^n$. Consider the task to decide whether $p\ ^s q$ or not. $p\ ^s q$ is eqivalent to $\bigwedge_{m=1}^{n} L(m)$, where $L(m) =$ $((\ \{i_1,\dots,i_m\} = \{j_1,\dots,j_m\} = \{1,\dots,m\})\ \&\ (q(i_1)< \dots <q(i_m))\ \&$ $(\ p(j_1)< \dots <p(j_m)))\Rightarrow(\ \forall k = 1,\dots,m)(\ q(i_k) \leq p(j_k))$. Use the following procedure

```
      PROC ASSIGN (n,p,q,k);
   L: DO m = 1 TO n ;
   B: IF not L(m) THEN BEGIN; k = ´NO´; GO TO E; END;
      END L;
      k = ´YES´;
   E: END ASSIGN;
```

The complexity of the step "B" , i.e. of decision "L(m) is false" can be found as follows: From the previous step $q(1),\dots,q(m-1)$, $p(1),\dots,p(m-1)$ are ordered. To construct the chain $p(j_1)<\dots<p(j_m)$ we need at most m-1 comparisons; similarly for q. Hence for decision

"L(m) is false" we need at most $3m-2$ comparisons ($2(m-1)$ for the above mentioned chains and m for $q(i_k) \leq p(j_k)$, $k = 1,\dots,m$.)

<u>Conclusion</u>. For $p \overset{s}{<} q$ we need at most $3\sum_{m=1}^{n} m = 3\binom{n+1}{2}$ comparisons.

4.3. <u>Fact</u>. For $p \leq^{swd} q$ we need at most $2n$ comparisons.

4.4 If a probability distribution on $\underline{P}^n$ is given, then the average number of comparisons for $p \leq^{s} q$ is at most

$$Z(n) = 3 \sum_{m=1}^{n-1} m \, P(\bigwedge_{i=1}^{n-1} L(i)) \ ;$$

hence the average number (expected number) depends heavily on the underlying distribution. We present here some values under the hypothesis of independence (i.e. under uniform distribution on $\underline{P}^n$); see Tab.1 :

TABLE 1

Expected number of comparisons

n	6	8	10	12	14
swd	4.06	3.98	3.96	3.95	3.86
wd	2.44	2.29	2.19	2.17	2.08
s	16.17	17.80	18.81	19.64	19.85

In all cases S.E. ≤ 0.03

We present estimates obtained by random sampling the number of samples for each n was 5 000 (for n=6 10 000).

4.5 We used a program PRELACE which for given n generates all $(n!)^2$ pairs of permutations and evaluates whether $p <^{s} q$, $p \leq^{wd} q$ or $p \leq^{swd} q$. The output contains: (i) frequencies of pairs satisfying relations $p <^{s} q$, $p \leq^{wd} q$ and $p \leq^{swd} q$; (ii) 2 X 2 X 2 frequency table containing frequencies of simultaneous occurencies of comparable pairs w.r.t. $<^{s}$, $\leq^{wd}$, $\leq^{swd}$.(iii) For each ordering it computed the number of pairs for that the algorithm for " $p <^{s} q$? ", " $p \leq^{wd} q$? " or " $p \leq^{swd} q$? " respectively halts exactly after k comparisons ($k = 1,2,\dots$).

Due to the exponential complexity of this program we can expect results in reasonable time for $n \leq 7$.

4.6 The second used program was RGP (random generation of permutations), which for a given n generates random samples of permutations and computes statistics as in (i) - (iii) above.

Results of point (ii),(iii) are presented in the table 2. For n=3,4,5 it contains exact values, for n=6,8 results obtained on 10 000 random samples, for n=10 on 5 000 random samples.

TABLE 2

Summary statistics

n	3	%	4	%	5	%	6	%	8	%	10	%
swd	13	36.1	83	14.4	725	5.0	$1.1\ 10^4$	2.1	$1.4\ 10^6$	0.1	$1.1\ 10^{10}$	0.08
wd	15	41.7	134	23.3	1740	12.1	$4.5\ 10^4$	8.7	$3.4\ 10^7$	2.1	$4.2\ 10^{10}$	0.3
s	19	52.8	213	37.0	3781	26.3	$1.1\ 10^5$	20.1	$1.8\ 10^8$	11.1	$8.8\ 10^{11}$	6.4
1	4	21.1	80	37.6	2099	55.5	$6.3\ 10^4$	59.0	$1.5\ 10^8$	80.4	$8.4\ 10^{11}$	95.2
2	0	0	1	0.5	58	1.5	$1.9\ 10^3$	1.8	$2.7\ 10^6$	1.5	0	0
3	2	10.5	50	23.5	975	25.8	$3.2\ 10^4$	30.4	$2.9\ 10^7$	16.1	$3.6\ 10^{10}$	3.5

1 - s&¬wd, 2 - wd&¬s, 3 - wd&s&¬swd . Percentages in first three rows in $(n!)^2$, in the second three rows in the nuber of s-comparable pairs. Note that wd&s&swd ≡ swd.

4.7 <u>Theorem</u>. Assume the uniform distribution on $\underline{P}^n$. Let $Z^{wd}(n)$ and $Z^{swd}(n)$ are expected numbers of steps for $\leq^{wd}$ and $\leq^{swd}$ respectively. Then

$$\text{(i) } \limsup \frac{Z^{wd}(n)}{\log_2(n)} < +\infty \quad \text{and (ii) } \limsup \frac{Z^{swd}(n)}{\log_2(n)} < +\infty$$

Proof . For expected complexity we have here $Z^a(n) \leq 2 \sum_{m=1}^{n} P(\bigwedge_{i=1}^{m-1} L^a(i))$, where for a = wd , $L^a(i) \equiv |p(i) - i| \geq |q(i) - i|$, and for a = swd, $L^a(i) \equiv (L^{wd}(i)\ \&\ |\operatorname{sign}(p(i)-i) - \operatorname{sign}(q(i)-i)| \leq 1)$. Let $f : N \to N$ and $1 \leq f(i) < i$. Then define

$$(4) \quad T^a(n) = \frac{Z^a(n)}{2} = \sum_{m=1}^{f(n)} P(\bigwedge_{i=1}^{m} L^a(i)) + \sum_{m=f(n)+1}^{n} P(\bigwedge_{i=1}^{m} L^a(i)).$$

Denote $R(n,m) = (\bigwedge_{i=1}^{m} L^a(i))$. Clearly

$$(5) \quad T^a(n) \leq f(n)\ (1 - R(n,f(n))) + (n-f(n))\ R(n,f(n)) \text{ and hence}$$

$$(6) \quad T^a(n) \leq f(n) + n\ R(n,f(n)) = Q^a(n).$$

<u>Lemma</u>. Let $f^2(n)/n \to 0$ then $\lim_{n\to\infty} (2^{-f(n)}/R(n,f(n))) \in (0,+\infty)$, i.e. $2^{-f(n)}$ and $R(n,f(n))$ are strongly asymptotically equivalent.

Proof of the lemma: Denote by (*p) the expression

$(\exists i, 1 \leq i \leq f(n))(p(i) \leq i)$ and similarly for $(*q)$. We have, for $n \to +\infty$

$$P((*p)) \leq \frac{1 + 2 + \ldots + f(n)}{n} \leq \frac{f(n)^2}{n} \to 0.$$

Similarly for $P((*q))$. On the other hand, from $\neg(*p)$ and $\neg(*q)$ it follows that for $i=1,\ldots,f(n)$ we have $L^{wd}(i) \equiv L^{swd}(i) \equiv (p(i) \geq q(i))$. Denote by $L(i,n)$ the inequality $q(i) \leq p(i)$. It is clear that the values $P(\bigwedge_{i=1}^{f(n)} L(n,i))$ and $2^{-f(n)}$ are under assumption $f^2(n)/n \to 0$ asymptotically equivalent.

Now we apply our lemma to expression (6). If $f^2(n)/n \to 0$, then $Q^a(n)$ is asymptotically equivalent to $W(n) = f(n) + n\,2^{-f(n)}$. Put $f(n) = \text{int} \log_2 n$ (integer part of $\log_2 n$). Then we obtain $W(n) = \text{int} \log_2 n + n\,2^{-\text{int} \log_2(n)}$. So we have the upper bound for the expected complexity: $W(n) \leq \log_2 n + 2$.

CONCLUSIONS

The properties of the ordering $<^s$ described in the previous sections lead us to the conclusion that reasonable definitions of correlational quantifiers (decision rules based on permutations) should be based on this ordering. Moreover, all reasonable statistics for testing hypothesis of independence versus positive dependence are monotone with rèspect to $<^s$.

From the theoretical point of view, the only concurent of $<^s$ is the ordering $<^w$ studied by Yanagimoto and Okamoto (1969); but this ordering is not self dual and hence it is not appropriate for our symmetric concept of dependence.

From the computational point of view it seems to be useful to work with the ordering $\leq^{swd}$, which is related to $<^s$. Hence theoretically the concept of correlational quantifiers should be based on $<^s$ and $\leq^{swd}$ should be used in practical applications of deduction rules. From this point of view, definitions, theorems and procedures of Hájek and Havránek (1977a) based on the ordering $\leq^d$ are to be inspected. Note here that all theorems based on the ordering $\leq^d$ in the just mentioned book remain true, since they use only very general features of such orderings.

REFERENCES

Daniels H.E. (1944) : The relation between measures of correlation in the universe of sample permutations. Biometrika 33(1943-1946), 129-135.

Flaschmeyer J. (1972) : Kombinatorik. VEB Deutscher Verlag der Wissenschaften, Berlin (1972).

Hájek J., Šidák Z. (1967) : Theory of rank tests. Academia, Prague and Academic Press, New York (1967).

Hájek P., Havránek T. (1977a) : Mechanizing hypothesis formation - mathematical foundations of a general theory. Universitext. Springer - Verlag, Heidelberg (1977).

(1977b) : On generation of inductive hypotheses. International Journal of Man-Machine Studies 9 (1977), 415-438.

Havránek T. (1976) : Problems of logical foundations of effective statistical inference (Problémy logických základů efektivní statistické inference - in czech). Thesis. Charles University - Department of Mathematics, Prague (1976).

(1978) : Enumeration calculi and rank methods. International Journal of Man-Machine Studies (in print).

Havránek T., Vosáhlo J. (1978) : A GUHA procedure with correlational quantifiers. International Journal of Man-Machine Studies (in print).

Kendall M.G. (1948) : Rank correlation methods. Griffin.,London (1958), second edition (1955).

Knuth D.E. (1968) : The art of computer programming.Vol. 1. - Fundamental algorithms. Addison / Wesley, Reading (1968).

Leathers B.L. (1977) : Computing paired comparison measures for ordinal data. Applied Statistics (submitted).

Savage I.R. (1957) : Contributions to the theory of rank order statistics - the "trend" case. Annals of

Mathematical Statistics 28 (1957), 968 - 977.

Yanagimoto T., Okamoto M. (1969) : Partial orderings of permutations Annals of the Institute of Statistical Mathematics 21 (1969), 489-506.

Czechoslovak Academy of Sciences
Center of Biomathematics
Budějovická 1083
142 20 Praha 4 - Krč
Czechoslovakia

DISTRIBUTION OF STUDENT'S RATIO BASED ON HALF-GAUSSIAN POPULATION

A.K.M. Sirajul Hoq (Mosul), Mir M. Ali (London), and James G.C. Templeton (Toronto)

ABSTRACT

For a sample of size n drawn from a half-Gaussian population with location parameter μ and scale parameter σ, let $T = \sqrt{n}(\overline{X}-\mu)/S$ where $\overline{X}$ is the sample mean and S, the sample standard deviation given by $(n-1)S^2 = \Sigma(X_i-\overline{X})^2$. The tail probability $G_n(t) = \Pr\{T \geq t\}$ and the density $f_n(t) = -dG_n(t)/dt$ are obtained for values of $t \geq \{(n-1)(n-2)/2\}^{\frac{1}{2}}$. For the special cases n = 2,3,4 explicit expressions for $G_n(t)$ and $f_n(t)$ are obtained. For small n and large t, asymptotic tail probabilities for T with Gaussian, half-Gaussian and exponential parent distributions are compared.

INTRODUCTION

In performing a t-test, the relevant portion of the distribution of the t-statistic is the tail probabilities. A t-test, in general, is called for when the sample size is small. Hence the exact tail probabilities, (approximations often not being accurate) of the t-distribution for small samples when the underlying sample is assumed to arise from a specified non-normal population would enable one to ascertain whether a test valid under the normality assumption would also be robust under the specified non-normality assumption.

The present work, while arising out of problems in robustness studies, is in itself a very specialized and restricted mathematical problem. Hence it has not been felt worth while to survey the literature in robustness of tests. However, two excellent reports, one by Posten and Hatch (1966) and the other, an annotated bibliography of robustness studies by Govindarajulu and Leslie (1970) survey many of the works that bear on the question of robustness. Recent studies of the distribution of the t-statistic for non-normal parent distributions include Sansing (1976) on tail probabilities and densities for double exponential parent, Shenton and Bowman (1977) on moments of t for exponential parent, Hoq, Ali and Templeton (1977) on tail probabilities and densities for exponential parent, and other works cited by these authors.

THE PROBLEM

A sample $X_1,X_2,\ldots,X_n$ of size n is drawn from a half-Gaussian population having the probability density function $f(x) = (2/\pi)^{\frac{1}{2}}\sigma^{-1}\cdot\exp[-\{(x-\mu)/\sigma\}^2/2]$ for $x \geq \mu$, zero otherwise. The sample has mean $\bar{X} = \Sigma X_i/n$ and standard deviation S given by $(n-1)S^2 = \Sigma(X_i-\bar{X})^2$. Student's ratio T is defined by $T = \sqrt{n}(\bar{X}-\mu)/S$. The problem is the evaluation of $G(t) = \Pr\{T\geq t\}$ for large enough values of t, $t \geq \left((n-1)(n-2)/2\right)^{\frac{1}{2}}$ being considered in the present work. Since T is invariant under changes in the parameters μ and σ, it suffices to take $\mu = 0$ and $\sigma = 1$. Hence we restate the problem as follows.

Let $X_1,X_2,\ldots,X_n$ be a sample from the population having probability density function $f(x) = (2/\pi)^{\frac{1}{2}}\exp(-x^2/2)$ for $x \geq 0$, zero otherwise. It is readily seen that our problem is the evaluation of

$$G(t)=G_n(t) = (2/\pi)^{n/2}\int_\Gamma \exp(-\Sigma x_i^2/2)\,\Pi\, dx_i \tag{1}$$

where the domain $\Gamma = \{x_1,\ldots,x_n \mid \sqrt{n}\,\bar{x}/s \geq t \text{ and } x_i \geq 0 \text{ for } i = i,\ldots,n\}$ with $\bar{x} = \Sigma x_i/n$ and $(n-1)s^2 = \Sigma(x_i-\bar{x})^2$.

We make the transformations (T1), (T2), (T3), defined below, in succession.

(T1) Order $x_1,x_2,\ldots,x_n$ in ascending order of magnitude $0 \leq u_1 \leq u_2 \ldots \leq u_n$, say, then the ordered transformation $(T1): (x_1,\ldots,x_n) \to (u_1\ldots,u_n)$ has Jacobian equal to n! The domain Γ is changed to Γ_1, $\Gamma_1 = \{u_1,u_2,\ldots,u_n \mid \sqrt{n}\,\bar{u}/\left(\Sigma(u_i-\bar{u})^2\right)^{\frac{1}{2}} \geq t/\sqrt{n-1},\ u_1 \geq 0,\ u_i-u_{i-1} \geq 0$ for $i = 2,\ldots,n\}$.

(T2) This is followed by the orthogonal Helmert's transformation $(T2): (u_1,\ldots,u_n) \to (y_1,\ldots,y_n)$ having Jacobian equal to 1 and defined by $y_1 = (u_1 + u_2 + \ldots + u_n)/\sqrt{n} = \sqrt{n}\,\bar{u}$ and $y_i = \left(-(n-i+1)u_{i-1}+u_i+\ldots+u_n\right)/\left((n-i+2)(n-i+1)\right)^{\frac{1}{2}}$ for $i = 2,\ldots,n$; say $Y = HU$ in obvious matrix notation, so that $U = H^TY$. The domain Γ_1 is now transformed to Γ_2 given by $\Gamma_2: y_1/(y_2^2+\ldots+y_n^2)^{\frac{1}{2}} \geq t/\sqrt{n-1}$, $y_1 \geq \sqrt{n-1}\, y_2$ and $(n-i+2)^{\frac{1}{2}} y_i \geq (n-i)^{\frac{1}{2}} y_{i+1}$ for $i = 2,\ldots,n-1$; and $y_n \geq 0$.

(T3) Finally the generalized polar transformation (T3): $(y_1,\ldots,y_n) \to (r,\theta_1,\ldots,\theta_{n-1})$ with $0 \leq r < \infty$, $0 \leq \theta_i \leq \pi$ for $i = 1,\ldots,\theta_{n-2}$; and $0 \leq \theta_{n-1} \leq 2\pi$ given by $y_1 = r\cos\theta_1$, $y_i = r\cos\theta_i \prod_{j=1}^{i-1} \sin\theta_j$ for $i = 2,\ldots,n-1$ and $y_n = r\sin\theta_1 \ldots \sin\theta_{n-1}$ has the Jacobian $r^{n-1} J(\theta_1,\ldots,\theta_{n-1})$ where we define

$$(2) \qquad J(\theta_i,\ldots,\theta_{n-1}) = (\sin\theta_i)^{n-i-1}(\sin\theta_{i+1})^{n-i-2} \ldots \sin\theta_{n-2}\,.$$

The domain Γ_2 is transformed into $\Gamma_3 = \{D_1 \cap D_2,\ 0 \leq r < \infty\}$ with

$$(3) \qquad D_1 = \{(\theta_1,\ldots,\theta_{n-1}) \mid 0 \leq \theta_i \leq \pi/2; \quad i = 1,\ldots,n-1\}$$

and

$$D_2 = \{(\theta_1,\ldots,\theta_{n-1}) \mid \cot\theta_1 \geq t/\sqrt{n-1},\quad \cos\theta_2 \leq (\cot\theta_1)/\sqrt{n-1}\,,$$

$$(n-i)^{\frac{1}{2}}\cos\theta_{i+1} \leq (n-i+2)^{\frac{1}{2}}\cot\theta_i \quad \text{for } i = 2,\ldots,n-2; \theta_{n-1} \leq \pi/3\}.$$

In arriving at Γ_3, we see that straightforward substitution in $y_1 \geq \sqrt{n-1}\, y_2$ and $(n-i+2)^{\frac{1}{2}} y_i \geq (n-i)^{\frac{1}{2}} y_{i+1}$ for $i = 2,\ldots,n-1$ and $y_n \geq 0$ yields $\sqrt{n-1}\sin\theta_1\cos\theta_2 \leq \cos\theta_1$, $(n-i)^{\frac{1}{2}}\sin\theta_1 \ldots \sin\theta_i \cos\theta_{i+1} \leq (n-i+2)^{\frac{1}{2}}\sin\theta_1 \ldots \sin\theta_{i-1}\cos\theta_i$ for $i = 2,\ldots,n-2$ and $\sin\theta_1 \ldots \sin\theta_{n-1} \geq 0$. The last inequality with $0 \leq \theta_i \leq \pi$, for $i = 1,\ldots,n-2$ gives $0 \leq \theta_{n-1} \leq \pi$, and the other inequalities give $0 \leq \theta_i \leq \pi/2$ for $i = 1,\ldots,n-1$.

These three transformations applied to (1) result in

$$(2/\pi)^{-n/2} G_n(t) = \int_\Gamma \exp(-\tfrac{1}{2}\Sigma x_i^2)\,\Pi dx_i = n!\int_{\Gamma_1} \exp(-\tfrac{1}{2}U^T U)\,\Pi du_i$$

$$= n!\int_{\Gamma_2} \exp(-\tfrac{1}{2}Y^T H H^T Y)\,\Pi dy_i$$

$$= n!\int_{\Gamma_3} e^{-r^2/2} r^{n-1} (\sin\theta_1)^{n-2} J(\theta_2,\ldots,\theta_{n-1}) dr \Pi d\theta_i, \text{ since } HH^T = I$$

$$= n!\left[\int_{r=0}^{\infty} r^{n-1} e^{-r^2/2} dr\right] \int_{D_1 \cap D_2} (\sin\theta_1)^{n-2} J(\theta_2,\ldots,\theta_{n-1}) \Pi d\theta_i$$

$$= n! 2^{\frac{n-2}{2}} \Gamma(n/2) \int_{D_1 \cap D_2} (\sin\theta_1)^{n-2} J(\theta_2,\ldots,\theta_{n-1}) \Pi d\theta_i$$

$$(4) \quad G_n(t) = n! \pi^{-n/2} 2^{n-1} \Gamma(n/2) \int_{D_1 \cap D_2} (\sin\theta_1)^{n-2} J(\theta_2,\ldots,\theta_{n-1}) \prod_{i=1}^{n-1} d\theta_i .$$

We remark at this point that the domain Γ, in terms of n-dimensional geometry, can be shown to be a part of a cone with apex at the origin, axis as the equiangular line to the co-ordinate axes and semi-vertical angle arc cot $(t/\sqrt{n-1})$ contained in the positive orthant. For $t \geq n-1$ it can be shown that the half-cone lies wholly inside the orthant and in this case the evaluation of the integral in (1) is easily accomplished. This case was studied by Hotelling (1961) for exponential parent population. For smaller values of t, the problem however is non-trivial and far from easy. In general, depending on the value of t, the following configurations arise: the cone intersects the (n-k)-faces of the orthant, but does not intersect lower-dimensional faces for $k = 1,\ldots,n-1$. In what follows we will not use any geometrical intuition or argument, so we will not enter into any detailed geometrical discourse.

EVALUATION OF $G_n(t)$ AND $f_n(t)$ FOR $t \geq [(n-1)(n-2)/2]^{\frac{1}{2}}$

Before we proceed to the evaluation of (4), we state a lemma which will be needed later. A proof is given by Ali, Hoq and Templeton (1977).

Lemma: $$(n-j+2)! \int_{C_j} J(\theta_j,\ldots,\theta_{n-1}) \prod_{i=j}^{n-1} d\theta_i = 2\pi^{(n-j+1)/2} [\Gamma((n-j+1)/2)]^{-1}$$

where $C_j = \{(\theta_j,\ldots,\theta_{n-1}) \mid 0 \leq \theta_i \leq \text{arc cot}\left((n-i)^{\frac{1}{2}}(n-i+2)^{-\frac{1}{2}}\cos\theta_{i+1}\right)$ for $2 \leq j \leq i \leq n-2$ and $0 \leq \theta_{n-1} \leq \pi/3\}$.

<u>Case 1.</u> The range of t considered is $t \geq n-1$. In this case $D_1 \cap D_2$ appearing in the expression for $G_n(t)$ in (4) can after some elementary

calculations be shown to be

$$D_1 \cap D_2 = \{\theta_1 \mid 0 \le \theta_1 \le \text{arc cot}\,(t/\sqrt{n-1})\} \cap C_2$$

where C_2 is defined in the above lemma. Hence from (4) we obtain

$$[n!\pi^{-n/2}2^{n-1}\Gamma(n/2)]^{-1}G_n(t) = \int_{\theta_1=0}^{\text{arc cot}\,(t/\sqrt{n-1})} (\sin\theta_1)^{n-2}d\theta_1 \cdot$$

$$\int_{C_2} J(\theta_2,\ldots,\theta_{n-1}) \prod_{i=2}^{n-1} d\theta_i$$

$$= \frac{2\pi^{(n-1)/2}}{n!\Gamma[(n-1)/2]} \int_{\theta_1=0}^{\text{arc cot}\,(t/\sqrt{n-1})} (\sin\theta_1)^{n-2}d\theta_1,$$

using the lemma

$$= \frac{2\pi^{(n-1)/2}}{n!\Gamma[(n-1)/2]} \int_{u=0}^{(n-1)/(n-1+t^2)} (\tfrac{1}{2})u^{\left(\frac{n-1}{2}-1\right)}(1-u)^{\frac{1}{2}-1}du$$

setting $u = \sin^2\theta_1$

whence

(5) $$G_n(t) = 2^{n-1}I_{\left(\frac{n-1}{n-1+t^2}\right)}\left(\frac{n-1}{2},\frac{1}{2}\right)$$

where $I_x(a,b) = [B(a,b)]^{-1}\int_0^x u^{a-1}(1-u)^{b-1}du$ is the incomplete Beta function. The corresponding density $f_n(t) = -dG_n(t)/dt$ for $t \ge n-1$ is given by

(6) $$f_n(t) = \frac{2^n(n-1)^{(n-1)/2}}{B\left(\frac{n-1}{2},\frac{1}{2}\right)\left(n-1+t^2\right)^{n/2}} \triangleq r_n(t)\ .$$

It may be noted that the rth moment of T about the origin will exist if $\int_{n-1}^{\infty} t^r f_n(t)dt$, which is the contribution to that moment from Case 1, is finite. From formula (6), $f_n(t) = K_n(n-1+t^2)^{-n/2} = O(t^{-n})$ for large t, where K_n is a constant independent of t. It follows that the rth moment will exist for $r \le n-2$ only.

Case 2. The range of t considered is $((n-1)(n-2)/2)^{1/2} \le t \le n-1$, so that $\text{arc cot}\,(n-1)^{1/2} \le \theta_1 \le \text{arc cot}\,(t/\sqrt{n-1})$. In this case the domain $D_1 \cap D_2$ appearing in (4) and defined in (3) can be written as $A_1 \cup A_2$, where A_1 and A_2 are disjoint sets defined below, with C_2, C_3 defined as in the above lemma.

$$A_1 = \{\theta_1 | 0 \le \theta_1 \le \text{arc cot}\,\sqrt{n-1}\} \cap C_2$$

$$\begin{aligned} A_2 = \{\theta_1, \theta_2 | &\text{arc cot}\,\sqrt{n-1} \le \theta_1 \le \text{arc cot}\,(t/\sqrt{n-1}), \\ &\text{arc cos}\,(\cot\theta_1/\sqrt{n-1}) \le \theta_2 \le \text{arc cot}\,(((n-2)/n)^{1/2} \cos\theta_3)\} \cap C_3 \\ = \{\theta_1 | &\text{arc cot}\,\sqrt{n-1} \le \theta_1 \le \text{arc cot}\,(t/\sqrt{n-1})\} \cap C_2 \\ -\{\theta_1, \theta_2 | &\text{arc cot}\,\sqrt{n-1} \le \theta_1 \le \text{arc cot}\,(t/\sqrt{n-1}), \\ &0 \le \theta_2 \le \text{arc cos}\,(\cot\theta_1/\sqrt{n-1})\} \cap C_3 \end{aligned}$$

so that

$$\begin{aligned} D_1 \cap D_2 &= A_1 \cup A_2 \\ &= \{\theta_1 | 0 \le \theta_1 \le \text{arc cot}\,(t/\sqrt{n-1})\} \cap C_2 \\ &\quad - \{\theta_1, \theta_2 | \text{arc cot}\,\sqrt{n-1} \le \theta_1 \le \text{arc cot}\,(t/\sqrt{n-1}), \\ &\quad 0 \le \theta_2 \le \text{arc cos}\,(\cot\theta_1/\sqrt{n-1})\} \cap C_3 . \end{aligned}$$

To obtain $G_n(t)$ in Case 2, we evaluate first

$$C_n(t) = \int_{\theta_1 = \text{arc cot}\,\sqrt{n-1}}^{\text{arc cot}\,(t/\sqrt{n-1})} \int_{\theta_2 = 0}^{\text{arc cos}\,(\cot\theta_1/\sqrt{n-1})} (\sin\theta_1)^{n-2} (\sin\theta_2)^{n-3} d\theta_1 d\theta_2 .$$

Setting $u = (n-1)^{-1/2} \cot\theta_1$ and $v = \cos\theta_2$, we have

(7) $$C_n(t) = \sqrt{n-1} \int_{u=t/(n-1)}^{1} \int_{v=u}^{1} \left[1+(n-1)u^2\right]^{-n/2} \left[1-v^2\right]^{+(n-4)/2} du\, dv$$

(8) $$G_n(t) = 2^{n-1} I_{\left(\frac{n-1}{n-1+t^2}\right)} \left(\frac{n-1}{2}, \frac{1}{2}\right) - n! \frac{2^{n-1}}{\pi^{n/2}} \Gamma(n/2) \left[\int_{C_3} J(\theta_3, ..\theta_{n-1}) \Pi d\theta_i\right] C_n(t)$$

$$= 2^{n-1} I_{\left(\frac{n-1}{n-1+t^2}\right)} \left(\frac{n-1}{2}, \frac{1}{2}\right) - \frac{n(n-2)2^{n-1}}{\pi} C_n(t), \text{ using the Lemma.}$$

$C_n(t)$, and hence $G_n(t)$ in Case 2, are difficult to evaluate in closed form for general values of n. The density $f_n(t)$ for Case 2 can however be expressed simply, using $r_n(t)$ defined above, as

(9) $$f_n(t) = r_n(t) \left[1 - (n/2) I_{1-\left(\frac{t}{n-1}\right)^2} \left(\frac{n-2}{2}, \frac{1}{2}\right)\right].$$

EVALUATION OF $G_n(t)$ AND $f_n(t)$ FOR $n = 2,3,4$.

For $n = 2,3,4$, formulas (5), (6), (7), (8) and (9) can be evaluated explicitly with the following results:

For $n = 2$, we have a truncated Cauchy distribution;

$$G_2(t) = (4/\pi) \text{arc cot}\, t, \text{ for } t \geq 1,$$

$$f_2(t) = (4/\pi)(1+t^2)^{-1}, \text{ for } t \geq 1.$$

For $n = 3$,

$$G_3(t) = 4\left(1 - \frac{t}{(2+t^2)^{\frac{1}{2}}}\right) - \frac{6}{\pi} \text{arc cos}\left(\frac{t^2-1}{3}\right) + \frac{12t\, \text{arc cos}\,(t/2)}{\pi(2+t^2)^{\frac{1}{2}}}, \text{ for } 1 \leq t \leq 2$$

$$= 4\left[1 - t/(2+t^2)^{\frac{1}{2}}\right], \quad \text{for } t \geq 2,$$

$$f_3(t) = \frac{24}{\pi} \cdot \frac{(\pi/3) - \text{arc cos}\,(t/2)}{(2+t^2)^{3/2}}\ , \quad 1 \le t \le 2\ ,$$

$$= 8/(2+t^2)^{3/2}, \quad t \ge 2\ .$$

For $n = 4$,

$$G_4(t) = \frac{16}{\pi}\left[\text{arc tan}\,(t/\sqrt{3}) + \frac{\sqrt{3}(2+t)}{3+t^2} - \frac{\pi}{6} - \frac{2}{\sqrt{3}}\right], \quad \sqrt{3} \le t \le 3\ ,$$

$$= \frac{16}{\pi}\left[\text{arc cot}\,(t/\sqrt{3}) - \frac{\sqrt{3}t}{3+t^2}\right]\ , \quad t \ge 3\ ,$$

$$f_4(t) = -\frac{dG_4(t)}{dt} = \frac{32\sqrt{3}\,(2t-3)}{\pi(3+t^2)^2}, \quad \sqrt{3} \le t \le 3\ ,$$

$$= \frac{96\sqrt{3}}{\pi(3+t^2)^2}\ , \quad t \ge 3\ .$$

From the above densities we can compute the first moment $E_3(T)$ for $n = 3$. Using the substitution $t = 2\cos\theta$, we find

$$E_3(T) = \int_1^\infty t\, f_3(t)\, dt = (4\sqrt{6}/\pi)F((2/3)^{\frac{1}{2}}, \pi/3) = 3.701\ ,$$

where $F(k,\phi) = \int_0^\phi (1 - k^2\sin^2\theta)^{-\frac{1}{2}}\, d\theta$ is an incomplete elliptic integral of the first kind.

COMPARISON OF TAILS OF DISTRIBUTION OF STUDENT'S RATIO FOR GAUSSIAN, HALF-GAUSSIAN, AND EXPONENTIAL PARENT DISTRIBUTIONS

For Gaussian parent, we have [Abramowitz and Stegun, formula 26.7.1]

$$2G(t) = P\left[|T| \ge t\right] = I_x\left(\frac{\nu}{2}, \frac{1}{2}\right)$$

where $x = \nu/(\nu+t^2)$ and $\nu = n-1$.

For half-Gaussian parent, in Case 1 ($t \ge n-1$), we have

$$G_n(t) = 2^{n-1} I_x\left(\frac{\nu}{2}, \frac{1}{2}\right)$$

with x and ν defined as above. Therefore the half-Gaussian parent leads to a T-distribution with upper tail exactly 2^n times as heavy as the upper tail of the T-distribution with Gaussian parent (2^{n-1} times as heavy as both tails).

Still in Case 1, we may compare tails for Gaussian and exponential parents for small sample sizes using an asymptotic formula for large t [Abramowitz and Stegun, formula 26.7.7]. We have $G_n(t) \approx a_\nu/t^\nu + b_\nu/t^{\nu+1}$, where $\nu = n-1$ and a_ν, b_ν are given constants. Results given in the following table show that for sample sizes $n = 2(1)6$ and $t \ge n-1$, the half-Gaussian parent gives the heaviest upper tail and the Gaussian the lightest, with the exponential between the other two.

TABLE 1

Upper Tails for Distribution of T-statistic with Gaussian, Half-Gaussian and Exponential Parent Distributions.

Table shows $G_n(t) = P[T \ge t]$ for sample size n and $t \ge n-1$.

n	Range	Exponential (exact)	Half-Gaussian (asymptotic, large t)	Gaussian (asymptotic, large t)
2	$t \ge 1$	t^{-1}	$1.2732\,t^{-1}$	$0.3183\,t^{-1}$
3	$t \ge 2$	$2.4184\,t^{-2}$	$3.9928\,t^{-2} + 0.4144\,t^{-3}$	$0.4991\,t^{-2} + 0.0518\,t^{-3}$
4	$t \ge 3$	$8.1621\,t^{-3}$	$17.7504\,t^{-3} - 0.7360\,t^{-4}$	$1.1094\,t^{-3} - 0.0460\,t^{-4}$
5	$t \ge 3$	$33.8981\,t^{-4}$	$99.011\,t^{-4} - 88.19\,t^{-5}$	$3.0941\,t^{-4} - 2.756\,t^{-5}$
6	$t \ge 5$	$163.4749\,t^{-5}$	$636.7\,t^{-5} - 899.2\,t^{-6}$	$9.948\,t^{-5} - 14.05\,t^{-6}$

Similar comparisons for Case 2 are more complicated, and have not been attempted.

The exact ratio of the densities for T with half-Gaussian and with exponential parent, $f_n^{HG}(t)$ and $f_n^{EXP}(t)$ respectively, has been found in Case 1, and is given by

$$f_n^{EXP}(t)/f_n^{HG}(t) = n^{-n/2}\pi^{(n-1)/2}\Gamma\left(\frac{n+1}{2}\right)[1 + (n-1)/t^2]^{n/2} \quad \text{for } t \ge n-1 .$$

This ratio is a decreasing function of t for given n. Setting $t = n-1$ and using a Stirling expansion for the Gamma function gives

$$f_n^{EXP}(n-1)/f_n^{HG}(n-1) = \pi^{\frac{1}{2}}(2e/\pi)^{-(n-1)/2}[1 + (6n)^{-1} + O(n^{-2})] .$$

Computer programs for computing $G_n(t)$ and $f_n(t)$, for half-Gaussian parent and also for exponential parent, have been prepared and are available on request.

ACKNOWLEDGEMENTS

The financial support of the National Research Council of Canada is acknowledged. We also wish to thank Gertrude Ip for computer programming, and the anonymous referee of another paper for the comment that led to this paper.

REFERENCES

Abramowitz M. and Stegun I.A. (1964): Handbook of Mathematical Functions with Formulas, Graphs and Mathematical Tables. U.S. Government Printing Office, 1964, (also reprinted by Dover, New York, 1965). (For correction to formula 26.7.1, see Mathematics of Computation 19 (1965), 705.)

Ali M.M. and Richards W.A. (1975): On a conjecture concerning the common content of an N-cube and a diagonal cylinder. Ann. Inst. Stat. Math. 27 (1975), 281-287.

Ali M.M., Hoq A.K.M.S. and Templeton J.G.C. (1977): Common content of a regular hypersimplex and a concentric hypersphere. Working Paper 77-005 issued by the Department of Industrial Engineering, University of Toronto, Toronto, Canada 1977.

Govindarajulu Z. and Leslie R.T. (1970): Annotated bibliography on robustness studies. Technical Report 7 issued by the Department of Statistics, University of Kentucky, Lexington, Kentucky, U.S.A. 1970.

Hoq A.K.M.S., Ali M.M. and Templeton J.G.C. (1977): Distribution of Student's ratio based on exponential population. Working Paper 77-001 issued by the Department of Industrial Engineering, University of Toronto, Toronto, Canada 1977.

Hotelling H. (1961): The behavior of some standard statistical tests under non-standard conditions. In: Proc. of the Fourth Berkeley Symposium on Mathematical Statistics and Probability, Berkeley 1960. University of California Press, Berkeley and Los Angeles 1961.

Posten H.O. and Hatch L.O. (1966): Robustness of the Student-procedures: a survey. Research Report 24 issued by the Department of Statistics, University of Connecticut, Storrs, Connecticut, U.S.A. 1966.

Sansing R.C. (1976): The t-statistic for a double exponential distribution. SIAM J. Appl. Math. 31 (1976), 634-645.

Shenton L.R. and Bowman K.O. (1977): A new algorithm for summing divergent series, Part 3: Applications. J. Comput. Appl. Math. 3 (1977), 35-51.

A.K.M. Sirajul Hoq

Department of Mathematics
College of Science
University of Mosul
Mosul, Iraq

Mir M. Ali

Department of Mathematics
University of Western Ontario
London, Ontario N6A 5B9, Canada

James G.C. Templeton

Department of Industrial Engineering
University of Toronto
Toronto M5S 1A4, Canada

LEAST SQUARES AND MAXIMUM LIKELIHOOD ESTIMATION OF FUNCTIONAL RELATIONS

Hans-Peter Höschel

Berlin

ABSTRACT

The problem considered is that of estimating a functional relation among real variables given observed random variate values if the covariance of observations is assumed to be known.

The model stated in the present paper contains a certain class of functional relations which include simple explicit and implicit ones and, moreover, those formed of quite arbitrary surfaces (manifolds). It is established the equivalence of the method of maximum likelihood under normality and that of a generalized least squares method. Generalized least squares estimators are contained almost surely in the set of solutions of a certain nonlinear system of equations which is in principle accessible to iterative numerical methods.

INTRODUCTION

The fitting of a surface when the observed variables are subject to error is a statistical problem with a long history. Papers on nonlinear functional relations have appeared rarely, some of them may be found in Dolby (1972), Dolby and Lipton (1972) and Villegas (1969). It seems that there arise some difficulties if one wants to apply the method given there to problems with more than two variables or to implicit relations.

In the present paper we will propose another approach. We will start with a slightly modified notion of a functional relation (FUR) including arbitrary differentiable manifolds which are of practical interest.

Then we will regard the connection between maximum likelihood estimators (MLE) and generalized least squares estimators (GLSE).

Secondly, it seems that up till now the question of identifiability neither for linear nor nonlinear relations has been investigated. Some FUR has always been stated for investigations including a rather large set of parameters. But what happens if with positive probability the estimators which has been obtained for a concrete model determines more than one of the surfaces of the model ? Such a situation seems to be likely undesirable. Moreover, we do not know how many points we will need at least to identify a surface of the model.

We will obtain a nonlinear system of equations which gives GLSE and is in principle directly accessible to iterative numerical methods for arbitrary dimensions of the observed variates.

It is a pleasurable duty for the author to thank O.Bunke, R.Sulanke, B.Martin and P.Wintgen for helpful discussions.

FUNCTIONAL RELATIONS

The initial point for the investigation of a FUR in a concrete practical problem always is some causal-effect-structure which influences the observed variates directly or indirectly and causes a dependence of a certain typ between them.

Any FUR includes at first some structural model $\mathcal{F}$ which describes our knowledge or conjecture about the type of dependence between the variates. $\mathcal{F}$ is a known class of submanifolds in R^p. We always assume that the model is an adequate one, that the "true" relation is contained in the structural model $\mathcal{F}$ (cp. Humak (1977), p.9).

Furthermore the FUR includes an observational model. In this paper we will consider a simple form only. Modifications which include, for instance, replicated observations or transformations from the "structural space" into the"observational space" will be excluded. An example for such more complicated models is the observations of planets. An appropriate structural model in this case is the Kepler model including all ellipses having one focus in the sun. But we will assume only that there is a fixed but unknown "true" manifold $F \in \mathcal{F}$ and n fixed but unknown points $\mu_i \in F$, $i = 1, \dots, n$, lying on this manifold. Then we are given n observations $y_i \in R^p$

of these points which may be correlated. In this sense we have a design (column vector) $\mu = (\mu_1, \ldots, \mu_n) \in F^n$ and an observation (vector) $y = (y_1, \ldots, y_n) \in R^{pn}$ with

$$y = \mu + \varepsilon, \quad \mu \in \mathcal{F}^{(n)}$$

where we have

1. The parameter set $\mathcal{F}^{(n)}$ for μ

$\mathcal{F}^{(n)}$ is the set of $\mu \in R^{pn}$ for which exists some manifold F in $\mathcal{F}$ with $\mu \in F^n$ and there is no other manifold in $\mathcal{F}$ with this property.

Hence $\mathcal{F}^{(n)}$ is the set of designs which determine uniquely a manifold in $\mathcal{F}$, socalled identifying designs. (Remark: $\mathcal{F}^{(n)}$ might be a void set.)

2. The observational error ε

$\varepsilon = (\varepsilon_1, \ldots, \varepsilon_n)$ is a random vector with zero mean and regular covariance Ω : $E\varepsilon = 0 \in R^{pn}$, $C(\varepsilon) = \Omega \in \mathfrak{M}^{>}_{pn}$.

3. The estimation problem

The problem is to estimate the parameter function

$$F(\mu) := \{ F \in \mathcal{F} : \mu \in F^n \}$$

which by definition is uniquely determined if $\mu \in \mathcal{F}^{(n)}$.

MLE AND GLSE

The connection between MLE and GLSE was considered by Dolby (1972). Now we will repeat very briefly another approach to this problem which was considered by Nussbaum (1976) and Höschel (1978).

Suppose we are given a family of identifiable probability distributions depending on a parameter $\xi \in \Xi$ and that the likelihood function is $l_y(\xi)$ given a fixed observation $y \in \mathcal{Y}$. Furthermore, we have a fixed measurable map

$$g : (\Xi, \mathcal{B}_\Xi) \longrightarrow (\Theta, \mathcal{B}_\Theta).$$

Then $\vartheta = g(\xi) \in \Theta$ is a parameter function depending on the parameter $\xi \in \Xi$.

Definition of MLE: Any solution $\hat{\vartheta} \in \Theta$ of the maximization problem

$$\sup_{\xi \in g^{-1}(\hat{\vartheta})} l_y(\xi) = \sup_{\vartheta \in g(\Xi)} \sup_{\xi \in g^{-1}(\vartheta)} l_y(\xi)$$

is called ML-solution (MLS) for $\vartheta = g(\xi)$ on $\Theta = g(\Xi)$. Measurable solutions $\hat{\vartheta} : (\mathcal{Y}, \mathcal{B}_y) \longrightarrow (\Theta, \mathcal{B}_\Theta)$ are called MLE.⌋

For MLE an important invariance principle is valid (cp. Zehna(1966))
<u>Invariance principle for MLE:</u> If $\hat{\xi} = \hat{\xi}(y)$ is a MLS on Ξ then $\hat{\vartheta}(y) := g(\hat{\xi}(y))$ is a MLS for $\vartheta \in \Theta$, where

$$\ell_y(\hat{\xi}) = \sup_{\vartheta \in g(\Xi)} \sup_{\xi \in g^{-1}(\vartheta)} \ell_y(\xi)$$
$$= \sup_{\xi \in g^{-1}(\hat{\vartheta})} \ell_y(\xi) \ \lrcorner$$

(Remark: The spaces regarded do not need to be Euclidian ones as often required (cp. e.g. Zacks (1971), 5.1) and g does not need to be bijective.)

Now let us consider a FUR under normal error distribution. The unknown parameter is $\mu \in \mathcal{F}^{(n)} =: \Xi$. The parameter function to estimate is $g(\xi) := F(\mu)$. The likelihood function is:

$$\ell_y(\mu) \propto \exp - | y - \mu|^2_{\Omega^{-1}} / 2 \ , \quad \mu \in \mathcal{F}^{(n)} .$$

Hence we obtain as the criterion for the determination of the MLS on $\mathcal{F}^{(n)}$ the functional

$$k_y(\mu) := | y - \mu |^2_{\Omega^{-1}}$$

which is nothing but a generalized least squares criterion.
<u>Definition of GLSE:</u> Let $\mathcal{M} \subset \Xi$ be a fixed subset. A generalized least squares solution (GLS) for μ on $\mathcal{M}$ is any $\hat{\mu} \in \mathcal{M}$ with

$$k_y(\hat{\mu}) = \min_{\mu \in \mathcal{M}} k_y(\mu) .$$

$\hat{\vartheta} = g(\hat{\mu})$ is called GLS for $\vartheta = g(\mu)$ on $\mathcal{M}$. Measurable solutions will be called GLSE. $\lrcorner$

Therefore we have proved the
<u>Equivalence statement:</u> MLS under normality and GLS are equivalent. $\lrcorner$

As shown in Joshi (1976) the principles of sufficiency and conditionality imply the principle of likelihood. Together with GLSE we therefore have a "chain" of inference principles which are quite tightly connected.

BASIC ASSUMPTIONS FOR FUR

Now let us give the suppositions which will restrict the FUR regarded in this paper. We will assume that $\mathcal{F}$ is a disjoint union $\bigcup_{b \in B} F_b$ of structural manifolds $F_b \subset R^p$ where B is some parameter set. We suppose that B and each F_b are closed smooth differentiable submanifolds in Euclidian spaces without singularities. For many practical applications it is sufficient to assume that the deformation of F_b after changing $b \in B$ differentially is an infinitesimal one. All this can be described explicitly in a functional form by the following assumption.

<u>1.Assumption:</u> $\mathcal{F}$ is a closed differentiable and smooth submanifold in R^p without singularities. The parameter manifold is described locally by

$$B := \{ b \in R^s : h(b) = 0 \} , \quad h : R^s \longrightarrow R^r , r < s ,$$

and we have locally as well

$$F_b = \{ u \in R^p : f(b, u) = 0 \} , \quad f(b, .): R^p \longrightarrow R^q, q < p ,$$

where the functional matrices $D_u f = [D_{u_1} f, \dots, D_{u_p} f]$ and $D_b h$ are of full row rank.⌋

Remarks: The functions f, h and the numbers q and r may change for different neighborhoods. Since $\mathcal{F}$ is a differentiable manifold there are only countable many of such local representations. The assumption is nothing but the statement that F_b is a fibre of the fibre bundle $\mathcal{F}$ with basis B (cp. Dieudonné (1976), 16.12./13.). $\mathcal{F}$ is closed if there is only a finite number of such local representations which are used in the assumption (cp. Dieudonné(1976), 16.8.9.1°.). If there is one such representation only then we obtain a usual FUR.

<u>Example:</u> In the plane we obtain all parabolas and pairs of parallel straight lines from

$$f(b, u) := \beta_0 + \beta_1 x + \beta_2 y + \beta_3 x^2 + 2\beta_4 xy + \beta_5 y^2 ,$$

and

$$h(b) := \beta_3 \beta_5 - \beta_4^2 ,$$

where $u := (x,y)$, $b = (\beta_0, \dots, \beta_5)$. Therefore B is a closed 5-dimensional submanifold in R^6 .⌋

Now we will introduce an assumption which is related to the identifiability problem. The following situation might occur.

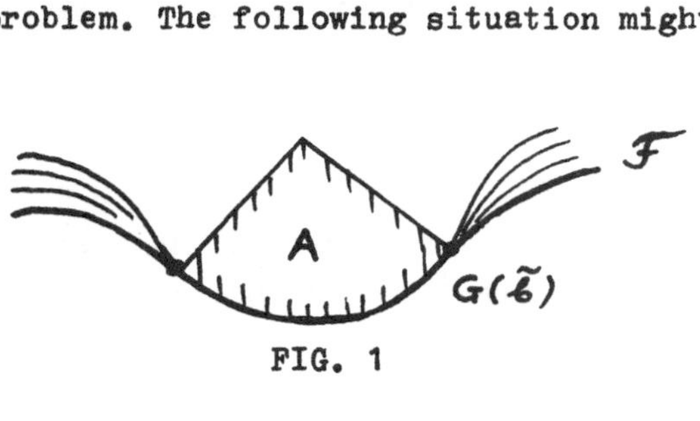

FIG. 1

Here the intersection $G(\tilde{b}) := F_{b_1} \cap F_{b_2}$, $\tilde{b} = (b_1, b_2)$, $b_1 \neq b_2$, is a subset contained in a submanifold with the same dimension as F_{b_1} or F_{b_2} . On $G(\tilde{b})$ the tangent spaces of F_{b_1} and F_{b_2} are equal.

For all observations y_i contained in the marked area A the distance to all manifolds from $\mathcal{F}$ is the same. Thus for all observations $y \in R^{pn}$ which are in the set $A^n \subset R^{pn}$ we can not estimate uniquely manifolds from $\mathcal{F}$ using the least squares principle. However, A^n has a positive Lebesgue measure. Therefore we want that such situations occur sufficiently seldom.

To describe this we will need the denotations

$$B^{(2)} := \{ \tilde{b} = (b_1, b_2) \in B \times B: \; b_1 \neq b_2 \} ,$$
$$H := g_{(n)}^{-1}(0) = \cup_{\tilde{b} \in B^{(2)}} G(\tilde{b})$$

where

$$g_{(n)}(\tilde{b}, v) := \begin{bmatrix} f(b_1, u_1) \\ f(b_2, u_1) \\ \vdots \end{bmatrix} : \quad B^{(2)} \times R^{pn} \longrightarrow R^{2qn}, \; v := (u_1, \dots, u_n).$$

If we write $f_{ij} = f(b_i, u_j)$ then

$$D g_{(n)} = \begin{bmatrix} D_b f_{11} & ,0 & ,D_u f_{11} & ,0 & ,\dots \\ 0 & ,D_b f_{21} & ,D_u f_{21} & ,0 & ,\dots \\ D_b f_{12} & ,0 & ,0 & ,D_u f_{12} & ,\dots \\ 0 & ,D_b f_{22} & ,0 & ,D_u f_{22} & ,\dots \\ \cdot & \cdot & \cdot & \cdot & \dots \end{bmatrix} .$$

Of course, a lower bound for the minimal rank of $D g_{(n)}$ we obtain from the 1.assumption by $r(D g_{(n)}) \geqq r(D f_{(n)}) = nq$. From the example given in figure 1 we could see that we have full nonidentifiability if the rank of $D g_{(n)}$ is minimal. Nearly that situation we also are if $r(D g_{(n)})$ is "nearly" nq. We will describe the fact that such situations are "sufficiently seldom" by the following

<u>2.assumption:</u> The set S_n of points $(\tilde{b}, v) \in B^{(2)} \times R^{pn}$ for which $r(D g_{(n)}) \leqq nq + 2(s - r)$ is contained in a countable union of C^1-submanifolds in R^{2s+np} with a codimension not less than $nq + 2(s - r)$.⌋

<u>Examples:</u> 1. A linear relation in R^p is given by

$$0 = Mu , \quad M \in \mathfrak{M}_{q \times p} , \quad r(M) = q , \quad u \in R^p .$$

The matrices M form a C^∞-submanifold B in R^{pq} with a dimension $q(p-q)$ (cp. Dieudonné (1976), 16.11.9.). Therefore " $b_1 \neq b_2$ " means that the row spaces of M_1 and M_2 are different. Hence we have $r(D g_{(1)}) = r[M_1', M_2']' \geqq q + 1$. Thus for $n > 2q(p-q)$ we have

$$r(D g_{(n)}) \geqq n(q+1) > nq + 2q(p-q)$$

and therefore $S_n = \phi$.⌋

2. Take all unit balls in R^3 with centers on a nonsingular C^∞-curve.

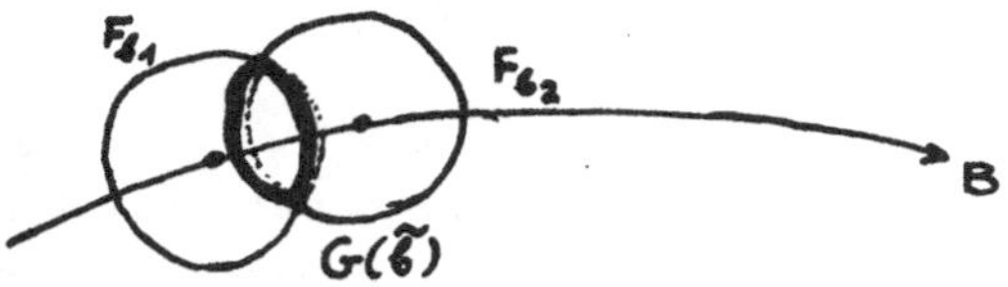

Then again for $b_1 \neq b_2$ the tangent spaces of F_{b_1} and F_{b_2} on $G(\tilde{b})$ are different.⌋

3. Let us consider all quadratic FUR of the form

$$0 = z + \alpha + \beta x^2 , \quad u := (x, z) , \quad b := (\alpha, \beta).$$

(More general FUR will be treated in the following example 4.).
Then $D g_{(n)}$ is of the form

$$\begin{bmatrix} 1 & x_1^2 & 0 & 0 & 2\beta_1 x_1 & 1 & 0 & \cdots \\ 0 & 0 & 1 & x_1^2 & 2\beta_2 x_2 & 1 & 0 & \cdots \\ \cdot & \cdot & \cdot & \cdot & \cdot & \cdot & \cdot & \cdots \end{bmatrix}$$

where $\tilde{b} = (\alpha_1, \beta_1, \alpha_2, \beta_2)$. If this matrix drops rank by t then there are at least t points $u_{i_\nu} = (x_{i_\nu}, z_{i_\nu})$, $\nu = 1, \ldots, t$, for which

$$\tilde{f}_{i_\nu}(\tilde{b}, v) := x_{i_\nu}(\beta_1 - \beta_2) = 0 .$$

The derivational vector of $\tilde{f}_{i_\nu}$ is:

$$D\tilde{f}_{i_\nu} = [\, 0 \,, x_{2i_\nu} \,, 0 \,, -x_{i_\nu} \,, \ldots \,, \beta_1 - \beta_2 , \, 0 \,, \ldots \,]$$

Then S_n is the set of zeros of $g_{(n)}$ and $\tilde{f}_{i_\nu}$. Thus it is a subset of the zeros of that system of equations where f_{2i_ν} is replaced by $\tilde{f}_{i_\nu}$. Then we can see that on the open subset in $B^{(2)} \times R^{pn}$ where $\beta_1 \neq \beta_2$ we have a regular functional matrix for the corresponding system of equations. Therefore the solutions are in a C^1-submanifold of R^{2s+np} with a codimension $2nq$. Hence for $nq = n > 2s = 4$ we have obtained the desired statement. On the other hand for $\beta_1 = \beta_2$ we have $\alpha_1 \neq \alpha_2$ on $B^{(2)}$. But then we also have $G(\tilde{b}) = \phi$ and therefore $S_n \subset V$, $\dim V < n(p-q)$.⌋

4. There is some hope to prove the 2.assumption for a large class $\mathcal{F}$ by application of the multijet transversality theorem (cp.Golubitsky and Guillemin (1973),p.57) since in this theorem it is stated that for an open and dense subset of functions the desired assertion is valid if the subset of functional matrices under consideration is itself a submanifold.

But in fact we are able to prove the assumption for an important class of structural models. Let us consider in generalization of example 3.

$$f(b,u) := -z + \bar{f}(b,x)$$

where $z \in R^q$, $x \in R^{p-q}$, $u = (x, z)$, $\bar{f} = (\bar{f}_1, \dots, \bar{f}_q)$.
Then we obtain a decomposition of R^{p-q} into such point sets where at least one of the partial derivatives of arbitrary order $D_u v(\bar{f}_j(b_1, x) - \bar{f}_j(b_2, x))$, v as multiindex, does not vanish and, secondly, that set where all this derivatives are zero.

If n is large enough we can obtain by a finite process similar to that of the foregoing example on sets of the first type a functional matrix of the desired rank since $\bar{f}_j(b_1, x) - \bar{f}_j(b_2, x)$ does not depend on z. But on the second set the functions $\bar{f}_j(b_1, .)$ and $\bar{f}_j(b_2, .)$ would agree on an open neighborhood of each point of the set. This set, however, will be void if we suppose the following assumption which is valid very often in practice.

Assumption: For each point $\tilde{b} = (b_1, b_2) \in B^{(2)}$ there is no point in $G(\tilde{b})$ such that the manifolds F_{b_1} and F_{b_2} agree on an open neighborhood of this point.⌋

GENERALIZED LEAST SQUARES SOLUTIONS

A GLS for μ on $\mathcal{F}^{(n)}$ we obtain from

$$k_y(\hat{\mu}) = \min_{\mu \in \mathcal{F}^{(n)}} k_y(\mu).$$

Then a GLS for $F \in \mathcal{F}$ is given by $\hat{F} = F(\hat{\mu})$.

However, $\mathcal{F}^{(n)}$ is a relatively complicated set and moreover we do not know whether it is closed or not to assure the existence of a solution $\hat{\mu}$. Therefore we have to try another approach. Let us replace the set $\mathcal{F}^{(n)}$ by the enlarged set

$$\mathcal{F}^n := \bigcup_{F \in \mathcal{F}} F^n .$$

Then we can formulate the enlarged GLS problem

$$k_y(\hat{\mu}) = \min_{\mu \in \mathcal{F}^n} k_y(\mu).$$

This problem is somewhat easier to handle, since $\mathcal{F}^n$ can be described explicitly locally by $f_{(n)}^{-1}(0) \cap h^{-1}(0)$.

On the other hand any GLS $\hat{\mu}$ on $\mathcal{F}^n$ might be nonidentifiable for the structural paramter $F \in \mathcal{F}$. But the happy-end is given by the following assertion.

Lemma: Let $\mathcal{F}$ be a structural model of the type described above. If the number of observations is larger than two times the (maximal) dimension of the parameter manifold then the set of observations $y \in R^{pn}$ which give nonidentifying GLS on $\mathcal{F}^n$ has Lebesgue measure zero in R^{pn} :

$$\lambda^{pn}\{y : \hat{\mu}(y) \in \mathcal{F}^n \setminus \mathcal{F}^{(n)}\} = 0 \;, \quad n > 2(s-r) .$$

Proof: A GLS on $\mathcal{F}^n$ is a solution of

$$k_y(\hat{\mu}) = \min_{0=h(b),\, 0=f_{(n)}(b,\mu)} k_y(\mu) .$$

Now $\mathcal{F}^n$ is closed since $\mathcal{F}$ is closed. Therefore the minimum is attained and at least one GLS $\hat{\mu}$ exists.

Now let us assume for a moment that there is only one local representation of $\mathcal{F}$. Since $\mathcal{F}$ is a smooth manifold without singularities we can apply the method of Lagrangian multipliers (cp. Dieudonné (1976), 16.20.) to obtain a system of equations the solutions of which contain all GLS. The resulting system of equations we will call normal equations in remembering the usual regression theory. Then we need the derivative

$$D k_y = [D_b k , D_v k] = [0' , 2(y-\mu)'\Omega^{-1}]$$

with $0 \in R^s$ and A' as the transposed of a matrix A and

$$D f_{(n)} = [D_b f_{(n)}, D_v f_{(n)}] = \begin{bmatrix} D_b f(b, u_1) & , D_u f(b, u_1), & \dots, & 0 \\ \vdots & \vdots & \ddots & \vdots \\ D_b f(b, u_n) , & 0 & ,\dots, & D_u f(b, u_n) \end{bmatrix}$$

Using the Lagrangian multipliers $(\lambda, \varkappa) \in R^{qn+r}$ we obtain the normal equation

$$0 = D' k_y + D' f_{(n)} \cdot (\lambda, \varkappa) \in R^{s+np}$$

or explicitly

$$\begin{aligned} y &= \mu + 2\Omega D_v' f_{(n)} \cdot \lambda =: l(b, \mu, \lambda) , \\ 0 &= D_b' f_{(n)} \cdot \lambda + D_b' h \cdot \varkappa , \\ 0 &= f_{(n)}(b, \mu) , \\ 0 &= h(b) . \end{aligned}$$

We assume now that there is any solution $\hat{\mu}$ which is nonidentfying. Thus there are at least two different $b_1 \neq b_2$ in B with $\hat{\mu}(y) \in G(\tilde{b})$. Or conversely, the set Y of "nonidentifyable"

y's we obtain from

$$Y = \tilde{\ell}(K) \ , \quad K := H \times R^{qn} \ ,$$

$$\tilde{\ell}(\tilde{\ell}, \mu, \lambda) := \ell(P_1(\tilde{\ell}), \mu, \lambda),$$

where P_1 is the projection of $\tilde{\ell}$ onto its first part ℓ_1 and H is defined above.

If we could show that H is contained in a countable union of C^1-manifolds V_ν each with a dimension less than $n(p-q)$ then

$$\dim(V_\nu \times R^{qn}) = \dim V_\nu + \dim R^{qn} < nq \ .$$

But then $\tilde{\ell}(V_\nu \times R^{qn})$ is of measure zero in R^{pn} since $\tilde{\ell}$ is continuously differentiable (cp. Sulanke and Wintgen (1972), III.1. Satz 2.). Therefore Y would be contained in a countable union of zero sets and thus be a zero set as well.

Now

$$H = g_{(n)}^{-1}(0) = g_{(n)}^{-1}(0) \cap (S_n \cup \bar{S}_n)$$

where S_n is defined in the 2.assumption and $\bar{S}_n$ is the set complement in $B^{(2)} \times R^{pn}$. $\bar{S}_n$ is open since the rank function is upper semicontinuous. For each point of $\bar{S}_n$ there is an open neighborhood $U(\mu)$ for which there are at least $qn + 2(s-r) + 1$ coordinate functions of $g_{(n)}$ the vector of which will be denoted by $g^{(\mu)}$. Then we have on U

$$r(D\, g^{(\mu)}) = qn + 2(s-r) + 1 \ .$$

Since $\mathcal{F}^n$ is a differentiable manifold and $\bar{S}_n \subset \mathcal{F}^n$ there is a countable covering $\{U_\nu\}_{1,2,\dots}$ of $\bar{S}_n$ by such neighborhoods $U_\nu := U(\mu_\nu)$. We denote by g_ν the restriction of $g^{(\mu)}$ on U_ν. Then we have

$$V_\nu := g_\nu^{-1}(0) \cap U_\nu \ .$$

Then V_ν is a C^1-submanifold of $B^{(2)} \times R^{pn}$ with a dimension

$$\dim V_\nu = 2(s-r) + pn - (qn + 2(s-r) + 1)$$

$$= n(p-q) - 1 \ ,$$

(cp. Dieudonné (1976), 16.8.9. and 16.8.7.). Thus we have

$$g_{(n)}^{-1}(0) \cap \bar{S}_n \subset \bigcup_\nu (g_\nu^{-1}(0) \cap U_\nu) = \bigcup_\nu V_\nu \ .$$

Secondly by assumption 2. we also have $S_n \subset V_0$ with $\dim V_0$ less than $n(p-q)$ which completes the proof.

Let us show briefly how to tackle the general case when $\mathcal{F}$ has more than one local representation. Since $\mathcal{F}$ is a C^∞-manifold there is at least a countable number of such local representations. However, for each of the resulting "local" normal equations the above assertion is valid and, thereafter, we have to take the countable union of the corresponding zero sets of y's . For the number $s - r$ we have to take its maximum over all local representations.⌋

<u>Corollary:</u> If $n > 2(s - r)$ then the set of nonidentifying designs is of Lebesgue measure zero in R^{pn} .

Proof: This set is a subset of $\mathcal{F}^n \setminus \mathcal{F}^{(n)}$ namely that one for which the observation y and the corresponding GLS coincide.⌋

(Remark: This assertion was shown for polynomial models with $q = 1$ directly in Höschel and Zink (1977).).

We do not know up till now whether the GLS $\hat{\mu}$ are uniquely determined or not. But if they are then from the uniqueness of $s(\hat{\mu})$ and the then following obvious continuity of $s(\hat{\mu})$ in y we obtain the GLS as GLSE.

But we are able to prove this for the simplest special case only, namely for linear FUR with a covariance of the form $\Omega = \Lambda \otimes \Sigma$ (cp. Höschel (1978), after 4.1.). It can be assumed that this will be true also for a large class of FUR which have been considered in this paper.

Moreover the question which structural models of practical interest satisfy the 2.assumption requires further investigations.

REFERENCES

Dieudonné J. (1976): Grundzüge der modernen Analysis 3, VEB Deutscher Verlag der Wissenschaften, Berlin 1976, (Transl. of the 2nd french edition).

Dolby G.R. (1972): Generalized Least Squares and Maximum Likelihood Estimation of Non-linear Functional Relationships, J.Royal Statist.Soc.B 34(1972), 393-400.

Dolby G.R. and Lipton S. (1972): Maximum likelihood estimation of the general nonlinear functional relationship with replicated observations and correlated errors, Biometrika 59(1972),121-129.

Golubitsky M. and Guillemin V. (1973): Stable mappings and their singularities, Springer-Verlag, New York Inc., 1973.

Höschel H.-P. (1978): Generalized Least Squares Estimators of Linear Functional Relations with known Error Covariance, Math.Operationsforsch.und Statistik, Ser. Statistics 2(1978)No.1,

Höschel H.-P. and Zink T. (1977): Consequences of the Veronese Transformation from Algebraic Geometry for the Design of Polynomial Functional Relations and Polynomial Regression with Errors in Variables, Paper presented at the 3rd International Summer School on Problems of Model Choice and Parameter Estimation in Regression Analysis, Mühlhausen, 1977.

Humak K.M.S. (1977): Statistische Methoden der Modellbildung, Band I, Statistische Inferenz für lineare Parameter, Akademie-Verlag, Berlin 1977.

Joshi V.M. (1976): A Note on Birnbaums Theory of the Likelihood Principle, Journ.Amer.Stat.Ass. 71(1976), 345-346.

Nussbaum M. (1976): Maximum Likelihood and Least Squares Estimation of Linear Functional Relationships, Math.Operationsforsch.u.Statist. 7(1976),23-49.

Sulanke R. and Wintgen P. (1972): Differentialgeometrie und Faserbündel, VEB Deutscher Verlag der Wissenschaften, Berlin 1972.

Villegas C. (1969): On the Least Squares Estimation of Non-linear Relations, Ann.Math.Stat. 40(1969), 462-466.

Zacks S. (1971): The Theory of Statistical Inference, Wiley & Sons, Inc. New York, 1971.

Zehna P.W. (1966): Invariance of Maximum Likelihood Estimation, Ann.Math.Statist. 37(1966), 755.

Sektion Mathematik
Humboldt-Universität
PSF 1297
DDR-1086 Berlin
G.D.R.

О НЕКОТОРЫХ СВОЙСТВАХ ГАУССОВСКИХ ПРОЦЕССОВ, СВЯЗАННЫХ С НЕПАРАМЕТРИЧЕСКОЙ ОЦЕНКОЙ ПЛОТНОСТИ РАСПРЕДЕЛЕНИЯ

Ш.А.Хашимов

Ташкент, СССР

АННОТАЦИЯ

Изучаются предельные поведения гауссовских процессов, связанных с непараметрической оценкой неизвестной плотности распределения.

Пусть $X_1, X_2, \ldots, X_n$ – независимые одинаково распределенные случайные величины с функцией распределения $F(x)$ и плотностью распределения $f(x)$. Пусть $f(x) \in L_2(a,b)$ и $\{\varphi_j(x)\}_0^\infty$ – полная ортонормированная система функций на $L_2(a,b)$, $-\infty \le a < b \le \infty$. Тогда

$$f(x) = \sum_{j=0}^{\infty} a_j \varphi_j(x),$$

где

$$a_j = \int_a^b \varphi_j(x) f(x)\, dx, \qquad \sum_{j=0}^{\infty} a_j^2 < \infty$$

Предположим, что

(I) $|\varphi_j(x)| < C < \infty$ (C не зависит от x и j)

В качестве оценки неизвестной плотности распределения $f(x)$ рассмотрим статистику вида

(2) $$f_n(x) = \sum_{j=0}^{q(n)} a_{jn} \varphi_j(x),$$

где

$$a_{jn} = \int_a^b \varphi_j(x)\, dF_n(x), \quad q(n) \to \infty$$

и $F_n(x)$ - эмпирическая функция распределения выборки $X_1, X_2, \ldots, X_n$.

Статистика вида (2) впервые была предложена Ченцовым Н.Н. Дальнейшие свойства $f_n(x)$ были изучены многими авторами.

Из теоремы 3 работы Комлоша, Майора, Тушнади следует, что с вероятностью единица

$$(3) \qquad F_n(x) = F(x) + \frac{1}{n} B_n(F(x)) + \frac{\ln n}{n} Z_n(x),$$

где $B_n(x)$ - последовательность броуновских мостов и

$\sup_x |Z_n(x)| = Z_n$ удовлетворяет соотношению

$P\{Z_n > \lambda + y\} < \kappa n^{-\mu y}$, где λ, κ, μ - положительные абсолютные константы.

Тогда используя (2) и (3), находим

$$f_n(x) - f(x) = b_n(x) + \frac{1}{\sqrt{n}} U_n(x) + \delta_n(x),$$

где

$$b_n(x) = M f_n(x) - f(x),$$

$$U_n(x) = \sum_{j=0}^{q(n)} \varphi_j(x) \int_a^b \varphi_j'(x)\, B_n(F(x))\, dx,$$

$$\delta_n(x) = \frac{\ln n}{n} \sum_{j=0}^{q(n)} \varphi_j(x) \int_a^b \varphi_j'(x)\, Z_n(x)\, dx.$$

Некоторые свойства величины $\delta_n(x)$ изучены в работе Хашимова Ш.А. Там же доказано, что $U_n(x)$ является гауссовским процессом с $M U_n(x) = 0$ и корреляционной функцией

$$r_{q(n)}(x,y) = \int_a^b K_{q(n)}(x,t)\, K_{q(n)}(y,t)\, f(t)\, dt - M f_n(x)\, M f_n(y),$$

где

$$\mathcal{K}_{q(n)}(x,t)=\sum_{j=0}^{q(n)} \varphi_j(x)\,\varphi_j(t),$$

$$\mathcal{K}_{q(n)}(y,t)=\sum_{j=0}^{q(n)} \varphi_j(y)\,\varphi_j(t).$$

Обозначим

$$\sup_{a<x<b} \mathcal{K}^2_{q(n)}(x,t)=\Psi^2_{q(n)}(t), \quad t\in(a,b)$$

$$\int_a^b \Psi_{q(n)}(t)\, f(t)\,dt = I_{q(n)},$$

$$p_{1n}(\varepsilon)=\sup_{|x-y|\le\varepsilon} \left\{\sum_{j=0}^{q(n)} \left|\varphi_j(x)-\varphi_j(y)\right|\right\}^{1/2},$$

$$p_{2n}(\varepsilon)=\int_0^\varepsilon \left\{\ln(1/u)\right\}^{1/2} d\,p_{1n}(\varepsilon),$$

$$\gamma_n(\varepsilon)=\max\left\{\frac{p_{1n}(\varepsilon)}{I^{1/2}_{q(n)}},\ \frac{p_{2n}(\varepsilon)}{I^{1/2}_{q(n)}\{\ln q(n)\}^{1/2}}\right\}$$

Пусть $\mathcal{J}$ некоторый интервал из (a,b).
Справедливы следующие
Теорема I.

Пусть выполнено (I), $\sup_{x\in\mathcal{J}} f(x)=C_0<\infty$. Пусть, кроме того $\lim_{n\to\infty}\gamma_n(\varepsilon)=A(\varepsilon)$ и $A(\varepsilon)\to 0$ при $\varepsilon\to 0$

Тогда, если

$$\lim_{n\to\infty} \frac{q(n)^{(1+a)} \ln q(n)}{I_{q(n)}} = 0 , \quad \frac{1}{4} < a < 1,$$

то по вероятности

$$\lim_{n\to\infty} \left(I_{q(n)} \ln q(n)\right)^{-1/2} \sup_{x\in \mathcal{Y}} U_n(x) = \frac{\sqrt{2}}{4} ,$$

$$\lim_{n\to\infty} \left(I_{q(n)} \ln q(n)\right)^{-1/2} \inf_{x\in \mathcal{Y}} U_n(x) = -\frac{\sqrt{2}}{4} .$$

Теорема 2.

Пусть выполнено (I) и $\sup_{x\in \mathcal{Y}} f(x) = C_0 < \infty$. Пусть, кроме того $\lim_{n\to\infty} \gamma_n(\varepsilon) = A(\varepsilon)$ и $A(\varepsilon) \to 0$ при $\varepsilon \to 0$. Тогда по вероятности

$$\lim_{n\to\infty} \left(I_{q(n)} \ln q(n)\right)^{-1/2} \sup_{x\in \mathcal{Y}} |U_n(x)| \le \sqrt{2\varepsilon_0} ,$$

где ε_0 - некоторое положительное число.

Теорема 3.

Пусть выполнено (I) и $\sup_{x\in \mathcal{Y}} f(x) = C_0 < \infty$. Пусть, кроме того $\lim_{n\to\infty} \gamma_n(\varepsilon) = A(\varepsilon)$, $A(\varepsilon) \to 0$ при $\varepsilon \to 0$

Тогда с вероятностью единица

$$\lim_{n\to\infty} \left\{I_{q(n)} \ln q(n)\right\}^{-1/2} \sup_{x\in \mathcal{Y}} |U_n(x)| \le \sqrt{2\varepsilon_1} ,$$

где ε_1 - такое положительное число, что $\sum_n q^{-\varepsilon_1}(n) < 0$.

Отметим, что подобные результаты для оценок неизвестной плотности вероятности, построенных при помощи "весовых" функций были получены Сильверманом.

ЛИТЕРАТУРА

Ченцов Н.Н.: Оценка неизвестной плотности распределения по наблюдениям. ДАН СССР, т.147 (1962), № I, 45–48.

Komlos I., Major P. and Tusnady G. An approfimation of partial sums of independent rv's and the sample df I. Z. Wahrscheinlichkeitstheorie und Verw. Gebiete 32 (1975), 111-131

Хашимов Ш.А.: О некоторых свойствах непараметрических оценок плотности вероятности. ДАН УзССР (1977), № II

Silverman B.W. On a Gaussian process related to multivariate probability density estimation. Math. Proc. Cambridge Phil. Soc. 80 (1976), N 1, 135-144.

Институт математики им.В,И.Романовского АН УзССР
Ташкент, туп. Астрономический, II

ON THE COMPACTNESS OF VECTOR-VALUED TRANSITION MEASURES

Albrecht Irle

Münster

ABSTRACT

A space of vector-valued transition measures is introduced and certain weak topologies are defined on this space. Some compactness results are given which also generalize and unify known results for the scalar case. An application in the context of decision theory is given.

1. INTRODUCTION

In a statistical decision problem with sample space $(\Omega,\mathcal{A})$ and terminal decision space $(S,\mathcal{S})$ decision functions for the statistician can be considered as mappings from Ω to the set of all probability measures on $(S,\mathcal{S})$ satisfying some measurability condition, i.e. as transition measures. Results on compactness of sets of transition measures thus are useful for existence theorems in general decision theory and appear scatteredly in papers on this subject, see f.e. LeCam (1964), Farrell (1967).

In this paper we introduce vector-valued transition measures, i.e. mappings from Ω to a certain space of vector-valued measures on $(S,\mathcal{S})$, thus generalizing the notion of transition measure. Vector-valued transition measures may be useful in describing the behaviour of certain physical systems, see part 5(ii).
In part 2 we state some technicalities on "generalized" transition measures using the well-known relationship between transition measures

and appropriate linear operators.

Part 3 and part 4 contain the basic statements on compactness in the space of vector-valued transition measures supplied with a certain weak topology, which also generalize and unify known statements for the case of (scalar-valued) transition measures. Part 3 is concerned with the situation that $(S,\mathcal{S})$ is a metrizable topological space with the σ-algebra of Borel sets, whereas in part 4 the case of an arbitrary measurable space $(S,\mathcal{S})$ is considered. Part 5 finally contains some remarks on applying the foregoing results.

2. PRELIMINARIES

Let $(\Omega,\mathcal{A},\mu)$ be a complete probability space; thus there exists a lifting λ on $L_\infty = L_\infty(\Omega,\mathcal{A},\mu)$ in the sense of Ionescu Tulcea, A. and C. (1969), ch. III. Mappings on Ω which agree μ-a.e. shall be identified in the following, but we will not distinguish in our notations between equivalence classes and their representatives.

Let V be an ordered normed vector space with positive cone V^+ and norm $\| \ \|_V$. We define the set $\mathcal{T}$ of *generalized transition measures* - compare f.e. LeCam (1964) - as the set of all mappings δ from Ω to V^* (the space of linear bounded functionals on V) satisfying the following *conditions:*

(i) the mapping $f*\delta: \Omega \longrightarrow \mathbb{R}$ defined by $f*\delta(\omega) = \delta(\omega)(f)$ belongs to L_∞ for all $f\in V$.

(ii) $\|\delta\|_T := \sup\{\|f*\delta\|_\infty : \|f\|_V \le 1\} < \infty$, where $\| \ \|_\infty$ denotes the norm in L_∞.

(iii) $\delta(\omega)$ belongs to $(V^*)^+$ for μ-a.a. ω, i.e. for μ-a.a. ω $\delta(\omega)(f) \ge 0$ holds for all $f\in V^+$.

A weak topology is defined on $\mathcal{T}$ in the following way:

(2.1) For $f\in V$, $g\in L_1 = L_1(\Omega,\mathcal{A},\mu)$, $\varepsilon > 0$, and $\delta_o \in \mathcal{T}$ consider the set $\{\delta : \delta\in\mathcal{T}, |\int (f*\delta - f*\delta_o) g \, d\mu| < \varepsilon\}$. Then there exists a unique topology on $\mathcal{T}$ for which the system of all sets of the above form is a subbase of the neighbourhood system for any δ_o. It is obvious that $\mathcal{T}$ is Hausdorff for this topology if there exists a countable subset of V which separates the points of $(V^*)^+$. In the following we will consider $\mathcal{T}$ as a topological space supplied with the above defined topology.

2.2 *Lemma:* *(i)* A set $T' \subset T$ is relatively compact if and only if $\{\|\delta\|_T : \delta \in T'\}$ is bounded.
(ii) If for any $\alpha > 0$ V has a countable subset which is convergence - determining for the topology $\sigma(\{v^* : v^* \in (V^*)^+, \|v^*\| \leq \alpha\}, V)$, then any relatively compact subset of T is sequentially compact. (Here and in the following $\sigma(E,F)$ denotes the weakest topology on E such that all elements of $F \subset \mathbb{R}^E$ are continuous.)

Proof: Let P be the set of all positive linear bounded operators φ from V to L_∞ with $\|\varphi\|_P = \sup\{\|\varphi(f)\|_\infty : \|f\|_V \leq 1\}$; consider P as topological space supplied with the weak operator topology defined analogously to *(2.1)*. It is a standard fact that a subset $P' \subset P$ is relatively compact for this topology if and only if P' is norm-bounded. Any element of T can obviously be considered as element of P, and so the proof of *(i)* is reduced to the following statement: for any $\varphi \in P$ there exists $\delta \in T$ such that $\varphi(f) = f * \delta$ μ-a.s. for any $f \in V$. Let λ be a lifting on L_∞ and for $\omega \in \Omega$ let π_ω denote the projection with respect to ω from $\mathbb{R}^\Omega$ to $\mathbb{R}$. For $\varphi \in P$ the mapping $\pi_\omega \circ \lambda \circ \varphi$ belongs to $(V^*)^+$ and it is easily seen that δ defined by $\delta(\omega) = \pi_\omega \circ \lambda \circ \varphi$ has the desired properties.
(ii) This follows by a diagonal argument from the fact, that any norm-bounded subset of L_∞ is sequentially $\sigma(L_\infty, L_1)$ - compact, see Nölle and Plachky (1967).

3. VECTOR-VALUED TRANSITION MEASURES ON A RADON SPACE

For a normal topological space S and a Banach space X let $V = C_{tb}(X)$ be the space of all totally bounded continuous functions from S to X. Then V^* is equal to the space of all additive set functions m on S with values in X^*, which are regular with bounded variation $|m|$ - i.e. $V^* = rba(X^*)$, see de Korvin and Easton (1971). Let $bca(X^*)$ denote the σ-additive set functions on S which have bounded variation. Let us remark that certain facts on vector measures which will be used in the following without further references can be found in Dinculeanu (1967). The following *assumptions* are made:

(i) X is a Banach lattice.

(ii) S is metrizable and has the Radon property, i.e. any $\mu \in bca(\mathbb{R})$ is tight, which means that for every $\varepsilon > 0$ there exists a compact subset K of S such that $|\mu|(S-K) \leq \varepsilon$.

We remark that any polish space S has the Radon property, but that there are non-Polish spaces with this property. The metrizability implies that $bca(X^*) \subset rba(X^*)$ holds. $C_{tb}(X)$ is a Banach lattice with generating positive cone $C_{tb}(X)^+$. We now define the set of *vector-valued transition measures:*

$$M := \{\delta : \delta \in T,\ \delta(\omega) \in bca(X^*)^+ \text{ for } \mu\text{-a.a. } \omega \in \Omega\}.$$

(3.1) Proposition: Let $m \in rba(X^*)$. Then m is σ-additive if and only if m_x is σ-additive for all $x \in X^+$, where m_x is the real-valued set function $m_x = m(\ )(x)$.

Proof: If m is σ-additive then obviously m_x is σ-additive for all $x \in X^+$, so it remains to prove the converse.

Assume m_x σ-additive for all $x \in X^+$; thus m_x is σ-additive for all $x \in X$ and by the Radon property of S m_x is tight for all $x \in X$. Let $\varepsilon > 0$ be given. Then there exist a finite family $(U_i)_{i \in I}$ of disjoint sets $U_i \in S$ and $x_i \in X$, $\|x_i\| \le 1$ for $i \in I$, with $|m|(S) \le |\sum_{i \in I} m(U_i)(x_i)| + \frac{\varepsilon}{2}$. By the tightness of the individual m_{x_i} there exist compact $K_i \subset S$ with $|m_{x_i}|(S - K_i) \le \frac{\varepsilon}{2|I|}$. Thus for $K := \bigcup_{i \in I} K_i$ - which is a compact subset of S, since I is finite:

$$|m|(S-K) = |m|(S) - |m|(K) \le \frac{\varepsilon}{2} + \sum_{i \in I} |m(U_i)(x_i)| - \sum_{i \in I} |m|(K \cap U_i)$$

$$\le \frac{\varepsilon}{2} + \sum_{i \in I} (|m(U_i)(x_i)| - |m(K \cap U_i)(x_i)|)$$

$$\le \frac{\varepsilon}{2} + \sum_{i \in I} |m_{x_i}|(U_i - (K \cap U_i)) \le \varepsilon.$$

Thus $|m|$ is tight. Furthermore $|m|$ is regular, since m is regular. From this and the well-known fact that any regular positive bounded additive set function on the Borel-σ-algebra of a compact space is σ-additive we infer that $|m|$ is σ-additive and thus that m is σ-additive.

So - in the situation considered here - weak* σ-additivity and σ-additivity are the same concepts. Let us remark that the Orlicz-Pettis theorem only yields that weak σ-additivity and σ-additivity are equivalent.

(3.2) Lemma: *(i)* For any compact $K \subset S$, $\delta \in T$ and $x \in X^+$ the mapping $I_K * \delta_x$ is measurable and ≥ 0 μ-a.s. .
(ii) For any $B \in S$, $\delta \in M$ and $x \in X^+$ the mapping $I_B * \delta_x$ is measurable and ≥ 0 μ-a.s..

Proof: *(i)* Let ρ be any metric for S inducing the given topology and for K compact let $K_n = \{s: \rho(s,t) < \frac{1}{n}$ for some $t \in K\}$. Then for every open $G \supset K$ we have $K_n \subset G$ for some n. It thus follows by regularity that for $\delta \in T$, $\omega \in \Omega$ $\lim_n |\delta(\omega)|(K_n - K) = 0$ holds. Choose real-valued continuous f_n with $0 \leq f_n \leq 1$, $f_n \equiv 1$ on K and $f_n \equiv 0$ on $S - K_n$. For any $x \in X^+$ it then follows $f_n x \in C_{tb}(X)^+$ and $I_K * \delta_x = \lim_{n \to \infty} f_n * \delta_x = \lim_{n \to \infty} (f_n x) * \delta$ is measurable and ≥ 0 μ-a.s. .
(ii) is proved analogously.

For $\delta \in M$, $x \in X^+$ and $g \in L_1$, $g \geq 0$ μ-a.s., define an element of $bca(\mathbb{R})^+$ by $\delta(g,x)(U) := \int (I_U * \delta_x) g \, d\mu$, $\delta(g,x)(S) \leq \|\delta\|_T \|g\|_1$.
In the following consider $bca(\mathbb{R})^+$ with the $\sigma(bca(\mathbb{R})^+, C_b(\mathbb{R}))$-topology, where $C_b(\mathbb{R})$ denotes the space of real valued bounded continuous functions on S.

(3.3) Theorem: Assume that X is separable. Then the following statements are *equivalent* for $M' \subset M$:

(i) M' is relatively compact.

(ii) $\{\delta(g,x): \delta \in M'\}$ is relatively compact in $bca(\mathbb{R})^+$ for any $g \in L_1, g \geq 0$, and $x \in X^+$.

If M' is norm-bounded, then (i) and (ii) are equivalent to (iii):

(iii) $\{\delta(I_\Omega, x): \delta \in M'\}$ is relatively compact in $bca(\mathbb{R})^+$ for any $x \in X^+$.

Proof: *(i)$\Longrightarrow$(ii)* This follows at once from the fact that the mapping $\delta \longrightarrow \delta(g,x)$ is a continuous mapping from M to $bca(\mathbb{R})^+$ for any $g \in L_1$, $g \geq 0$, $x \in X^+$.

(ii)$\Longrightarrow$(i): Assume that (ii) holds; then for any $x \in X$, $g \in L_1$ and $h \in C_b(\mathbb{R})$ the set $\{|\int ((xh) * \delta) g d\mu| : \delta \in M'\}$ is bounded. Since the linear space generated by $\{xh: x \in X, h \in C_b(\mathbb{R})\}$ is dense in $C_{tb}(X)$ for the uniform norm, the above boundedness holds for any $f \in C_{tb}(X)$ also. This implies that for any $f \in C_{tb}(X)$ the set $\{f * \delta : \delta \in M'\}$ is norm-bounded in L_∞ and by the uniform boundedness principle M' is norm-bounded. Having established this we will only use (iii) for the rest of the proof which will also show the equivalence of condition (iii) with (i) - resp. (ii) - for a norm-bounded M'.

Let $(\delta_\alpha)_\alpha$ be a net in M'. Since M' is norm-bounded, there are a subnet $(\delta_\beta)_\beta$ of $(\delta_\alpha)_\alpha$ and $\delta\in T$ with $\delta = \lim_\beta \delta_\beta$ - using (2.2). Let $x\in X^+$. We will show in the following that $\delta(\omega)_x$ is σ-additive for μ-a.a. ω. By (iii) we obtain a subnet $(\delta_\gamma)_\gamma$ of $(\delta_\beta)_\beta$ and $\nu\in bca(\mathbb{R})^+$ (depending on x in general) with $\nu = \lim_\gamma \delta_\gamma(I_\Omega, x)$. For any $h\in C_b(\mathbb{R})$ we have

$$\int h\, d\nu = \lim_\gamma \int h\, d\delta_\gamma(I_\Omega, x) = \lim_\gamma \int (x\cdot h)*\delta_\gamma d\mu$$
$$= \int (xh)*\delta d\mu = \int h*\delta_x\, d\mu.$$

Using (3.2) it follows that for any compact $K\subset S$ $\nu(K) = \int I_K*\delta_x\, d\mu$ and also $\nu(S-K) = \int I_{S-K}*\delta_x\, d\mu$ holds. Since ν is tight, there exists a non-decreasing sequence of compacts K_n with $\lim_n \nu(S-K_n) = 0$. Since $I_{S-K_n}*\delta_x \geq 0$ μ-a.s. for any n it follows that $\lim_n \delta(\omega)_x(S-K_n) = 0$ for μ-a.a.ω and thus $\delta(\omega)_x$ is tight for μ-a.a. ω. As in (3.1) this implies that $\delta(\omega)_x$ is σ-additive for μ-a.a.ω - the μ-zero set N_x of those ω for which $\delta(\omega)_x$ is not σ-additive depending on x in general. Now let $\{x_k: k\in \mathbb{N}\}$ be a dense subset of X^+, $N:= \bigcup_k N_{x_k}$ with $\mu(N) = 0$. Let $\omega\in\Omega-N$. For any $k\in\mathbb{N}$ there exists a non-decreasing sequence of compact $K_n^{(k)}$ with $\lim_n \delta(\omega)_{x_k}(S-K_n^{(k)}) = 0$. Let $K_n := \bigcup_{j\leq n}\bigcup_{k\leq n} K_j^{(k)}$: then K_n is compact and $\lim_{n\to\infty} \delta(\omega)_{x_k}(S-K_n) = 0$ for all $k\in\mathbb{N}$. Since $\{x_k: k\in\mathbb{N}\}$ is dense in X^+, this obviously implies $\lim_{n\to\infty} \delta(\omega)_x(S-K_n) = 0$ for all $x\in X^+$ and thus $\delta(\omega)_x$ is σ-additive for all $x\in X^+$. Using (3.1) we obtain that $\delta(\omega)$ is σ-additive for all $\omega\in\Omega$ -N which proves (i).

(3.4) *Corollary:* Assume that X and D are separable. Then the following statements are *equivalent* for $M'\subset M$:

(i) M' is relatively compact in M.

(ii) M' is sequentially compact in M.

Proof: *(i)$\Longrightarrow$(ii)* As in (2.1) it is obviously enough to show that for any $\alpha>0$ there exists a countable subset of $C_{tb}(X)$ which is convergence determining for the topology $\sigma(\{m: m\in bca(X^*)^+, \|m\|\leq\alpha\}, C_{tb}(X))$. Since D is separable there exists a countable convergence-determining subset $\{h_k: k\in\mathbb{N}\}$ of $C_b(\mathbb{R})$ for $\sigma(bca(\mathbb{R})^+, C_b(\mathbb{R}))$ - compare Parthasarathy (1967), p. 43-44. Let $\{x_k: k\in\mathbb{N}\}$ be a dense subset of X. It follows easily that $\{x_j h_k: j,k\in\mathbb{N}\}\subset C_{tb}(X)$ has the disired properties.

(ii) $\Longrightarrow$ *(i)*: If M' is sequentially compact in M, then for any $g \in L_1$, $g \geq 0$, $x \in X^+$ the set $\{\delta(g,x): \delta \in M'\}$ is sequentially compact in $bca(\mathbb{R})^+$ and thus relatively compact, since the topology $\sigma(bca(\mathbb{R})^+, C_b(\mathbb{R}))$ is metrizable. This implies (i) by theorem (3.3).

For a subset H of $bca(X^*)^+$ let $M(H) := \{\delta: \delta \in M, \delta(\omega) \in H$ for μ-a.a.$\omega\}$. If (Ω, A, μ) is non-atomic then obviously $\{\delta(I_\Omega, x): \delta \in M(H)\}$ contains the convex hull of $H_x := \{m_x : m \in H\}$ for any $x \in X$.

(3.5) Corollary: Let X be separable. For a convex subset H of $bca(X^*)^+$ are *equivalent:*

(i) $M(H)$ is relatively compact in M.

(ii) H_x is relatively compact in $bca(\mathbb{R})^+$ for any $x \in X^+$.

Proof: '*(i)* $\Longrightarrow$ *(ii)*' follows from (3.3). If (ii) holds, then $M(H)$ is norm-bounded and furthermore an easy application of the Hahn-Banach theorem yields that $\{\delta(I_\Omega, x): \delta \in M(H)\}$ is contained in the closure of H_x in $bca(\mathbb{R})^+$ for any $x \in X^+$. Then (3.3) implies (i).

(3.6) Proposition: Let X and D be separable, H a norm-bounded convex subset of $bca(X^*)^+$. Then $M(H)$ is closed in M *if and only if* H is closed in $bca(X^*)^+$.

Proof: It is obvious that H is closed if $M(H)$ is closed, so we will only prove the converse.

Let Q be the linear space generated by $\{x_j h_k: j,k \in \mathbb{N}\}$ - see the proof of (3.4) - and consider on $bca(X^*)$ the topology $\sigma(bca(X^*), Q)$. Then $\sigma(bca(X^*), Q)$ has the Hausdorff property and is locally convex. It easily seen that H is closed with respect to the topology $\sigma(bca(X^*)^+, Q)$. Let Q' be a countable dense subset in Q for the uniform norm. Consider $\delta \in M - M(H)$. Let $A = \{\omega : \delta(\omega) \notin H\}$. By the Hahn-Banach theorem applied with respect to $\sigma(bca(X^*)^+, Q)$ we obtain - for any $\omega \in A$ - $q_\omega \in Q$ with $\sup\{\int q_\omega dm : m \in H\} < \int q_\omega d\delta(\omega)$. Since H is norm-bounded we may assume $q_\omega \in Q'$ for all $\omega \in A$. Now for $A(n,q) := \{\omega : \sup\{\int q dm : m \in H\} \leq \frac{1}{n} + \int q d\delta(\omega)\}$ we have $A(n,q) \in A$ and $A = \bigcup_{n \in \mathbb{N}} \bigcup_{q \in Q'} A(n,q)$. Since $\delta \in M - M(H)$ this implies that there exist $n \in \mathbb{N}$, $q \in Q'$ with $\mu(A(n,q)) > 0$ and thus for any $\delta' \in M(H)$ $\int (q * \delta') I_{A(n,q)} d\mu \leq \frac{1}{n} \mu(A(n,q)) + \int (q * \delta) I_{A(n,q)} d\mu$, which proves that δ does not belong to the closure of $M(H)$.

4. VECTOR-VALUED TRANSITION MEASURES ON AN ARBITRARY MEASURABLE SPACE

For a measurable space $(S,\mathcal{S})$ and a Banach space X let $V:=\mathcal{B}(X)$ denote the space of all bounded totally measurable functions from S to X. Then V^* is equal to the space of all additive set functions on $\mathcal{S}$ with values in X^* which have finite variation $|m|$, i.e. $V^* = ba(X^*)$ - see Dinculeanu (1967), p.147. Assume that X is a Banach lattice. Then $\mathcal{B}(X)$ is a Banach lattice with generating positive cone $\mathcal{B}(X)^+$.

Again we define the set of *vector-valued transition measures* as

$$\mathcal{M}=\{\delta:\ \delta\in\mathcal{T},\ \delta(\omega)\in bca(X^*) \text{ for } \mu\text{-a.a. } \omega\in\Omega\}.$$

Let us consider for the following $bca(X^*)$ with the topology $\sigma(bca(X^*),\mathcal{B}(X))$.

(4.1) *Theorem:* For $H\subset bca(X^*)^+$ are *equivalent:*

(i) $\mathcal{M}(H)$ is relatively compact in $\mathcal{M}$.

(ii) H is relatively compact in $bca(X^*)^+$.

Proof: That (i) implies (ii) obvious, thus assume (ii); then H is norm-bounded, thus $\mathcal{M}(H)$ is norm-bounded. For a net $(\delta_\alpha)_\alpha$ in $\mathcal{M}(H)$ there exist a subnet $(\delta_\beta)_\beta$ and $\delta'\in\mathcal{T}$ with $\delta' = \lim_\beta \delta_\beta$ by (2.2).
Define $\delta\in\mathcal{T}$ by $\int f\, d\delta(\omega) := \lambda(f*\delta')(\omega)$, where λ is a lifting as in (2.2); then obviously $\delta = \lim_\beta \delta_\beta$ holds.
Now let $(A_n)_{n\in\mathbb{N}}$ be a sequence in $\mathcal{A}$, $A_n \downarrow \emptyset$. If $\varepsilon_n := \sup\{|m|(A_n): m\in H\}$, then for any $f\in\mathcal{B}(X)^+$, $\|f\|\le 1$, and any $m\in H$ $0\le\int f\cdot I_{A_n}\, dm \le \varepsilon_n$ holds.
This implies $0\le (f\cdot I_{A_n})*\delta' \le \varepsilon_n$ μ-a.s. and by the properties of a lifting:
$0\le\int f\cdot I_{A_n}\, d\delta(\omega)\le\varepsilon_n$ for all $\omega\in\Omega$ and thus $|\delta(\omega)|(A_n)\le\varepsilon_n$. Since H is compact in $bca(X^*)^+$, it follows that $\lim_n \varepsilon_n = 0$, see Batt (1974).
But this implies that $\delta(\omega)$ is σ-additive for all $\omega\in\Omega$.

5. APPLICATIONS

Let us scetch some possibilities for applications.

(i) Assume $X = \mathbb{R}$. Then $\mathcal{M}_1 := \{\delta: \delta\in\mathcal{M}, \delta(\omega)(S) = 1\}$ is a closed subset of $\mathcal{M}$ and can be interpreted as the set of decision functions in a statistical decision problem. Now any lower semi-continuous function

ℓ on S is the limit of an increasing sequence of continuous functions, thus for any $\ell \geq 0$ the mapping on $M\delta \longrightarrow \int(\ell*\delta)g\,d\mu$ is sequentially lower-semicontinuous for any $g \in L_1$. Now if $M' \subset M_1$ fulfills condition (3.3) (iii) and if S is separable, then M' is sequentially compact by (3.3) and (3.4). From this one can conclude the existence of Bayes- and minimax decision functions in rather general settings of decision theory.

(ii) Consider a physical system the behaviour of which can be described by a vector measure, see f.e. Kluvánek and Knowles (1975), ch. V,IX. Assuming further that this behaviour additionally depends on a random quantity ω (f.e. temperature) we may describe it by a transition measure $\delta(\omega)$. In the case that we only have approximate knowledge of the system we can also try to represent it by a certain set M' of transition measures. (3.3) and (4.1) then give conditions for the relative compactness of the set of possible behaviours of the system, which then yield f.e. the existence of extremal points and 'least favourable' behaviours.

REFERENCES

Batt, J. (1974): On weak compactness in spaces of vector-valued measures and Bochner-integrable functions in connection with the Radon-Nikodym property of Banach spaces. Rev. Roum. Math. Pures Appl. 19 (1974), 285-304.

de Korvin, A. and R.J. Easton (1971): Some representation theorems. Rocky Mountain J. Math. 1(1971), 561-573.

Dinculeanu, N. (1967): Vector Measures. Pergamon Press, Oxford 1967.

Farrell, R.H. (1967): Weak limits of sequences of Bayes procedures in estimation theory. Proc. Fifth Berkeley Symposium Math. Stat. Prob., Vol. 1(1967), 83-112.

Ionescu Tulcea, A. and C. Ionescu Tulcea (1969): Topics in the Theory of Lifting. Springer-Verlag, Berlin 1969.

Kluvánek, I. and G. Knowles (1975): Vector Measures and Control Systems. North-Holland Mathematics Studies 2o, Amsterdam 1975.

LeCam, L. (1964): Sufficiency and approximate sufficiency. Ann. Math. Stat. 35(1964), 1419-1456.

Nölle, G. and D. Plachky (1967): Zur schwachen Folgenkompaktheit von Testfunktionen. Z. Wahrscheinlichkeitstheorie verw. Geb. 8(1967), 182-184.

Parthasarathy, K.R. (1967): Probability Measures on Metric Spaces. Academic Press, New York 1967.

Institut für Mathematische Statistik der
Westfälischen Wilhelms-Universität Münster
Roxeler Straße 64
D-44 Münster
Germany

АНАЛИТИЧЕСКИ ИЗМЕРИМЫЕ СТРАТЕГИИ В УПРАВЛЯЕМЫХ СКАЧКООБРАЗНЫХ МАРКОВСКИХ ПРОЦЕССАХ

А.А.Юшкевич

Москва

АННОТАЦИЯ

Рассматриваются управляемые скачкообразные марковские процессы с борелевскими пространствами состояний и управлений и стратегиями, зависящими от прошлого. Основные результаты: 1.Переход от борелевских стратегий к более широкому классу А-стратегий не меняет оценки модели. 2.В моделях отрицательного типа существуют марковские равномерно ε-оптимальные стратегии. 3.В однородных моделях с дисконтированием существуют стационарные равномерно ε-оптимальные А-стратегии.

1.Рассматриваемая модель Z задается следующими элементами: 1/пространством состояний X - борелевским пространством; 2/пространством управлений A - борелевским пространством; 3/проекцией j - измеримым отображением A на X /слой $A(x)=j^{-1}(x)$ - это множество управлений, допустимых в состоянии x /; 4/мерой скачков $q(t,a,\Gamma)$ - мерой в пространстве X, измеримо зависящей от (t,a) ; 5/скоростью дохода $r(t,a)$ - измеримой функцией от (t,a) / $0\leq t<\infty$, $x\in X$, $a\in A$, $\Gamma\in\mathcal{B}(X)$, на оси t измеримыми считаются борелевские множества/. Предполагается, что соответствие $x\to A(x)$ допускает измеримый выбор, что мера $q(t,a,\cdot)$ сосредоточена на множестве $X\setminus j(a)$, и что

$$q(t,a,X)\leq K<\infty.$$

Если q не зависит от t и $r(t,a)=r(a)e^{-\alpha t}$, то модель называется однородной.

Пространство Ω всех траекторий - это совокупность непрерывных слева кусочно-постоянных и не имеющих точек накопления скачков функций $x(t)$ $(0<t<\infty)$ со значениями в X , пространство H всех историй - это множество всевозможных сужений $x_0^s(t)$ $(0<t\leq s)$ траекторий на конечных промежутках $(0,s]$. Траектория задается конечной или счетной последовательностью $[0\xi_0\tau_1\xi_1\tau_2\xi_2\ldots]$, история - конечным набором $[0\xi_0\tau_1\ldots\xi_n s]$, где $0<\tau_1<\tau_2<\ldots$ - моменты скачков, $\xi_0\neq\xi_1\neq\xi_2\neq\ldots$ - значения $x(t)$ между скачками. Вводя с помощью этих представлений измеримые структуры в Ω и H , превращаем их в борелевские пространства.

Стратегией называется измеримое отображение π пространства H в A , удовлетворяющее условию $\pi[x_0^s]\in A(x(s))$. Марковская стратегия задается измеримым отображением φ пространства $(0,\infty)\times X$ в A , удовлетворяющим условию $\varphi(t,x)\in A(x)$, по формуле $\varphi[x_0^s]=\varphi(s,x(s))$. Если $\varphi(t,x)=\varphi(x)$, марковская стратегия называется стационарной. Начальным распределением называется вероятностная мера μ на X . Формулы

/I/ $$P_\mu^\pi\{\xi_0\in\Gamma\}=\mu(\Gamma),$$

$$P_\mu^\pi\{\tau_{n+1}<u,\ \xi_{n+1}\in\Gamma \mid \xi_0\tau_1\xi_1\ldots\tau_n\xi_n\}=$$

/2/

$$=\begin{cases}\int\limits_{\tau_n}^{u} q(t,\pi[x_0^t],\Gamma)e^{-\int\limits_{\tau_n}^{t} q(s,\pi[x_0^s],X)ds}dt & \text{при } u>\tau_n, \\ 0 & \text{при } u\leq\tau_n\end{cases} \quad (\text{п.н.} P_\mu^\pi),$$

где $x_0^s=[0\xi_0\tau_1\ldots\xi_n]$, позволяют сопоставить произвольной паре (μ,π) распределение вероятностей P_μ^π в Ω. Соответствующее математическое ожидание обозначим E_μ^π /если $\mu(\Gamma)=1_\Gamma(x)$, то будем писать P_x^π и E_x^π /. Оценками стратегии π и модели Z называются функции

/3/ $$w(x,\pi)=E_x^\pi\int\limits_0^\infty r(t,\pi[x_0^t])dt, \quad v(x)=\sup_\pi w(x,\pi) \quad (x\in X).$$

Предполагается, что модель суммируема сверху или снизу, т.е. $w(x,\pi)$ не обращается в $\infty-\infty$ и либо $w(x,\pi)<+\infty$ при всех x и π, либо $w(x,\pi)>-\infty$ при всех x и π. Стратегия π называется ε-оптимальной, если $w(x,\pi)\geq v(x)-\varepsilon$ $(x\in X)$, и называется /п.н.μ/ ε-оптимальной, если это верно при почти всех x относительно меры μ.

Проблема заключается в отыскании марковских /если можно - стационарных/ ε-оптимальных стратегий.

2. В такой постановке борелевские модели изучены в работе Юшкевича/1977/. Пусть $v(t,x)=v_t(x)$ - оценка модели Z_t, полученной из Z сокращением промежутка управления с $(0,\infty)$ до (t,∞). Частичная редукция к модели с дискретным временем показала, что функция $v(t,x)$ - аналитическая, т.е. множества $\{(t,x): v(t,x)>C\}$ - аналитические. Этот факт вместе с использованием немарковских стратегий позволил вывести уравнение оптимальности и с его помощью установить следующую формулу: если $\|v_t\|$ ограничена на любом

ограниченном множестве значений t, то для любой марковской стратегии φ

$$v(x)-w(x,\varphi)=\int_0^\infty E_x^\varphi\{T_t v_t(x(t))-T_t^\varphi v_t(x(t))\}dt+æ, \tag{4}$$

где всегда $æ \geqslant 0$, в случае сходимости интеграла в /4/

$$æ=\lim_{t\to\infty} E_x^\varphi v_t(x(t))$$

и операторы T_t и T_t^φ определены формулами

$$T_t f(x)=\sup_{a\in A(x)}\left[r(t,a)+\int_X [f(y)-f(x)]\,q(t,a,dy)\right],$$

$$(x\in X).$$

$$T_t^\varphi f(x)=r(t,\varphi(t,x))+\int_X [f(y)-f(x)]\,q(t,\varphi(t,x),dy)$$

С помощью /4/ при определенных условиях было установлено существование марковских и стационарных /п.н. μ/ ε-оптимальных стратегий. Нетрудно убедиться, что для любой стратегии π оценка $w(x,\pi)$ - измеримая /т.е.борелевская/ функция x. С другой стороны, тривиальная модификация известного примера Блекуэлла/1965/ приводит к модели, в которой $v(x)=1$ на неборелевском аналитическом множестве $\mathcal{D}$ и $v(x)=0$ на $X\setminus\mathcal{D}$. Поэтому в приведенных выше формулировках нельзя отказаться от оговорки /п.н. μ/.

В случае дискретного времени в сходной ситуации Блекуэлл, Фридман и Оркин /1974/ показали, что расширение класса стратегий с борелевски до аналитически измеримых отоброжений позволяет строить равномерно ε-оптимальные стратегии. В настоящем сообщении аналогичные результаты получаются для скачкообразных моделей.

3. Пусть $\mathcal{B}(M)$, как выше, обозначает σ-алгебру борелевских множеств борелевского пространства M, а $\mathcal{A}(M)$ - минимальную σ-алгебру, содержащую все аналитические множества в M. Назовем А-стратегией измеримое отображение π пространства

историй $\{H,\mathcal{A}(H)\}$ в пространство управлений $\{A,\mathcal{B}(A)\}$, удовлетворяющее условию $\pi[x_0^t]\in A(x(t))$. Марковская А-стратегия задается измеримым отображением φ пространства $\{M,\mathcal{A}(M)\}$, $M=(0,\infty)\times X$, в $\{A,\mathcal{B}(A)\}$, удовлетворяющим условию $\varphi(t,x)\in\in A(x)$ по формуле $\varphi[x_0^t]=\varphi(t,x(t))$ /нетрудно убедиться, что при этом φ будет А-стратегией/.

Предложение I. При любом начальном распределении μ и любой А-стратегии π формулы /I/-/2/ однозначно задают меру P_μ^π в пространстве Ω.

Обозначим H_n пространство историй $[0\xi_0\tau_1\xi_1\ldots\tau_n\xi_n s]$ длины n $(n=0,1,2,\ldots)$ и сопоставим начальному распределению μ и А-стратегии π меры в пространствах H и $(0,\infty)\times X$ формулами

$$\Pi_\mu^\pi\{\Gamma\times dt\}=P_\mu^\pi(\Gamma)dt \qquad (\Gamma\in\mathcal{B}(\xi_0\tau_1\ldots\xi_n),\ \Gamma\times dt\in H_n),$$

$$\Phi_\mu^\pi\{dt\times\Delta\}=P_\mu^\pi\{x(t)\in\Delta\}dt \qquad (\Delta\in\mathcal{B}(X)).$$

Можно доказать следующие леммы.

Лемма I. Если для А-стратегий π и ρ имеем $\pi=\rho$ /п.н. Π_μ^π/, то $P_x^\pi=P_x^\rho$ /п.н. μ/ и $w(x,\pi)=w(x,\rho)$/п.н. μ/.

Лемма 2. Если для марковских А-стратегий φ и ψ имеем $\varphi=\psi$ /п.н. Φ_μ^π/, то $P_x^\varphi=P_x^\psi$ /п.н. μ/ и $w(x,\varphi)=$ $=w(x,\psi)$ /п.н. μ/.

Из леммы I и Возможности приблизить А-измеримую функцию почти всюду равной ей В-измеримой функцией легко выводится

Теорема I. В суммируемой модели Z оценка $w(x,\pi)$ осмыслена для любой А-стратегии π, а оценка $v(x)$ модели Z совпадает с супремумом $w(x,\pi)$ по классу всех А-стратегий π.

С помощью леммы 2 нетрудно распространить формулу /4/ на А-стратегии:

Л е м м а 3. Если $\|v_t\|$ ограничена на любом ограниченном множестве значений t, то для любой марковской А-стратегии справедливы равенства /4/-/5/.

С л е д с т в и е I. В условиях леммы 3 для оптимальности марковской А-стратегии φ необходимо и достаточно, чтобы при каждом x

$$T_t v_t(x) = T_t^{\varphi} v_t(x) \quad \text{/при почти всех } t\text{ /}$$

и

$$\varliminf_{t\to\infty} E_x^{\varphi} v_t(x(t)) \leq 0.$$

Используя лемму 3 и теоремы об аналитически измеримом выборе для аналитических множеств, можно получить результаты о существовании ε-оптимальных стратегий.

Т е о р е м а 2. Если $\|v_t\|$ ограничена на каждом ограниченном множестве изменения t и для любой марковской стратегии φ справедливо /6/, то для любого $\varepsilon > 0$ существует марковская ε-оптимальная А-стратегия.

Т е о р е м а 3. Если модель Z однородна, $\alpha > 0$ и $\|v\| < \infty$, то для любого $\varepsilon > 0$ существует стационарная ε-оптимальная А-стратегия.

Замечание. В однородной модели с $\alpha > 0$ для конечности $\|v\|$ необходимо и достаточно, чтобы функция $L(x) = \sup_{a \in A(x)} r(a)$ была ограничена.

ЛИТЕРАТУРА

Blackwell D. (1965) : Discounted dynamic programming. Ann. Math. Statistics 36 (1965), 226-235.

Blackwell D., Freedman D., Orkin M. (1974) : The optimal reward operator in dynamic programming. Ann. Probability 2 (1974), No.5, 926-941.

Юшкевич А.А. /1977/ : Управляемые скачкообразные марковские процессы. Доклады Академии наук СССР 233 /1977/, № 2, 304-307.

Московский институт инженеров транспорта

Москва, ул. Образцова 15

СССР

ON SAMPLE FUNCTIONS OF MULTIPARAMETER STOCHASTIC PROCESSES WITH INDEPENDENT INCREMENTS

Nijolė Kalinauskaitė

Vilnius

ABSTRACT

Some statements concerning the sample function behavior near origin and the γ-variation of multiparameter stochastic processes with independent increments are presented.

INTRODUCTION

Let $\xi(t),\ t=(t_1, t_2, \dots, t_s) \in R_s^+$ be a separable homogeneous weakly continuous random process with multidimensional time parameter the increments of which in non-overlapping s-dimensional rectangles are independent. Such a random process is defined (see Straf (1972)) with the help of characteristic functions

$$E e^{iz\xi(t)} = \exp\{-\prod_{j=1}^{s} t_j \cdot \psi(z)\}$$

and the independence of increments on non-overlapping intervals. Here

$$\psi(z) = ita + \int_{-\infty}^{+\infty} \left(e^{izx} - 1 - \frac{izx}{1+x^2}\right) \frac{1+x^2}{x^2} \, d\,\Theta(x) .$$

The integrand is defined for $x=0$ by continuity and is therefore equal to $-\frac{1}{2}t^2$ if $x=0$. Suppose $a=0$. The Lévy canonical representation of $\psi(z)$ is

$$\psi(z) = -\frac{\sigma^2 z^2}{2} + \int_{+0}^{+\infty} \left(e^{izu} - 1 - \frac{izu}{1+u^2}\right) \nu(du) +$$

$$+\int_{-\infty}^{-0}\left(e^{izu}-1-\frac{izu}{1+u^2}\right)\nu(du)$$

where

$$\int_{|u|\leq 1} u^2\nu(du) < \infty \quad \text{and} \quad \sigma^2 \geqslant 0.$$

By $\Delta_{[a,b]}\xi$ denote an increment of the process ξ on a rectangle $[a,b]$ with walls parallel to the coordinate planes. Such an interval can be described by means of two its vertices $a=(a_1,\dots,a_s)$ and $b=(b_1,\dots,b_s)$ with $a_j \leq b_j$, $1\leq j\leq s$

The increment

$$\Delta_{[a,b]}\xi = \xi(b)-\sum_{k=1}^{s}\xi(a_k)+\sum_{k<\ell}\xi(a_{k,\ell})-\dots+(-1)^s\xi(a)$$

where $\xi(a_{k,\ell,\dots,p})$ denotes $\xi(c_1,\dots c_s)$ with $c_k, c_\ell,\dots,c_p$ equal to b_k, $b_\ell,\dots,b_p$, respectively, and the remaining c_j equal to a_j. In other words $\Delta_{[a,b]}\xi$ is the sum of values of the function $\xi(t,\omega)$ taken at the vertices of the rectangle, where ξ has opposite signs for every pair of vertices sharing a common edge.

Many important results concerning various properties of sample functions of infinitely divisible random processes ($s=1$) have been obtained. In the case $s\geqslant 2$ the multiparameter Wiener process (in our notation it is the case $\theta(x)\equiv 0, \sigma^2>0$) has been investigated at most. The law of the iterated logarithm, the modulus of continuity, strong and weak ψ -variations, some questions concerned with Hausdorff dimension were discussed for the multiparameter Wiener process. Our aim is to investigate similar properties of sample functions in the case of multiparameter infinitely divisible processes with independent increments. Some of our statements will be presented below.

ON THE BEHAVIOR OF SAMPLE FUNCTIONS NEAR ORIGIN

We use the following notations:

$$\|t\| = \prod_{j=1}^{s} t_j, \quad |t| = \sqrt{t_1^2+t_2^2+\dots+t_s^2}\ .$$

By β denote the Blumenthal-Getoor index, of the cumulant $\psi(z)$, namely,

$$\beta = \inf\{\alpha > 0 : \int_{|x| \le 1} |x|^{\alpha} \nu(dx) < \infty\}.$$

As the distribution of the random variable $\xi(t)$ depends on $\|t\|$ it is natural to compare the sample functions of the process $\xi(t)$ with the functions in $\|t\|$.

Let $g_1(u)$ be a continuous monotone increasing positive function in one variable of regular growth, satisfying the relation

$$\lim_{|u| \to 0} g_1(u) = 0 .$$

The function $g(t) \equiv g_1(\|t\|)$ is called an upper bound for the process ξ if

$$P\{\lim_{|t| \to 0} \frac{\xi(t)}{g(t)} = 0\} = 1 .$$

The function $g(t)$ is a lower bound of ξ if with probability one

$$\limsup_{|t| \to 0} \frac{\xi(t)}{g(t)} = \infty$$

The function $g(t)$ is an exact upper bound of the process if with probability one

$$\limsup_{|t| \to 0} \frac{\xi(t)}{g(t)} = const.$$

Let $P_c(t) = P\{|\xi(t)| > c\,g(t)\}$ where $c > 0$ is a constant.

Theorem 1. If the integral

$$J_1 = \int_{[0,1]} P_c(t) \frac{dt}{\|t\|} < \infty \qquad \forall c > 0$$

then the function $g(t)$ is an upper bound of the process $\xi(t), t \in R_s^+$

If this integral diverges for some $c > 0$, then the function $g(t)$ cannot be an upper bound of the process ξ.

In the case $s = 1$ such a statement has been proved in A. Khintchine (1939). We prove it for the multidimensional time parameter case following the scheme of reasoning indicated by A. Khinchine adding necessary generalizations.

As the integrand of the integral J_1 depends on $\|t\|$ by virtue of the equality presented in M.S. Klamkin (1976), the integral J_1 can be rewritten in the form

$$J_1 = \frac{1}{(s-1)!} \int_0^1 P_c(u) \lg^{s-1} \frac{1}{u} \frac{du}{u}$$

where $P_c(u) = P\{|\xi(u)| > g_1(u)\}$ corresponds to a one-dimensional time process $\xi(u)$ with the same function $\psi(z)$. Thus in general the problem cannot be reduced to the one-dimensional time parameter case.

If we have some more information on the process $\xi(t)$, we can obtain more explicit statements on the sample function behavior near origin.

If the process $\xi(t), t \in R_s^+$ is stable then the integral J_1 converges (or diverges) together with the integral

$$J_2 = \int_{[0,1]} \frac{dt}{c^\alpha g^\alpha(t)} .$$

Hence, for the stable multiparameter processes (if we consider their modulus) the exact upper bound does not exist.

Theorem 2. If the process ξ has no Gaussian component ($\sigma^2 = 0$) and its spectral function $\Theta(x)$, satisfies the condition

$$\int_{-\infty}^{\infty} \left| \lg \frac{1}{|z|} \right|^{s-1} d\Theta(z) < \infty \tag{A}$$

then the function

$$u(t) = \sqrt{\|t\| \lg\lg \frac{1}{\|t\|}}$$

is an upper bound of the process $\xi(t)$, $t \in R_s^+$.

For the case $s=1$ such a theorem was proved in A. Khinchine (1939).

Theorem 2 together with the local law of iterated logarithm for the multiparameter Wiener process yields the following.

Corrollary. The process $\xi(t), t \in R_s^+$ satisfies the local law of the iterated logarithm if it has a Gaussian component and the condition (A) is satisfied.

Theorem 3. If $\gamma > \beta$ then under the condition (A) with probability one

$$\|t\|^{-\frac{1}{\gamma}} |\xi(t)| \to 0 \quad as \quad |t| \to 0 .$$

If $\gamma < \beta$ then under the condition (A) with probability one

$$\limsup_{|t| \to 0} \|t\|^{-\frac{1}{\gamma}} |\xi(t)| = \infty .$$

For $s=1$ theorem 3 was proved in Blumenthal-Getoor (1961). It is clear that condition (A) is satisfied if $\beta < 2-\varepsilon$ where $\varepsilon > 0$ is arbitrarily small.

ON THE STRONG γ -VARIATION

Let $\pi = \{A_i\}$ be finite partition of s-dimensional cube $[0,1]$ into s-dimensional rectangles A_i . The strong γ -variation of $\xi(t)$ is

$$V_\gamma^s(\xi) = \sup_{\pi} \sum_i |\Delta_{A_i}\xi|^\gamma$$

where $\Delta_{A_i}\xi$ denotes the increment of the process $\xi(t)$ on a rectangle A_i and sup is taken over all partitions of $[0,1]$. It is not difficult to see that $V_\gamma^s(\xi)$ is a measurable function of ω (cf. Blumenthal-Getoor (1960), lemma 2.2) and hence a random variable. The following two statements are valid:

Theorem 4. If $\gamma < \beta$ then $V_\gamma^s(\xi) = \infty$ with probability one.

If $\beta < \gamma \leq 1$ then $V_\gamma^s(\xi)$ is finite with probability one.

Here β is a Blumenthal-Getoor index.

To prove the Theorem 4 of the method of Bochner (1955 Sec., 5.3) is employed. For the case $s=1$ such statements were considered in Blumenthal-Getoor (1961).

REFERENCES

Blumenthal R.M., Getoor R.K.(1960): Some properties of stable processes. Trans.Amer.Math.Soc., v. 95, No 4, 263-273.

Blumenthal R.M., Getoor R.K. (1961): Sample functions of stochastic processes with stationary independent increments. J.Math.Mech. v. 10, No 3, 493-516.

Bochner S. (1955): Harmonic analysis and the probability theory. 1955 Berkeley and Los Angeles.

Klamkin M.S.1976): On some multiple integrals. J.of Math.Analysis and Appl., v. 54, No 2, 476-479.

Orey S., Pruitt W.E. (1973): Sample functions of the N-parameter Wiener process. The Annals of Prob., v. 1, No 1, 138-163.

Paranjape S.R., Park C. (1973) : Laws of iterated logarithm of multiparameter Wiener processes. J.of Multivariate Analysis, v. 3, 132-136.

Straf H.L. (1972): Weak convergence of stochastic processes with several parameters. In: Proc.Sixth Berkeley Symp. on Probab.Theory, v. 2, 187-221.

Yoder L.(1974) : Variation of multiparameter Browninan motion. Proc.Amer.Math.Soc., v. 46, No 2, 302-309.

Zimmerman G.(1972): Some sample function properties of the two-parameter Gaussian process. Annals of Math.Stat., v. 43, 1235-1246.

Ченцов Н.Н.(1956): Винеровские случайные поля от нескольких параметров. ДАН СССР, т. 106, № 4.

Хинчин А.Я.(1939): О локальном росте однородных стохастических процессов без последействия. Известия АН СССР. 1939 № 5-6, 487-508.

Institute of Mathematics and Cybernetics
Academy of Sciences, Lithuanian SSR
K. Poželos 54
Vilnius 232600
USSR

AN APPROXIMATIVE SOLUTION OF A STOCHASTIC OPTIMIZATION PROBLEM

Vlasta Kaňková

Prague

ABSTRACT

Let us consider a stochastic optimization problem in which the optimum is sought with respect to the mathematical expectation. Assume the probability laws of the random variables to be unknown. Some estimates of the optimal solution and the optimal value of the optimalized function were given in [3] . These estimates were obtained using realizations of random variables having the same probability laws as those in the problem. In this paper we are going deeper in this direction. We investigate the rate of convergence for these estimates. It is shown that this rate is at least exponencial.

INTRODUCTION

Let $(\Omega, \mathcal{S}, P)$ be probability space where P is a complete probability measure,

$T(\omega) = [T_1(\omega), \ldots, T_s(\omega)]$ be random vector defined on $(\Omega, \mathcal{S}, P)$.

Let, further, $g(t,x)$ be a real valued, continuous function defined on $\mathcal{T} \times \mathcal{X}$ where $\mathcal{X} \subset E_r$ is a non-empty, compact, convex set and $\mathcal{T} \subset E_s$ satisfies the condition $P\{\omega : T(\omega) \in \mathcal{T}\} = 1$

(E_r denotes r-dimensional Euclidean space).

It is easy to see that $g(T,x)$ is for every $x \in \mathcal{X}$ a random variable defined on $(\Omega, \mathcal{S}, P)$. We shall introduce the stochastic optimitation problem as the problem of finding

(1) $$\max_{x \in \mathcal{X}} \mathcal{E}\, g(t,x)$$

($\mathcal{E}$ denotes the operator of the mathematical expectation).

Let random vectors $T_i(\omega) = [T_{i1}(\omega), \dots, T_{is}(\omega)]$, $i = 1,2,\dots$ (defined on $(\Omega, \mathcal{S}, P)$) satisfy the following condition

(i) $\{T_i\}_{i=1}^{\infty}$ is a sequence of independent s-dimensional random vectors having the same distribution function $F(t)$ as the random vector T.

We shall define the functions $U_i(t,\omega)$, $F_n(t,\omega)$, $t = [t_1, \dots, t_s] \in E_s$, $\omega \in \Omega$, $i = 1,2,\dots$, $n = 1,2,\dots$

$$U_i(t,\omega) = 1 \iff T_{ij} < t_j \quad \text{for all } j = 1,2,\dots,s$$
$$U_i(t,\omega) = 0 \iff T_{ij} \geq t_j \quad \text{for at least one } j \in \{1,2,\dots,s\}$$

(2) $$F_n(t,\omega) = \frac{1}{n} \sum_{i=1}^{n} U_i(t,\omega)$$

If $g(t,x)$ satisfies the condition,

(ii) $g(t,x)$ is a continuous, bounded function on $\mathcal{T} \times \mathcal{X}$,

we can formulate the theorem

Theorem 1: If T , T_i, $i = 1,2,\dots$, $g(t,x)$ satisfy the conditions (i), (ii), $F_n(t)$ the condition (2) then if we define for every $x \in \mathcal{X}$, $\omega \in \Omega$, $n = 1,2,\dots$

(3) $$I_n(x,\omega) = \int_{\mathcal{T}} g(t,x)\, dF_n(t,\omega)$$
$$I(x) = \int_{\mathcal{T}} g(t,x)\, dF(t),$$

it holds

$$P\{\omega : \max_{x \in \mathcal{X}} I_n(x,\omega) \xrightarrow[(n\to\infty)]{} \max_{x \in \mathcal{X}} I(x)\} = 1.$$

If, further, $g(t,x)$ is for every $t \in \mathcal{T}$ a strictly concave function of x and x_0 , $\hat{x}_n(\omega)$, $n = 1,2,\dots$ are points for which

(4) $$I_n(\hat{x}_n(\omega),\omega) = \max_{x \in \mathcal{X}} I_n(x,\omega)$$
$$I(x_0) = \max_{x \in \mathcal{X}} I(x)$$

then

$$P\{\omega : \hat{x}_n(\omega) \xrightarrow[(n\to\infty)]{} x_0\} = 1.$$

This Theorem was proved, in case $\mathcal{T} = E_s$ and sequence $\{T_i(\omega)\}_{i=1}^{\infty}$ ergodic, in [3]. The aim of this paper is to find an upper estimate of

$$P\{\omega : |\max_{x \in \mathcal{X}} I_n(x,\omega) - \max_{x \in \mathcal{X}} I(x)| \geq z\}$$

for $z \in E_1$, $n = 1,2,\dots$

THE MAIN RESULT

Let the function $g(t,x)$ satisfies the condition

(iii) $g(t,x)$ is for every $t \in \mathcal{T}$ a Lipschitz function of x with Lipschitz constant L not depending on t .

Let, further, the constants K , M satisfy the conditions

(5)
$$\sup_{x', x \in \mathcal{K}} \|x - x'\| < K$$
$$|g(t,x)| \leq M \quad \text{for } t \in \mathcal{T},\ x \in \mathcal{K}$$

with $\|\cdot\|$ denoting the Euclidean norm in E_r .

If we define a number $N(z)$ by

(6)
$$N(z) = \min\{n \text{ integrel} : n \geq \frac{3LK}{z}\}$$

then the following main our result is valid.

Theorem 2: If the conditions (i),(ii),(iii) are satisfied and if $I(x)$, $I_n(x,\omega)$, $n = 1,2,\ldots$ are defined by (3), $N(z)$ by (6) and M by (5) then

$$P\{\omega : |\max_{x \in \mathcal{K}} I_n(x,\omega) - \max_{x \in \mathcal{K}} I(x)| \geq z\} \leq 2[N(z)]^r \exp\{-\frac{nz^2}{18M^2}\}$$

for $z \in E_1$, $z > 0$, $n = 1,2,\ldots$

Proof: Let $z_0 > 0$, $z_0 \in E_1$ be arbitrary. From (6) it follows the existence of points $x_1, x_2, \ldots, x_{N_0}$, $N_0 = [N(z_0)]^r$ such that

$$\sup_{x \in \mathcal{K}} \inf_{x_i \in [x_1,\ldots,x_{N_0}]} |g(t,x) - g(t,x_i)| \leq \frac{z_0}{3} \quad \text{for all } t \in \mathcal{T}$$

so that

(7)
$$\sup_{x \in \mathcal{K}} \inf_{x_i \in [x_1,\ldots,x_{N_0}]} |I(x) - I(x_i)| \leq \frac{z_0}{3}$$

and

$$\sup_{x \in \mathcal{K}} \inf_{x_i \in [x_1,\ldots,x_{N_0}]} |I_n(x,\omega) - I_n(x_i,\omega)| \leq \frac{z_0}{3} \quad n = 1,2,\ldots,\ \omega \in \Omega$$

Since it, further, follows from the triangular inequality

$$|I_n(x,\omega) - I(x)| \leq |I_n(x,\omega) - I(x_i,\omega)| + |I_n(x_i,\omega) - I(x_i)| + |I(x_i) - I(x)|$$

for $x \in \mathcal{K}$, $x_i \in [x_1,\ldots,x_{N_0}]$, $n = 1,2,\ldots$

we get

(8)
$$P\{\omega : |\max_{x \in \mathcal{K}} I_n(x,\omega) - \max_{x \in \mathcal{K}} I(x)| \geq z_0\} \leq$$
$$\leq P\{\omega : |I_n(x,\omega) - I(x)| \geq z_0 \quad \text{for at least one } x \in \mathcal{K}\} \leq$$
$$\leq P\{\omega : |I_n(x,\omega) - I(x_i)| \geq \frac{z_0}{3} \quad \text{for at least one } x_i \in [x_1,\ldots,x_{N_0}]\}$$

Further, in [2] it is proved

$$P\{\omega : (I_n(x,\omega) - I(x)) > y\} \leq \exp\{-\frac{ny^2}{2M^2}\}$$

for all $y > 0$, $y \geq 0$, $x \in \mathcal{K}$.

Hence we can easily obtain

$$P\{\omega : |\max_{x\in\mathcal{K}} I_n(x,\omega) - \max_{x\in\mathcal{K}} I(x)| \geq z_0\} \leq$$
$$\leq 2[N(z_0)]^r \cdot \{-\frac{nz_0^2}{18M^2}\}$$

This completes the proof of Theorem 2.

In Theorem 2 it was assumed that $g(t,x)$ is for every $t\in\mathcal{T}$ a Lipchitz function of x. This is a rather strong assumption. We shall see that under rather general conditions it is sufficient to assume that $g(t,x)$ is, for every $t\in\mathcal{T}$, a concave function of x.

Let for $\varepsilon\geq 0, \mathcal{K}(\varepsilon)$ be defined by

(9) $\mathcal{K}(\varepsilon) = \mathcal{K} + B(\varepsilon) = \{u : u = x+v,\ x\in\mathcal{K},\ v\in B(\varepsilon)\}$,

where $B(\varepsilon)$ denotes ε-surroundings of $0\in E_r$.

If there is ε_0 such that $g(t,x)$ can be defined on $\mathcal{T}\times\mathcal{K}(\varepsilon_0)$ then the following Corollary holds.

Corollary: Let there exists $\varepsilon_0 > 0$ such that $g(t,x)$ is a continuous, bounded function $[|g(t,x)|\leq M]$ on $\mathcal{T}\times\mathcal{K}(\varepsilon_0)$. Let, further, $g(t,x)$ is for every $t\in\mathcal{T}$ a concave function of $x\in\mathcal{K}(\varepsilon)$,. If T, T_i, $i=1,2,\dots$ satisfy the condition (i), $I(x), I_n(x,\omega)$ $n=1,2,\dots$ the condition (3) then for every $z>0$, $z\in E$, it holds

$$P\{\omega : |\max_{x\in\mathcal{K}} I_n(x,\omega) - \max_{x\in\mathcal{K}} I(x)| \geq z\} \leq 2[\bar{N}(z)]^r \exp\{-\frac{nz^2}{18M^2}\}$$

where $\bar{N}(z) = \min\{n \text{ integrel} : n \geq \frac{6M}{\varepsilon_0}\frac{K}{z}\}$.

Proof: Obviously to prove the Corollary it is enough to prove that

(10) $|g(t,x) - g(t,x')| \leq \frac{2M}{\varepsilon_0}\|x-x'\|$

for every $x, x'\in\mathcal{K}$, $t\in\mathcal{T}$.

Let x, $x'\in\mathcal{K}$ be arbitrary. If we put
$\bar{x} = x + \frac{\varepsilon_0}{\|x-x'\|}(x-x')$, then $\bar{x}\in\mathcal{K}(\varepsilon_0)$
and $x = (1-\lambda)x' + \lambda\bar{x}$ where $\lambda = \frac{\|x-x'\|}{\varepsilon_0+\|x-x'\|} \in \langle 0,1\rangle$

As $g(t,x)$ is for every $t\in\mathcal{T}$ a concave function of x we get

$$g(t,x) \geq (1-\lambda)g(t,x') + \lambda g(t,\bar{x})$$
$$= g(t,x') + \lambda[g(t,\bar{x}) - g(t,x')] \qquad \text{for all } t\in\mathcal{T}$$

But from this it is easy to see

$$g(t,x) - g(t,x') \leq \lambda|g(t,x') - g(t,\bar{x})| \leq$$
$$\leq \frac{\|x-x'\|}{\varepsilon_0+\|x-x'\|}\cdot 2M \leq \frac{2M}{\varepsilon_0}\|x-x'\|$$

Thus (10) is proved.

Until now we have dealt with estimates of the optimal value

of the optimalized function. The rate of the convergence of the optimal solution will certainly depend on the form of the function $g(t,x)$ If the error of the approximating solution is measured by

$$|I(x_n(\omega)) - I(x_0)|$$

(where $x_n(\omega)$, x_0, $n = 1,2,\dots$ are defined by (4)) then another results analogical to Theorem 2 and Corollary, can be proved.

REFERENCES

[1] Dupačová J. (1976) : Experience in Stochastic Programming Models. IX International Symposium on Mathematical Programming, Budapest 1976

[2] Hoeffding W. (1963) : Probability Inequalities for Sums of Bounded Random Variables. Journal of the Americ. Statist. Ass. 58 (1963), No.301, 13-30.

[3] Kaňková V. (1974) : Optimum Solution of a Stochastic Optimization Problem with Unknown Parameters. In: Trans. of the Seventh Prague Conference, Prague 1974, Academia, Prague 1978.

[4] Rockafellar R. (1970) : Convex Analysis, Princeton Press New Jersey 1970

Czechoslovak Academy of Sciences
Institute of Information Theory and Automation
Pod vodárenskou věží 4
180 76 Praha 8 - Libeň
Czechoslovakia

WHEN ARE TWO SPECIAL LINEAR FORMS OF INDEPENDENT RANDOM VECTORS IDENTICALLY DISTRIBUTED ?

Leo Klebanov

Leningrad

ABSTRACT

Let X_i, $i=1,\dots,n$ be independent p_i-dimensional random vectors. Consider two linear functions

$$L_1 = A_1X_1+\dots+A_nX_n \quad , \qquad L_2 = B_1X_1+\dots+B_nX_n$$

where A_i, B_i are $(m \times p_i)$ matrices. In the paper the condition of identically distribution of L_1 and L_2 is studied. The results are applied to a characterization of the normal law and to the problem of characterization of probability distributions of random vectors given the joint distribution of linear functions of them.

The results provide a generalization of those concerning characterization of probability laws throught properties of linear functions.

INTRODUCTION AND NOTATIONS

Let $X_1, \dots, X_n$ be independent vector random variables, the dimension of X_i being p_i. Consider two linear forms

$$L_1 = A_1X_1+\dots+A_nX_n \; , \qquad L_2 = B_1X_1+\dots+B_nX_n$$

where A_j, B_j are $m\times p_j$ matrices of ranks a_j and b_j respectively. We investigate the problem of characterizing the distributions of random vectors X_i by the property of identically distribution of L_1 and L_2.

Introduce the product of matrices $E \odot G$ as defined by Khatri and Rao (1968). Let $E=(E_1 \vdots \cdots \vdots E_k)$ and $G=(G_1 \vdots \cdots \vdots G_k)$ be partitions of

E and G , then

$$E \odot G = (E_1 \otimes G_1 \vdots \dots \vdots E_k \otimes G_k)$$

where $\otimes$ denotes Kronecker product. We denote by $(E\odot)^2 E$ the product $(E\odot E)\odot E$. In the similar way $(E\odot)^r E$ is devined. The properties of such products were studied in Khatri and Rao (1968) and Khatri (1971).

Let $F_1, \dots, F_n, H_1, \dots, H_n$ be $m \times p_j$ ($j = 1, \dots, n$) matrices. We denote by I the set of all integer numbers i, $1 \le i \le n$. Let S_1, S_2 be two subsets of I. A family $(F_j)_{j \in J}$ ($J \subset S_1$) is called a determinating family for $(F_j)_{j \in S_1}$, $(H_j)_{j \in S_2}$ if for any $j \in S_1, s \in S_2$ there exist $i_1, i_2 \in J$ and matrices C_j, D_s such that $F_j = F_{i_1} C_j$, $H_s = F_{i_2} D_s$. (The families $(F_j)_{j \in S_1}$ and $(H_j)_{j \in S_2}$ are generated by the family $(F_j)_{j \in J}$). For any $j \in J$ let $I_j^{(1)} = \{s \in S_1 \mid F_s = F_j C_s\}$, $I_j^{(2)} = \{s \in S_2 \mid H_s = F_j D_s\}$.

THE MAIN THEOREM

We denote by $R(A)$ the rank of the matrix A.

Theorem A. Let L_1 and L_2 be identically distributed. Then the following propositions hold.

1) If there exist

$$J_1 = \{j_1^{(1)} < \dots < j_{r_1}^{(1)}\} \subset I, \qquad J_2 = \{j_1^{(2)} < \dots < j_{r_2}^{(2)}\} \subset I$$

such that

(1) $$R(A_i \vdots B_j) = a_i + b_j \quad \text{for } i \in J_1,\ j \in I,$$

(2) $$R(A_i \vdots A_j) = a_i + a_j \quad \text{for } i \in J_1,\ j \in I \setminus J_1,$$

(3) $$R(B_i \vdots A_j) = b_i + a_j \quad \text{for } i \in J_2,\ j \in I \setminus J_1,$$

(4) $$R(B_i \vdots B_j) = b_i + b_j \quad \text{for } i \in J_2,\ j \in I \setminus J_2.$$

then random vectors $A_i X_i$ ($i \in J_1$) and $B_j X_j$ ($j \in J_2$) have multivariate normal (possible degenerate) distributions.

2) Let families $(A_j)_{j \in I \setminus J_1}$, $(B_\ell)_{\ell \in I \setminus J_2}$ be generated by a family $(A_s)_{s \in J_3}$ where $J_3 = \{j_1^{(3)} < \dots < j_{r_3}^{(3)}\} \subset I \setminus J_1$. Let $A = (A_{j_1^{(3)}} \vdots \dots \vdots A_{j_{r_3}^{(3)}})$. If for some r, $2 < r < r_3 - 2$ (and for empty J_1, J_2: $1 \le r < r_3 - 2$) the following conditions hold

(5) $$R[(A\odot)^{r-1} A] < \sum_{j \in J_3} a_j^r, \qquad R[(A\odot)^r A] = \sum_{j \in J_3} a_j^{r+1},$$

then the difference of log characteristic functions (c.f.) of

$A_j \sum_{i\in I_j(k)} C_i X_i$ and $A_j \sum_{i\in I_j(k)} D_i X_i$ $(j\in \mathcal{J}_s)$ is a polynomial of degree $\leq \tau$ in any neighbourhood O^ρ of the origin where both log s of c.f. are well defined.

First we state two lemmas (for the proof see Khatri and Rao (1971))

Lemma 1. Let Φ_i be a continuous complex valued function of a real p_i -vector variable ($i=1,\dots,s$) defined in a neighbourhood of the origin in appropriate spaces, and F_i be matrix of the order $m\times p_i$. If

$$(6) \qquad \sum_{i=1}^{s} \Phi_i (F_i' t) = P_k(t)$$

where $P_k(t)$ is a polynomial of degree k , and if

$$R(F_i : F_j) = R(F_i) + R(F_j), \quad i=1,\dots,\rho\ ;\ j=\rho+1,\dots,s$$

then both $\sum_{i=1}^{\rho} \Phi_i (F_i' t)$ and $\sum_{i=\rho+1}^{s} \Phi_i (F_i' t)$ are polynomials in t of degree $\leq \max(k, s-2)$.

Lemma 2. Let Φ_i and F_i ($i=1,\dots,s$) be as in Lemma1 and $F = (F_1 \vdots \dots \vdots F_s)$. If for some τ , $k<\tau<s-2$ the relations

$$R[(F\odot)^{\tau-1} F] < \sum_{i=1}^{s} (R(F_i))^{\tau}$$

and

$$R[(F\odot)^{\tau} F] = \sum_{i=1}^{s} (R(F_i))^{\tau+1}$$

hold, then each Φ_i is a polynomial of degree $\leq \tau$.

Proof of Theorem. Let f_j be the c.f. of X_j ($j=1,\dots,n$). From the condition of identically distribution of L_1 and L_2 we have

$$(7) \qquad \prod_{j=1}^{n} f_j(A_j' t) = \prod_{j=1}^{n} f_j(B_j' t), \quad t\in R^m .$$

Let $\varphi_j = \log f_j$, $\varphi_{j+n} = -\log f_j$, $A_{j+n}' = B_j'$, $j=1,\dots,n$. Then in any neighbourhood O^ρ the equation (7) is equivalent to

$$(8) \qquad \sum_{j=1}^{2n} \varphi_j (A_j' t) = 0 .$$

From (1),(2) and Lemma 1 we can deduce, that

$$\sum_{j\in \mathcal{J}_1} \varphi_j (A_j' t) \quad \text{and} \quad \sum_{j\notin \mathcal{J}_1} \varphi_j (A_j' t)$$

are polynomials in t of a degree $\leq 2n-2$. Thus the relation

$$\prod_{j\in \mathcal{J}_1} f_j (A_j' t) = \exp\{P(t)\}$$

holds where P is a polynomial of a degree $\leq 2n-2$, i.e. P is a polynomial of degree ≤ 2 and the vector $A_j X_j$ ($j\in \mathcal{J}_1$) has a normal distribution (see Marcinkiewicz (1938) and Cramér (1936)).

From the relation $\sum_{j\in \mathcal{J}_1} \varphi_j (A_j' t) = P(t)$ and the equation (8) we get

$$\sum_{j\notin \mathcal{J}_1} \varphi_j(A_j' t) = -P(t)$$

where $P(t)$ is a polynomial of a degree ≤ 2. Now we obtain from (3), (4) and Lemma 1 that the function

$$\sum_{j\in \mathcal{J}_2'} \varphi_j(A_j' t)$$

is a polynomial in t. Here $\mathcal{J}_2' = \{j' \mid j' = j + n,\ j \in \mathcal{J}_2\}$.

Proceeding further for $\sum_{j\in\mathcal{J}_2'} \varphi_j(A_j' t)$ in the same wey as we did above for $\sum_{j\in\mathcal{J}_1} \varphi_j(A_j' t)$ we conclude that the vector $A_j X_j$ $(j \in \mathcal{J}_2')$ has a normal distribution, i.e. the vector $B_j X_j$ $(j \in \mathcal{J}_2)$ has a normal distribution. Moreover the relation

$$\sum_{j\in \mathcal{J}_1 \cup \mathcal{J}_2'} \varphi_j(A_j' t) = P_2(t) \tag{9}$$

holds where P_2 is a polynomial of a degree ≤ 2 and $P_2 \equiv 0$ if $\mathcal{J}_1$ and $\mathcal{J}_2$ are empty.

Let

$$\psi_j(A_j' t) = \sum_{i\in I_j^{(1)}} \varphi_i(C_i' A_j' t) + \sum_{i\in I_j^{(2)}} \varphi_{i+n}(\mathcal{D}_i' A_j' t), \quad j \in \mathcal{J}_3 .$$

Then the equation (9) can be rewritten in the form

$$\sum_{j\in\mathcal{J}_3} \psi_j(A_j' t) = P_2(t).$$

According to Lemma 2 the function $\psi_j(A_j' t)$ is a polynomial in t of a degree $\leq \tau$. Theorem A is proved.

COROLLARIES OF THEOREM A

Corollary 1. Let $\Xi = (x_1, \dots, x_n)$ be a random vector in R^n with independent components and

$$L_1 = \sum_{i=1}^n a_i x_i, \quad L_2 = \sum_{i=1}^n b_i x_i, \quad L_3 = \sum_{i=1}^n c_i x_i, \quad L_4 = \sum_{i=1}^n d_i x_i$$

be four linear forms. If the random vectors $Y = (L_1, L_2)$ and $Z = (L_3, L_4)$ are identically distributed, than all those x_j for which $a_j d_\ell \neq c_\ell b_j$ for all $\ell = 1, \dots, n$ have normal distributions.

Proof. Consider the matrices

$$A_i = \begin{pmatrix} a_i \\ b_i \end{pmatrix}, \qquad B_i = \begin{pmatrix} c_i \\ d_i \end{pmatrix}.$$

Then the conditions of identically distribution of Y and Z and of $\tilde{L}_1 = \sum_{i=1}^n A_i x_i$ and $\tilde{L}_2 = \sum_{i=1}^n B_i x_i$ are equivalent. Let for some j $a_j d_\ell \neq$

$\neq c_\ell b_j$ for all $\ell = 1,\dots,n$. Let $\mathcal{I}_1 = \{j' \mid A_{j'} = \lambda_{j'} \cdot A_j$ for some real $\lambda_{j'}\}$. Then we have the following relations

$$R(A_{j'} : B_\ell) = R(A_{j'}) + R(B_\ell)\,;\ j' \in \mathcal{I}_1,\ \ell \in I,$$

$$R(A_{j'} : A_s) = R(A_{j'}) + R(A_s),\ j' \in \mathcal{I}_1,\ s \in I \setminus \mathcal{I}_1.$$

Our statement follows now immediately from Theorem A.

Let Ξ, L_1, L_2 be as in Corollary 1. Suppose that L_1 and L_2 are independent and consider a vector $\Xi' = (x_1', \dots, x_n')$ such that Ξ and Ξ' are independent and identically distributed. Then the conditions of independence of L_1 and L_2 and of identically distribution (L_1, L_2) and (L_1, L_2') where $L_2' = \sum_{i=1}^n b_i x_i'$ are equivalent. From Corollary 1 we obtain now that those x_j, for which $a_j b_j \neq 0$, have normal distributions, i.e. we obtain Darmois-Skitovič theorem (see Darmois (1953) and Skitovič (1953)).

Corollary 2. Let Ξ, L_1, L_2, L_3 be as in Corollary 1. If the conditional distribution of L_1 given L_3 and the conditional distribution of L_2 given L_3 are equal then those x_j, for which $a_j c_\ell \neq b_\ell c_j$ for all $\ell = 1,\dots,n$ have normal distributions.

Proof. It is clear that if the conditional distributions of L_1 and L_2 given L_3 are equal then (L_1, L_3) and (L_2, L_3) are identically distributed. Our statement follows now from Corollary 1.

Corollary 2 is a generalization of a theorem by Heyde (1970). The result due to Heyde is a speciel case of Corollary 2 corresponding to $L_2 = -L_1$.

Corollary 3. Let Ξ, L_1, L_2 be again as in corollary 1. If L_1 and L_2 are symmetrically dependent, i.e.

$$P\{L_1 < t,\ L_2 < s\} = P\{L_1 < s,\ L < t\},$$

then those x_j, for which $a_j a_\ell \neq b_j b_\ell$ for all $\ell = 1,\dots,n$ have normal distributions.

Our statement follows from corollary 1 applied to the pairs (L_1, L_2) and (L_2, L_1).

Corollary 4. Let $X_1, \dots, X_n$ be independent p-dimensional random vectors, $L_1 = \sum_1^n A_j X_j$, $L_2 = \sum_1^n B_j X_j$, $L_3 = \sum_1^n C_j X_j$, $L_4 = \sum_1^n \mathcal{D}_j X_j$ where A_j, B_j, C_j, $\mathcal{D}_j$ are nonsingular (pxp) matrices. If for all $i,j = 1,\dots,n$ the matrix

$$\begin{pmatrix} A_i & C_j \\ B_i & \mathcal{D}_j \end{pmatrix}$$

is nonsingular then all vectors X_j , $j=1,\dots,n$ have normal distributions.

The result is established on the same lines as in the proof of corollary 1.

Corollary 4 is a generalization of Ghurye-Olkin (1962) theorem.

Corollary 5. Let $X_1,\dots,X_n$ be as in Theorem A, C_j be $m\times p_j$ matrices, $R(C_j)=p_j$, $j=1,\dots,n$ and

$$\mathcal{Z}=C_1X_1+C_2X_2+\dots+C_nX_n .$$

Suppose that the c.f. of $\mathcal{Z}$ does not vanish. Let $C=(C_1\vdots\cdots\vdots C_n)$ and for $r\le(n-2)$

$$R((C\odot)^{r-1}C)<\sum_{i=1}^{n}p_i^{r}\ , \quad R((C\odot)^{r}C)=\sum_{i=1}^{n}p_i^{r+1}.$$

Then the c.f. of X_i can be determined upto a multiplication by $\exp(P_r)$ where P_r is a polynomial of the degree at most r .

Proof. Let vectors $X_1',\dots,X_n'$ satisfy the conditions of Corollary 5. Then the linear forms

$$L_1=\sum_1^n C_iX_i \quad , \quad L_2=\sum_1^n C_iX_i'$$

are identically distributed and Corollary 5 follows from Theorem A applied to the case when $I=\{1,\dots,2n\}$, $\mathcal{J}_1,\mathcal{J}_2$ are empty, $\mathcal{J}_3=\{1,\dots,n\}$.

Corollary 5 coincides with Theorem 8 from Khatri and Rao (1971).

Theorem A generalizes also the results of Kotlarski (1971), Rao (1971) and many of results from Chapter 10 of Kagan, Linnik, Rao (1972)

The results of the paper were announced in Klebanov (1975).

REFERENCES

Cramér H. (1936) : Uber eine Eigenschaft der normalen Verteilungsfunktion. Math. Zs. 41 (1936), 405-414.

Darmois G. (1953) : Analyse générale des liaisons stochastiques. Rev. Inst. Intern. Statist. 21 (1953), 2-8.

Ghurye S.G., Olkin I. (1962) : A characterization of the multivariate normal distribution. Ann. Math. Statist. 33 (1962), No. 2, 533-541.

Heyde C.C. (1970) : Characterization of the normal law by the symmetry of a certain conditional distribution. Sankhyā A 32 (1970), No.1, 115-118.

Kagan A.M.,Linnik Yu.V.,Rao C.R. (1972) : Characterization problems of mathematical statistics. Russian edition, Moskow 1972. English edition, John Wiley, New York 1973.

Khatri C.G.,RAo C.R. (1968) : Solutions to some functional equations and their applications to characterization of probability distributions. Sankhyā A 30 (1968), 167-180.

Khatri C.G. (1971) : On characterization of gamma and normal distributions by solving some functional equations in vector variables, J. Multivariate Analysis 1 (1971), 70-89.

Khatri C.G.,Rao C.R. (1971) : Functional equations and characterization of probability laws through linear functions of random variables. Indian Stat. Inst., Discussion Paper (1971), No.68, 1-18.

Klebanov L.B. (1975) : On the condition of the identical distribution of linear forms in a speciel case. Teoriya Veroyatn. i yeye priminen. 20 (1975), No.3, 684-685. (in Russian).

Kotlarski I. (1971) : On a characterisation of probability distributions by the joint distribution of their linear functions. Sankhyā a 33 (1971), 73-80.

Marcinkiewicz I. (1938) : Sur une propriété de la loi de Gauss. Math. Zs. 44 (1938), No. 4-5, 622-638.

Rao C.R. (1971) : Characterization of probability laws through linear functions. Sankhyā A 33 (1971), 255-259.

Skitovič V.P. (1953) : On a property of normal distribution. Doklady Acad. Nauk SSSR 18 (1953), No.2, 217-219.

Leningrad Civil Engeneering Institute

II Krasnoarmeiskaja 4.,
Leningrad,
U.S.S.R.

ON THE ASYMPTOTIC SEPARABILITY OF TWO GAUSSIAN SEQUENCES

Nerutė Kligienė

Vilnius

ABSTRACT

The asymptotic behaviour of the error probability is investigated while discriminating two normal distributions $\mathcal{N}(\mu_{1N}^{(i)}, \Sigma_N^{(i)})$, $i=1,2$ using the likelihood ratio on the basis of an available realization $x_1, \ldots, x_N$ of the random sequence $\{X_t\}$ satisfying the regularity condition. The error probability is proved to decrease exponentially when N increases. The rate of convergence is expressed in terms of the known parameters of Gaussian distributions.

INTRODUCTION

Let $\{X_t, t=0,\pm 1,\ldots\}$ be a stationary sequence of random variables on the measurable space $(\mathcal{X}, \mathcal{B})$. Denote the measurable space corresponding to random variables $X_{st} = (X_s, X_{s+1}, \ldots, X_t)'$, $s \le t$; $s,t = 0, \pm 1, \ldots$ by $(\mathcal{X}_{st}, \mathcal{B}_{st})$. Let either $P^{(1)}$ or $P^{(2)}$ be the probability measure induced by the infinite sequence $\{X_t\}$ on $(\mathcal{X}, \mathcal{B})$ and $P_{1N}^{(1)}$, $P_{1N}^{(2)}$ be the restrictions of $P^{(1)}$ and $P^{(2)}$ on the $(\mathcal{X}_{1N}, \mathcal{B}_{1N})$. By H_1 and H_2 denote the respective statistical hypotheses occurring with a priori probabilities p and $q = 1-p$. There is no restriction to take $p=q=\frac{1}{2}$ when investigating the asymptotic behaviour of the probability of error. We shall study the maximum likelihood error probabilities

$$\text{(1)} \quad \varepsilon_N(H_1) = P\{T_N \le 0 \mid H_1\}, \quad \varepsilon_N(H_2) = P\{T_N > 0 \mid H_2\}$$

where

$$T_N = \log\left[\frac{dP_{1N}^{(1)}}{dx} \Big/ \frac{dP_{1N}^{(2)}}{dx}\right]$$

and $\mathcal{X}$ is a measure dominating $P_{1N}^{(1)}$ and $P_{1N}^{(2)}$ while discriminating H_1 and H_2 on the base of a growing number of the observed random variables $x_{1N}=(x_1,\dots,x_N)'$ on the sequence $\{X_t, t=0,\pm1,\dots\}$. The asymptotic rate according to results of A.Perez (1973) is given by

$$\begin{aligned}(2)\qquad &\lim_{N\to\infty}\frac{1}{N}\log\mathcal{E}_N(H_1)=\lim_{N\to\infty}\frac{1}{N}\log\mathcal{E}_N(H_2)=\\ &=\lim_{N\to\infty}\frac{1}{N}\log\left[\min_{0\le\alpha\le1}H_\alpha(P_{1N}^{(1)},P_{1N}^{(2)})\right]=h_{\alpha_0}(P^{(1)},P^{(2)}),\end{aligned}$$

where the function $H_\alpha(P_{1N}^{(1)},P_{1N}^{(2)})$ is defined by

$$(3)\qquad H_\alpha(P_{1N}^{(1)},P_{1N}^{(2)})=\int_{\mathcal{X}}\left(\frac{dP_{1N}^{(1)}}{dx}\right)^{\alpha}\left(\frac{dP_{1N}^{(2)}}{dx}\right)^{1-\alpha}dx$$

for real α, $0\le\alpha\le1$. The function $h_{\alpha_0}(P^{(1)},P^{(2)})$ is called minimum alpha-entropy rate of $P^{(1)}$ with respect to $P^{(2)}$ and general conditions on the pair $(P^{(1)},P^{(2)})$ were derived by A.Perez (1973) for the existence of the limits (2).

In the sequel we shall restrict ourselves with the study of two normal distributions for the sake of applicability of the general results mentioned above in evaluating the error probabilities while discriminating two statistical hypotheses.

THE ASYMPTOTIC BEHAVIOUR OF THE ERROR PROBABILITIES

Let $\{X_t, t=0,\pm1,\dots\}$ be a stationary sequence with a spectral density $f(\lambda)$ satisfying a regularity condition, i.e. there exist real constants $\{c_\nu\}$ and independent identically distributed random variables $\{\xi_t, t=0,\pm1,\dots\}$ with $\mathcal{E}\xi_t=0$, $\mathcal{E}\xi_t^2=\sigma^2$ such that

$$(4)\qquad X_t=\mu+\sum_{\nu=0}^{\infty}c_\nu\xi_{t-\nu},\quad c_0\equiv1.$$

Denote $\mu_{1N}=\mathcal{E}\{X_{1N}\}=(\mu,\dots,\mu)'$, $\Sigma_N=\{\mathcal{E}(X_{1N}-\mu_{1N})(X_{1N}-\mu_{1N})'\}$ and consider a normally distributed vector $X_{1N}\sim\mathcal{N}(\mu_{1N},\Sigma_N)$ with a components satisfying (4) and the observed value $x_{1N}=(x_1,\dots,x_N)$.

We shall evaluate the maximum likelihood error probabilities $\mathcal{E}_N(H_i)$, $i=1,2$ while discriminating the hypotheses

(5) $$H_i : X_{1N} \sim \mathcal{N}(\mu_{1N}^{(i)}, \Sigma_N^{(i)}) \quad i=1,2$$

on the base of a growing number of $x_1,\dots,x_N$, when $\Sigma_N^{(1)}, \Sigma_N^{(2)}$ are known non-singular matrices, $\mu_{1N}^{(1)}$ and $\mu_{1N}^{(2)}$ are known vectors.

Assume $\Sigma_N^{(i)}$ corresponds to the variance σ_i^2 in (4) and to the spectral density $f_i(\lambda)$. Denote matrix $R_N=[\Sigma_N^{(1)}]^{-1}\cdot\Sigma_N^{(2)}$, $r(\lambda)=\det(R_N-\lambda I)$ its characteristic polynomial, $g(\alpha)=\alpha^N\cdot r(\frac{\alpha-1}{\alpha})$ and define α_{0N}, $0\leq\alpha_{0N}\leq 1$, as a solution of the following equation

(6) $$g'(\alpha)/g(\alpha) = \log\det R_N .$$

Theorem 1. Let $\Sigma_N^{(1)}\neq\Sigma_N^{(2)}$ be known non-singular matrices, $\sigma_1^2>\sigma_2^2$. In testing hypotheses H_i (5) the error probabilities $\varepsilon_N(H_i)$, $i=1,2$ satisfy

(7) $$\lim_{N\to\infty}\frac{1}{N}\log\varepsilon_N(H_i)=\ln\left(\frac{\sigma_2}{\sigma_1}\right)^{\alpha_0}-\frac{1}{4\pi}\int_{-\pi}^{\pi}\log\left[\alpha_0-\frac{f_1(\lambda)}{f_1(\lambda)-f_2(\lambda)}\right]d\lambda,$$

where $\alpha_0=\lim_{N\to\infty}\alpha_{0N}$ and α_{0N} was defined as the solution of equation (6).

Proof. According to the assumptions (5), the densities of the random vector X_{1N} are

(8) $$\frac{dP_{1N}^{(i)}}{dx}=(2\pi)^{-\frac{N}{2}}\left(\det\Sigma_N^{(i)}\right)^{-\frac{1}{2}}\exp\left\{-\frac{1}{2}(x_{1N}-\mu_{1N}^{(i)})'[\Sigma_N^{(i)}]^{-1}(x_{1N}-\mu_{1N}^{(i)})\right\}$$

and in this case the function $H_\alpha(P_{1N}^{(1)},P_{1N}^{(2)})$ may be obtained explicitly by integrating expression (3) when $dP_{1N}^{(i)}/dx$ are given by (8). Consequently the expression

(9) $$\ln H_\alpha(P_{1N}^{(1)},P_{1N}^{(2)})=\frac{1}{2}\ln\frac{(\det\Sigma_N^{(1)})^{\alpha}(\det\Sigma_N^{(2)})^{1-\alpha}}{\det(\alpha\Sigma_N^{(1)}+(1-\alpha)\Sigma_N^{(2)})}-$$
$$-\frac{\alpha(\alpha-1)}{2}(\mu_{1N}^{(2)}-\mu_{1N}^{(1)})'[\alpha\Sigma_N^{(1)}+(1-\alpha)\Sigma_N^{(2)}]^{-1}(\mu_{1N}^{(2)}-\mu_{1N}^{(1)})$$

consists oftwo terms. The first term may be considered as the class separability due to the difference of covariance matrices, the second one may be considered as the class separability due to the mean difference when two covariance matrices are equal to each other and to $\alpha\Sigma_N^{(1)}+(1-\alpha)\Sigma_N^{(2)}$. Let us concentrate our attention on the case $\Sigma_N^{(1)}\neq\Sigma_N^{(2)}$ and $\mu_{1N}^{(1)}=\mu_{1N}^{(2)}$ because the case of equal covariance matrices and different means is investigated rather well.

Then

(10)
$$H_\alpha(P_{1N}^{(1)},P_{1N}^{(2)})=\frac{(\det\Sigma_N^{(1)})^{-\frac{\alpha}{2}}(\det\Sigma_N^{(2)})^{\frac{\alpha}{2}}}{\alpha^{\frac{N}{2}}\det\{[\Sigma_N^{(1)}]^{-1}\cdot\Sigma_N^{(2)}-\frac{\alpha-1}{\alpha}I\}^{\frac{1}{2}}}=$$
$$=\frac{(\det R_N)^{\frac{\alpha}{2}}}{\alpha^{\frac{N}{2}}\, r^{\frac{1}{2}}(\frac{\alpha-1}{\alpha})},$$

where $r(\lambda)=(-1)^N\lambda^N+r_1\lambda^{N-1}+\dots+r_{N-1}\lambda+r_N$ denotes the characteristic polynomial of the matrix $R_N=[\Sigma_N^{(1)}]^{-1}\cdot\Sigma_N^{(2)}$. Noticing that $\det R_N=r_N$, let us introduce the polynomial

(11)
$$g(\alpha)=\alpha^N\cdot r(\tfrac{\alpha-1}{\alpha})=(-1)^N(\alpha-1)^N+r_1(\alpha-1)^{N-1}\alpha+\dots$$
$$\dots+r_{N-1}(\alpha-1)\alpha^{N-1}+r_N\alpha^N=\prod_{j=1}^N g_0\cdot(\alpha-\alpha_{j,N}),$$

where $\alpha_{j,N}$, $j=1,2,\dots,N$ are zeros of the $g(\alpha)$.

Then we get

(12)
$$H_\alpha(P_{1N}^{(1)},P_{1N}^{(2)})=\left(\frac{r_N^\alpha}{g(\alpha)}\right)^{\frac{1}{2}}$$

or

(13)
$$\log H_\alpha(P_{1N}^{(1)},P_{1N}^{(2)})=\tfrac{1}{2}(\alpha\log r_N-\log g_0-\sum_{j=1}^N\log(\alpha-\alpha_{j,N}).$$

Let the value α_{0N}, $0\leq\alpha_{0N}\leq 1$ be the solution of the equation

(14)
$$\frac{d\log H_\alpha}{d\alpha}=\log r_N-\sum_{j=1}^N\frac{1}{\alpha-\alpha_{j,N}}=0$$

that minimizes $H_\alpha(P_{1N}^{(1)}, P_{1N}^{(2)})$ because

$$\frac{d^2 \log H_\alpha}{d\alpha^2} = \sum_{j=1}^{N} (\alpha - \alpha_{j,N})^{-2} > 0 .$$

The equation

(15) $$\sum_{j=1}^{N} (\alpha - \alpha_{j,N})^{-1} = \log r_N$$

is equivalent to equation (6) and it is not difficult to prove the existence of their solution α_{0N}, $0 \leq \alpha_{0N} \leq 1$ using the correspondence $\alpha_{j,N} = (1 - \lambda_{j,N})^{-1}$ between the eigenvalues $\lambda_{j,N}$, $j=1,2,\ldots,N$ of the matrix R_N and the zeros $\alpha_{j,N}$, $j=1,2,\ldots,N$ of the polynomial $g(\alpha)$.

The relationships (2), (6), (13) lead us to

(16) $$h_{\alpha_0}(P^{(1)}, P^{(2)}) = \lim_{N\to\infty} \frac{1}{N} \log H_{\alpha_{0N}}(P_{1N}^{(1)}, P_{1N}^{(2)}) = = \frac{1}{2} \lim_{N\to\infty} \frac{1}{N} [\alpha_{0N} \log \det R_N - \log g(\alpha_{0N})].$$

Define $\alpha_0 = \lim_{N\to\infty} \alpha_{0N}$, $0<\alpha_0<1$, then

(17) $$h_{\alpha_0}(P^{(1)}, P^{(2)}) = \frac{\alpha_0}{2} \lim_{N\to\infty} \frac{1}{N} \log \det R_N - \frac{1}{2} \lim_{N\to\infty} \log g(\alpha_{0N}).$$

Denote $\lambda_{j,N}^{(1)}, \lambda_{j,N}^{(2)}, j=1,2,\ldots,N$ the corresponding eigenvalues of the matrices $\Sigma_N^{(1)}$ and $\Sigma_N^{(2)}$, both Teoplitz, due to the stationarity of the sequence $\{X_t\}$. Using the asymptotic properties of the eigenvalues of Teoplitz matrices investigated in U. Grenander and G. Szego (1958) we derive

(18) $$\lim_{N\to\infty} \frac{1}{N} \log \det R_N = \lim_{N\to\infty} \frac{1}{N} \sum_{j=1}^{N} \log \frac{\lambda_{j,N}^{(2)}}{\lambda_{j,N}^{(1)}} = \frac{1}{2\pi} \int_{-\pi}^{\pi} \log \frac{f_2(\lambda)}{f_1(\lambda)} d\lambda.$$

The validity of (4) in addition to the results proved by A.N. Kolmogorof (1941) enables us to express the last integral by means of variances σ_1^2, σ_2^2 of the random variables $\{\xi_t\}$:

(19) $$\frac{1}{2\pi} \int_{-\pi}^{\pi} \log \frac{f_2(\lambda)}{f_1(\lambda)} d\lambda = \ln \frac{\sigma_2^2}{\sigma_1^2} .$$

Expressions (13),(17) lead us to

$$\alpha_{j,N} = (1-\lambda_{j,N})^{-1} = \lambda_{j,N}^{(1)} / (\lambda_{j,N}^{(1)} - \lambda_{j,N}^{(2)}), \tag{20}$$

$$\lim_{N\to\infty} \frac{1}{N} \log g(\alpha_{0N}) = \lim_{N\to\infty} \frac{1}{N} \sum_{j=1}^{N} \log\Big[\alpha_{0N} - \frac{\lambda_{j,N}^{(1)}}{\lambda_{j,N}^{(1)} - \lambda_{j,N}^{(2)}}\Big].$$

Due to the fact that the last expression is a continuous function of the eigenvalues $\lambda_{j,N}^{(1)}, \lambda_{j,N}^{(2)}$ on the appropriate interval, the statement of theorem 4.1 presented by R.M. Gray (1972) is valid, i.e.

$$\lim_{N\to\infty} \frac{1}{N} \log g(\alpha_{0N}) = \lim_{N\to\infty} \frac{1}{N} \sum_{j=1}^{N} \log(\alpha_{0N} - \alpha_{j,N}) =$$

$$= \lim_{N\to\infty} \frac{1}{N} \sum_{j=1}^{N} \log\Big[\alpha_{0N} - \frac{\lambda_{j,N}^{(1)}}{\lambda_{j,N}^{(1)} - \lambda_{j,N}^{(2)}}\Big] = \frac{1}{2\pi} \int_{-\pi}^{\pi} \log\Big[\alpha_0 - \frac{f_1(\lambda)}{f_1(\lambda) - f_2(\lambda)}\Big] d\lambda \tag{21}$$

Equations (16)-(21) give us the result

$$h_{\alpha_0}(P^{(1)}, P^{(2)}) = \ln\Big(\frac{\sigma_2}{\sigma_1}\Big)^{\alpha_0} - \frac{1}{4\pi} \int_{-\pi}^{\pi} \log\Big[\alpha_0 - \frac{f_1(\lambda)}{f_1(\lambda) - f_2(\lambda)}\Big] d\lambda \tag{22}$$

that in addition to (2) completes the proof of Theorem 1.

In concluding this paper let us remark that the asymptotic behaviour of the error probability $\varepsilon_N(H_i)$ is stipulated essentially by variances σ_1^2, σ_2^2 of the corresponding independent variables and spectral densities $f_1(\lambda), f_2(\lambda)$. The expression (7) is rather simple and may be useful in practice when it is necessary to evaluate the separability of two random processes in principle.

REFERENCES

Gray R.M. (1972) : On the asymptotic eigenvalue distribution of Teoplitz matrices. IEEE Trans. on Information Theory IT-18 (1972), No 6, 725-730.

Grenander U., Szego G. (1958) : Teoplitz forms and their applications. Univ. of Calif. Press, Berkeley, 1958.

Колмогоров А.Н. (I94I): Интерполирование и экстраполирование стационарных случайных последовательностей. Изв. АН СССР, сер. мат. 5(I94I), № 5, 3-I4.

Perez A. (1973) : Asymptotic discernibility of random processes. In: Proc. of the Prague Symposium on Asymptotic Statistics, 3-6 Sept. 1973, 311-322.

Lithuanian Academy of Sciences
Institute of Mathematics and Cybernetics

K. Poželos 54
232600 Vilnius
Lithuanian SSR, USSR

SOME REMARKS ON PROBABILITIES OVER FORMALIZED LANGUAGES

Ivan Kramosil

Prague

ABSTRACT

Some applications of probability theory, its methods and results in the field of automated problem solving and artificial intelligence in general need the possibility to ascribe probability values to well-formed formulas of a formalized language. First, a possibility is studied how to convert, using the model theory, such a formula into a random event in its classical set-theoretic sense. Second an, immediate definition of probability measure as a real-valued function over a formalized language is suggested. The obtained notions and assertions enable to formalize and develop some mathematical tools important when introducing the stochastic methods into the domain of automated problem solving (e.g., the notion of the so called randomized incidental phenomena).

1. A SEMANTICALLY-ORIENTED INTERPRETATION OF PROBABILITIES OVER FORMULAS

Since its origins in 18-th century the probability theory had passed a rather long way of developing and modifying before taking its first formalized and logically consistent form due to Kolmogorov (Kolmogorov (1933)). This conception, known also as axiomatic probability theory and considered, now, as the classic one is based on notions and methods known and borrowed from measure theory and measurable and integrable function theory. These fields of mathematics had achieved a great success just in the period preceding the mentioned Kolmogorov's basic work, so his intention and effort to profit of these branches, their methods and results, was quite natural.

In axiomatic probability theory random events are conceived as point sets. The construction starts from a nonempty set Ω and its points (elements) ω are called elementary random events (Ω itself is sometimes called the universum of elementary random events). Certain subsets of Ω are proclaimed axiomatically to be random events and they are supposed to form a σ-algebra (σ-field). In another words, random events are defined by specifying a σ-algebra $\mathcal{S}$ of subsets of Ω (i.e. $\mathcal{S} \subset \mathcal{P}(\Omega)$). Defining a probability measure P on $\mathcal{S}$ we have completed the construction of the so called probability space $\langle \Omega, \mathcal{S}, P \rangle$, the basic notion of axiomatic probability theory.

It is a matter of fact that in probability theory, as far as the author knows, the differences between a language of this theory and a corresponding necessary metalanguage are not studied in such a detail as in other branches of pure mathematics. Namely, when defining the σ-algebra $\mathcal{S}$ of random events, usually a combined theoretic and metatheoretic way is used. First, a generating class of random events is defined within the framework of a theory, second, this class is extended to the minimal covering σ-algebra (a metatheoretical construction). The random events by which the original class has been enriched are not, in general, definable in the theory in question itself, however, as they are, considered as sets, defined unambiguously on the metalevel, this circumstance is not considered by the classical probability theory as a weak point.

This approach, even if, maybe, justifiable and legitimate from the viewpoint of a practically oriented mathematician or statistician must be submitted to a serious criticism if we consider some systems of the so called artificial intelligence (AI) together with the possibility of such a system to use some probabilistically or statistically based procedures, e.g. when planning its future activity. An at least partial formal description of the environment in which an "intelligent" automaton is situated seems to be necessary for such a device to organize its autonomous interaction with this environment. A formalized language is an inevitable tool of such a description which means, returning back to the question of probabilistic and statistical applications in AI, that the device is able to perform statistical or probabilistic considerations only within the framework of a given formalized language. The borderlines between this language and its metalanguage are, from the point of view of an automaton, absolute and cannot be crossed. A particular consequence of this situation consists in the fact that the only way in which an AI system or device is able to define a random event and to handle it (e.g., combining it with some other random event, to ascribe a probability to the event or to estimate it on the ground of relative frequencies etc.) is to define the random event by a formula of the formalized language being at its disposal, i.e. to define it as the set of all elements of a universum which satisfy the defining formula. This gives that the descriptions of the type P (A) with P being a probability measure and A being a well-formed formula of the formalized language in question are unevitable in any mathematical paper dealing with stochastically oriented methods of AI. In this section we offer a formal justification of such expressions together with their semantics; an interesting particular case of the so called randomized incidental phenomena will be investigated in Section 2.

Definition 1. Let (,) and * be abstract symbols, by the set of types we shall call the minimal set τ of strings over $\{(\,)\,,\,*\}$ such that

(i) $* \in \tau$,

(ii) if $n \geq 1$ and $c_1, c_2, \ldots, c_n \in \tau$, then $(c_1, c_2, \ldots, c_n) \in \tau$.

The elements of τ are called types.

The notion of type is borrowed from simple type theory (see the description given in Mostowski (1948)). Symbols ( ) , and * are supposed to be different from all symbols in formalized languages introduced below.

In what follows we suppose that the logical connectives and quantifiers are the same in all the languages we shall investigate, the rules of forming well-formed formulas are also supposed to be given as well as a sequence of sequences of logical indeterminates of all types. Under these conditions any formalized language $\mathcal{L}$ is defined by a sequence $\{a_i\}_{i=1}^{n}$, $n \leq \omega$, where any a_i is either a constant symbol, or a function symbol or a relation symbol of certain type (here ω is the first infinite ordinal number).

Definition 2. Let $\mathcal{L} = \{a_i\}_{i=1}^{n}$, $n \leq \omega$, be a formalized language. Define a sequence $\Sigma = \{\sigma_i\}_{i=1}^{n}$ in this way:

(i) $\sigma_i \in \{0,1,2\} \times \tau$, $i = 1,2,\dots,n$,
(ii) if a_i is a constant symbol of type c, then $\sigma_i = \langle 0,c \rangle$,
(iii) if a_i is a function symbol of type c, then $\sigma_i = \langle 1,c \rangle$,
(iv) if a_i is a relation symbol of type c, then $\sigma_i = \langle 2,c \rangle$.

Then the sequence $\Sigma = \Sigma(\mathcal{L})$ is called <u>signature</u> of the language $\mathcal{L}$.

As can be easily seen, constants can be only of the type $*$, i.e. names of elements in an interpretation domain, so if $\sigma_i = \langle 0,c \rangle$, then $c = *$. Hence, the definition of signature could be simplified, but this is not necessary for our purposes.

Another notion necessary in order to build our interpretation of probabilities over formulas is that of a (semantic) relational structure. Any relational structure $\mathcal{R}$ is defined by, first, a nonempty set M and, second, by a sequence $\{b_i\}_{i=1}^n$, $n \leq \omega$, where any b_i is either an element of M, or a function of certain type defined in M or power sets of M with values in M or a power set of M, or a relation of certain type defined in M or power sets of M. Replacing in Definition 2 the words "constant symbol" by "element of M", "function symbol" by "function" and "relation symbol" by "relation" we obtain a definition of the signature of a relational structure $\mathcal{R}$. If, for a formalized language $\mathcal{L}$ and a relational structure $\mathcal{R}$, $\Sigma(\mathcal{L}) = \Sigma(\mathcal{R})$, then $\mathcal{L}$ can be interpreted in $\mathcal{R}$, or $\mathcal{R}$ can serve as a model of $\mathcal{L}$, simply ascribing a_i to b_i and vice versa, and any closed formula of $\mathcal{L}$ (i.e. containing no free indeterminates) becomes true of false in this interpretation or model. These notions of true and false formulas in $\mathcal{R}$ are defined in the usual way with which the reader is supposed to be familiar and the fact that a closed formula A from $\mathcal{L}$ is valid in the relational structure $\mathcal{R}$ is denoted by $\mathcal{R} \models A$.

Consider a formalized language $\mathcal{L}$ and a nonempty set Ω of relational structures such that, for any $\omega \in \Omega$, $\Sigma(\omega) = \Sigma(\mathcal{L})$. We use the symbols ω , Ω because of their common use in probability theory and let us hope that no misunderstanding of ω as the first infinite ordinal number can occur. Denote, for any closed formula $A \in \mathcal{L}$,

$$V(A) = \{\omega : \omega \in \Omega, \omega \models A\},$$

denote, also,

$S_o = \{V(A) : A \in \mathcal{L}, A \text{ closed}\}$.

Of course, V(A) and S_o depend also on Ω, however, this dependence is not important enough, in what follows, to be expressed explicitly.

<u>Theorem 1.</u> S_o is a set-theoretical algebra, but not generally a σ-algebra (σ-field).

<u>Proof.</u> The definition of satisfiability in a model gives that if not $\omega \models A$, $\omega \in \Omega$, then $\omega \models \neg A$. Hence, $\Omega - V(A) = V(\neg A) \in S_o$. Moreover, $\omega \models A \vee B$ iff $\omega \models A$ or $\omega \models B$, so $V(A) \cup V(B) = V(A \vee B) \in S_o$ and the first part of Theorem 1 is proved. $\mathcal{L}$, as well as S_o, are infinite but countable, while σ-algebras, as complete algebras, can be only finite or of cardinality at least of the continuum (Stone theorem), hence, with the exception of finite cases, S_o is not a σ-algebra. Q.E.D.

Let $\mathcal{S}$ be any σ-algebra containing S_o, let P be a probability measure on the measurable space $\langle \Omega, \mathcal{S} \rangle$. Now, the expression P(A), A being a closed formula from $\mathcal{L}$, can be understood as a conventional abbreviation of P(V(A)).

Let us remark that our interpretation of P(A) does not touch the classical probability in any way, as we have reduced probabilities over formulas to classical probabilities over random events as sets. Our way of reasoning was very close to that of Kripke (1963) and (1965) when he builds a semantic system for modal logics using the notion of "possible worlds" - our conception is, in fact, nothing else than a randomization of this" space of possible worlds".

2. AN APPLICATION TO AUTOMATED PROBLEM SOLVING: RANDOMIZED INCIDENTAL PHENOMENA

A need for a formal representation of the environment is almost absolute for each automaton-environment system or robot to be able to make "intelligent" decisions in this environment. A formal structure for such a representation can be based on the concept of the <u>image space</u>, cf. Štěpánková, Havel (1976). Mathematically, an image space I is a collection of formal theories with a common first-order language $\mathcal{L}$ and a common subset T_I of axioms (called the <u>core theory</u>

of I) representing the unchangeable facts about the environment (thus an "image" can be treated as an extension $T_I[A]$ of T_I by a "specific" axiom $A \in \mathcal{L}$) . In addition, a set Ξ of operators is given, each $\varphi \in \Xi$ presented by a pair $\langle C_\varphi, R_\varphi \rangle \in \mathcal{L}^2$ (the condition and result of φ, respectively). An operator φ is applicable in $T_I[A]$ iff $T_I[A] \vdash C_\varphi$, the outcome of such an application is then $T_I[R_\varphi]$. A problem in I is just a pair $\langle X, Y \rangle \in \mathcal{L} \times \mathcal{L}$ (the initial and the goal formulas, respectively).

Usually, formulas C_φ and R_φ describe only the important and essential properties which the environment is to satisfy before or satisfies after an application (execution) of φ (so C_φ and R_φ can be called attributes of φ). The desire of an exhaustive description of the two states of the environment (before and after) by C_φ and R_φ would lead to an enormous size of the set of operators. The notion of incidental phenomenon seems to offer a partial remedy. A pair $\langle A, B \rangle \in \mathcal{L} \times \mathcal{L}$, A, B closed, is called an incidental phenomenon with respect to an operator φ ($\langle A, B \rangle \in \mathrm{Inc}\,\varphi$, in symbols), if the simultaneous validity of C_φ and A followed by the execution of φ result in the simultaneous validity of R_φ and B. In the stage of planning an incidental phenomenon $\langle A, B \rangle \in \mathrm{Inc}\,\varphi$ represents a tool which can but also need not be used in a particular case and an intelligent choosing of appropriate incidental phenomena can significantly simplify the plan formation (hence, $\langle C_\varphi, R_\varphi \rangle$ can be seen as a "basic" or "trivial" incidental phenomenon with respect to φ, whose use is obligatory in any case when φ is considered). The notion of incidental phenomenon can be easily extended in the way enabling to define incidental phenomena with respect to a finite string of operators, not only to single operators themselves.

The more incidental phenomena are at our (or robot's) disposal, the easier is to choose an appropriate operator in a situation and, after all, the easier is to find a desired plan. This idea, together with the fact that the dependences in a real world are of a stochastic rather than of a deterministic type lead to a temptation to include a pair $\langle A, B \rangle$ into $\mathrm{Inc}\,\varphi$ also in case it is only "highly probable" that validity of C_φ and A followed by the execution of φ would result in the simultaneous validity of R_φ and B. Trying to formalize this idea we must ascribe to some (or all) pairs $\langle A, B \rangle \in \mathcal{L}^2$ the probability that if $C_\varphi \& A$ ($C_{\varphi_1} \& A$, resp.) holds, then the execution of φ ($\varphi_1 \ldots \varphi_n$, resp.) assures the validity of $R_\varphi \& B$

(R_{φ_n} &B, resp.). Denoting such a probability by $P_\varphi(A,B)$ ($P_{\varphi_1\dots\varphi_n}(A,B)$, resp.) we are able, using the ideas of Section 1, to convert this expression into an expression of classical probability theory with the usual set-theoretic semantics.

Denote, for a relational structure $\mathcal{R}$ such that $\Sigma(\mathcal{R}) = \Sigma(\mathcal{L})$ by $Tr(\mathcal{R})$ the set of all closed formulas from $\mathcal{L}$, which are valid in $\mathcal{R}$, i.e.

$$Tr(\mathcal{R}) = \{A : A \in \mathcal{L}, \text{ A closed}, \mathcal{R} \models A\}.$$

Denote, for a subset $\mathcal{L}' \subset \mathcal{L}$, by $Cn(\mathcal{L}')$ the set of all logical consequences of formulas from $\mathcal{L}'$ with respect to usual deduction rules. When mentioning the notion of operator at the beginning of this section we used the expression "presented" for describing the relation between an operator φ and the corresponding pair $\langle C_\varphi, R_\varphi \rangle$ of formulas. So we can now define what the operators <u>are</u>.

<u>Definition 3.</u> Let $\mathcal{L}$ be a formalized language, let Ω be a set of relational structures such that $\Sigma(\mathcal{L}) = \Sigma(\omega)$ for any $\omega \in \Omega$. <u>Operator φ presented by $\langle C_\varphi, R_\varphi \rangle \in \mathcal{L}^2$</u> is a partial mapping from Ω into Ω such that

(a) $\varphi(\omega)$ is defined iff $\omega \models C_\varphi$,

(b) if $\varphi(\omega)$ defined, then $\varphi(\omega) \models R_\varphi$.

<u>Theorem 2.</u> Let $\mathcal{L}$ be a formalized language, let Ω be a set of relational structures such that $\Sigma(\mathcal{L}) = \Sigma(\omega)$ for any $\omega \in \Omega$. Let Ξ be a set of operators over Ω such that there are, for each $\varphi \in \Xi$, three sets $\{C_\varphi\}$, $Out\,\varphi$, $In\,\varphi$ of formulas from $\mathcal{L}$ with the following property: for each $\omega \in \Omega$, $\varphi \in \Xi$, if $\varphi(\omega)$ defined, then

$$\varphi(\omega) \in \{\omega' : \omega' \in \Omega, Tr(\omega') = Cn[(Tr(\omega) - Out\,\varphi) \cup In\,\varphi]\}.$$

Then, for any $\varphi \in \Xi$, $A \in \mathcal{L}$, the inverse image of $V(A)$ with respect to φ is measurable with respect to any σ-algebra S_1 covering the algebra S_0 (recall that $S_0 = \{V(A) : A \in \mathcal{L}, \text{ A closed}\}$).

<u>Proof.</u> Clearly, it is sufficient to prove the assertion only in case when S_1 is the minimal σ-algebra over S_0. There are three cases to be considered separately.

(1) Let $A \in In\,\varphi$, then A is valid in $\varphi(\omega)$ in any case when it is defined, i.e., iff $\omega \models C_\varphi$. Hence,

$$\{\omega : \omega \in \Omega, \varphi(\omega) \in \Omega, \varphi(\omega) \models A\} = \{\omega : \omega \in \Omega, \omega \models C_\varphi\} = V(C_\varphi) \in S_0 \subset S_1$$

(2) Let $A \in Out\,\varphi - Cn(Tr(\omega))$, then A is not valid in $\varphi(\omega)$, if $\varphi(\omega)$ defined, so

$$\{\omega : \omega \in \Omega, \varphi(\omega) \in \Omega, \varphi(\omega) \models A\} = \emptyset = \{\omega : \omega \in \Omega, \omega \models A \& \neg A\} = V(A \& \neg A) \in S_0 \subset S.$$

(3) For any other $A \in \mathcal{L}$, A closed, A is valid in $\varphi(\omega)$ iff $\varphi(\omega)$ is defined and A was valid before, i.e. in ω, so

$$\{\omega : \omega \in \Omega, \varphi(\omega) \in \Omega, \varphi(\omega) \models A\} = \{\omega : \omega \in \Omega, \omega \models C_\varphi, \omega \models A\} =$$

$$= \{\omega : \omega \in \Omega, \omega \models C_\varphi \,\&\, A\} = V(C_\varphi \,\&\, A) \in S_0 \subset S_1,$$

which completes the proof. Q.E.D.

Corollary. Under the same conditions as in Theorem 2 and for each finite string $\varphi_1 \varphi_2 \dots \varphi_n$ of operators

$$\{\omega : \omega \in \Omega, \varphi_n(\varphi_{n-1}(\varphi_{n-2} \cdots (\varphi_1(\omega)) \dots)) \in \Omega, \varphi_n(\varphi_{n-1}(\varphi_{n-2} \cdots (\varphi_1(\omega)) \dots)) \models A\} \in S_1.$$

Proof. An easy inductive repeating of the same way of reasoning as in the proof of Theorem 2. Q.E.D.

Under the conditions of Theorem 2 we are able to express the probabilities $P_{\varphi_1\varphi_2\dots\varphi_n}(A,B)$ for randomized incidental phenomena in the form of conditional probabilities with the usual set-theoretic semantics, namely

$$P_{\varphi_1\varphi_2\dots\varphi_n}(A,B) = P(\{\omega : \omega \in \Omega, \varphi_n(\varphi_{n-1} \cdots \varphi_1(\omega) \dots) \in \Omega, \varphi_n(\varphi_{n-1} \cdots \varphi_1(\omega) \dots) \models B\} /$$
$$\{\omega : \omega \in \Omega, \omega \models C_\varphi \,\&\, A\}) =$$

$$= \frac{P(\{\omega : \omega \in \Omega, \varphi_n(\varphi_{n-1} \cdots \varphi_1(\omega) \dots) \in \Omega, \varphi_n(\varphi_{n-1} \cdots \varphi_1(\omega) \dots) \models B, \omega \models C_\varphi \,\&\, A\})}{P(\{\omega : \omega \in \Omega, \omega \models C_\varphi \,\&\, A\})} =$$

$$= \frac{P(\{\omega : \omega \in \Omega, \varphi_n(\varphi_{n-1} \cdots \varphi_1(\omega) \dots) \in \Omega, \varphi_n(\varphi_{n-1} \cdots \varphi_1(\omega) \dots) \models B\} \cap \{\omega : \omega \in \Omega, \omega \models C_\varphi \,\&\, A\})}{P(\{\omega : \omega \in \Omega, \omega \models C_\varphi \,\&\, A\})} =$$

$$= \frac{P(\{\omega : \omega \in \Omega, \varphi_n(\varphi_{n-1} \cdots \varphi_1(\omega) \dots) \in \Omega, \varphi_n(\varphi_{n-1} \cdots \varphi_1(\omega) \dots) \models B\} \cap V(C_\varphi \,\&\, A))}{P(V(C_\varphi \,\&\, A))},$$

if $P(V(C_\varphi \,\&\, A)) > 0$, the value being defined arbitrarily or not being defined otherwise. Theorem 2 assures the measurability of all the subsets of Ω in question.

3. PROBABILITY MEASURES DEFINED ON FORMALIZED LANGUAGES

As we have already mentioned, in any mathematical system or theory intending to serve as a useful tool in an AI system the dominating role is played by the syntax of this theory, not by its semantics which is not "accessible" for the AI system (environment-automaton system, robot). Namely, considering random events the main role will be played by formulas describing these events, not by the set-theoretic interpretation supposed to stand behind in the classical probability theory. We have suggested a possibility how to equip probabilities over formulas by semantics acceptable from the classical point of view, neverthless, there is also another possibility: to define probability measures not as set-theoretic functions, but as appropriate mappings from a formalized language $\mathcal{L}$ into the interval $\langle 0,1\rangle$ of reals.

Definition 4. Let $\mathcal{L}$ be a formalized language, let Ω be a set of relational structures such that $\Sigma(\mathcal{L}) = \Sigma(\omega)$ for each $\omega \in \Omega$. A model-based probability (measure) on $\mathcal{L}$ with respect to Ω is a mapping P from $\mathcal{L}$ into $\langle 0,1\rangle$ satisfying:

(1) $0 \leq P(A) \leq 1$, $P(A \,\&\, \neg A) = 0$, $P(A \vee \neg A) = 1$ for each $A \in \mathcal{L}$ (normality axiom).

(2) let $A_1, A_2, \ldots$ be a sequence of formulas from $\mathcal{L}$ such that

(i) there exists a formula $A \in \mathcal{L}$ containing just one free indeterminate not occuring freely in an A_j, $j = 1,2,\ldots$, and

(ii) for each $i = 1,2,\ldots$ there exists an individual constant e_i of $\mathcal{L}$ such that $i \neq j$ implies $e_i \neq e_j$, $i \neq j$ implies $(\Omega) \models (A(e_i) \rightarrow \neg A(e_j))$, and, for all $i = 1,2,\ldots$, $(\Omega) \models (A(e_i) \equiv A_i)$.

Then $P((\exists x)A(x)) = \sum_{i=1}^{\infty} P(A(e_i))$ (restricted σ-aditivity axiom).

Here $A(e_i)$ denotes the result of replacing of all free occurences of x in A by e_i, $(\Omega) \models A$ denotes "$\omega \models A$ for all $\omega \in \Omega$".

Definition 5. Let $\mathcal{L}$ be a formalized language, let $\vdash$ be a deducibility relation in $\mathcal{L}$ defined by a set of axioms (recursive subset of $\mathcal{L}$) and by deduction rules usual in mathematical logic. A deducibility-based probability (measure) on $\mathcal{L}$ with respect to $\vdash$ is a mapping P from $\mathcal{L}$ into $\langle 0,1\rangle$ satisfying (1) and (2) of Definition 4 with $(\Omega)\models$ replaced by $\vdash$.

Consider a model-or deducibility-based probability P on $\mathcal{L}$, let Ω' be a set of relational structures of the same signature as that of $\mathcal{L}$. P is called invariant with respect to Ω', if $(\Omega') \models (A_1 \equiv A_2)$ implies $P(A_1) = P(A_2)$, for each $A_1, A_2 \in \mathcal{L}$.

<u>Theorem 3.</u> Let P be a model-or deducibility-based probability over a formalized language $\mathcal{L}$, let Ω' be a set of relational structures of the same signature as that of $\mathcal{L}$. Then the invariance of P with respect to Ω' is a sufficient and necessary condition for the possibility to extend P to the minimal σ-algebra over the algebra $\{\{\omega:\omega\in\Omega',\omega\models A\}: A\in\mathcal{L}$, A closed$\}$.

<u>Proof.</u> Denote by S_1' the minimal σ-algebra from the assertion of Theorem 3, denote, for each $A\in\mathcal{L}$, $V_1(A)=\{\omega:\omega\in\Omega',\omega\models A\}$, and set $\tilde{P}(V_1(A))=P(A)$. Consider $V_1(A_1)$ and $V_1(A_2)$ such that $V_1(A_1)=V_1(A_2)$. We must be sure that in this case $\tilde{P}(V_1(A_1))=\tilde{P}(V_1(A_2))$. However, the supposed identity means that $\{\omega:\omega\in\Omega',\omega\models A_1\}=\{\omega:\omega\in\Omega',\omega\models A_2\}$, hence, for each $\omega\in\Omega$, $\omega'\models(A_1\equiv A_2)$. This is just the same as $(\Omega')\models(A_1\equiv A_2)$, hence, the invariance condition has been proved to be sufficient and necessary for extending P onto $S_o'=\{V_1(A): A\in\mathcal{L}$, A closed$\}$. The unique extension of P from S_o' onto S_1' follows immediately from a well-known theorem of measure theory (Halmos (1950), e.g.) Q.E.D.

The probability measures defined in this section seem to be similar to the so called <u>probabilistic logics</u>, however, a very important difference exists: a probabilistic logic is extensional, our probability measures are not. So, in our case, P(A) is not defined by $P(A_1)$, $P(A_2)$, ..., $P(A_n)$, $A_1, A_2, \ldots, A_n$ being the subformulas of A (as it is the case for probabilistic logics). So the more developed possibilities offered by probability theory when various types and degrees of stochastic dependence are to be expressed and preserved (cf. Gaines (1973) for a more detailed study and discussion on this problem). Some connections with modal logics should be also mentioned, not only because of the similar semantics of "possible worlds" but also for the reason that probability theory over a formalized language can be seen as a modal logic the degrees of modality being numerically quantified (cf. Feys (1965) for modal logics). It seems to be highly probable that the more detailed study of these problems would request to introduce and strictly distinguish two formal languages: $\mathcal{L}_1$, in which random events are defined as subsets of a universum, and $\mathcal{L}_2$, in which we should be able to ascribe probabilities to the random events definable in $\mathcal{L}_1$ and to speak about these probabilities. The inclusion $\mathcal{L}_1\subset\mathcal{L}_2$ seems to be quite natural, however, the possibility that $\mathcal{L}_1-\mathcal{L}_2\neq\emptyset$ can be also interesting and worth studying from theoretical as well as philosophical points of view (a worsened possibility of speaking about random events as a penalty for the possibility to speak about their probabilities).

In any case, the relations between theory and metatheory, language and metalanguage in probability theory deserve more detailed and more systematical investigations.

REFERENCES

Feys R. (1965): Modal Logics, Dunond, Paris 1965

Gaines B.R.(1973): Fuzzy Reasoning and the Logics of Uncertainty. Research Report, Man-Machine Systems Laboratory, Dept. of Electrical Eng., Univ. of Essex, Colchester, Essex 1973.

Halmos P.R. (1950): Measure Theory.D. van Nostrand Com., New York -Toronto-London 1950.

Kolmogorov A.N. (1933): Grundgriffe der Wahrscheinlichkeitsrechnung. Springer Verlag, Berlin 1933.

Kripke S.A. (1963): Semantical Analysis of Modal Logic I. Normal Modal Propositional Calculi. Zeitschrift für Math.Logik und Grundlagen der Math. 9 (1963), 67-96.

__________ (1965): Semantical Analysis of Modal Logic II. Non-Normal Modal Propositional Calculi. The Theory of Models. In: Proc. of The 1963 Internat.Symp at Berkeley. North Holland Publ.Comp., Amsterdam 1965, 206-220.

Mostowski A. (1948): Logika matematiczna. PWN, Warszawa 1948.

Štěpánková O., Havel I.M. (1976): A Logical Theory of Robot Problem Solving. Artificial Intelligence 7 (1976), 129-161.

Czechoslovak Academy of Sciences
Institute of Information Theory and Automation
Pod vodárenskou věží 4
182 08 Praha 8-Libeň
Czechoslovakia

ON VALUE OF INFORMATION IN GAME INTERRELATIONS OF ECONOMIC MODELS

Iosif A. Krass

Novosibirsk

ABSTRACT

A rather general dynamic model of a two-person game, each of them controlling an economical dynamic model, is described. The influence of different kinds of information available for each player on the structure of winning strategies is explored.

One of the most complex problems arising in the interrelation of economical systems is the problem connected with reciprocal informativeness of persons controlling systems' dynamics. As it is known, the influence of reciprocal informativeness appears more strongly in game interrelations. Beginning with the works by Isaaks, in the works by Lyapunov, Poletaev, Volokitin and others, and in the author's works as well, one and the same game model of antagonistic interrelations of economics with different suppositions of informativeness of both sides was explored. This game model in the usual classification is reduced to differential or difference game of quality.

We shall describe now the discrete version of the explored game. Each economical system is described by von Neumann-Gale model determined by technological mappings a_1, a_2, correspondingly. These mappings map $R^n_+ = \{x : x \in R^n, x \geq 0\}$ (space of commodities) on Ξ^n (set of all convex closed subsets in R^n_+) and define possible outputs of commodities in moment t with fixed outputs in moment ($t-1$).

Besides mappings $a_i\,(i=1,2)$, there are given two n -dimensional matrices with non-negative elements: S_1 - matrix of the first model relation upon the second one, and S_2 - matrix of the second model relation upon the first one, and also two non-negative vectors $\mathcal{E}_1, \mathcal{E}_2$ - vectors of minimal welfare.

If at some moment t a state $x_2(t)$ of the second model is such that one of the relations

$$x_2(t) \geq \mathcal{E}_2,\; a_2(x_2(t)) \cap (R^n_+ + \mathcal{E}_2) \neq \emptyset \tag{1}$$

is broken and the state of the first one satisfies the constrainings:

$$x_1(t) \geq \mathcal{E}_1,\; a_1(x_1(t)) \cap (R^n_+ + \mathcal{E}_1) \neq \emptyset, \tag{2}$$

then we shall say that the first model won the second one at the step t .(Or, rather, we are to say: "The player controlling the first model won the second player", but for brevity we shall omit the words "player controlling a model").

In analogy is defined the victory of the second model over the first one at step t , and at last the result is draw if both (1) and (2) relations are damaged.

The pair of vectors $(x_1(0), x_2(0))$ is called the initial state of game if these vectors satisfy relations (1), (2) at $t=0$, where $x_1(0)$ is the initial state of the first model and $x_2(0)$ - of the second one.

If $(x_1(t), x_2(t))$ is state of the game at moment t such that relations (1), (2) are satisfied, then the state of the game at moment $(t+1)$ is defined by relations

$$\begin{aligned} &x_1(t+1) + f_1(t-\tau_1, V_1(t-\tau_1)) + S_2 f_2(t-\tau_2, V_2(t-\tau_2)) \in a_1(x_1(t)) \\ &x_2(t+1) + f_2(t-\tau_2, V_2(t-\tau_1)) + S_1 f_1(t-\tau_1, V_1(t-\tau_1)) \in a_2(x_2(t)) \end{aligned} \tag{3}$$

Here $V_i(t)$ is the information about the state of game known to the i -th player at moment t (his informative set); τ_i - is the delay in getting the information and taking the decision; f_i - is the fun-ction mapping $T \times \mathcal{U}_i$ on R^n_+ , where T is the set of natural numbers (the range of definition of time), and $\mathcal{U}_i$ - the space of information of the i -th player ($V_i(t) \in \mathcal{U}_i, t \in T, i \in \{0,1\}$); $f_1(t, V_1(t))$ is the load taken from the output of the first model for the damaging of the second one; similarly, $f_2(t, V_2(t))$ is defined.

In fact, in includings (3) economical steps of each player connected with choosing of the fixed element from set $a_i(x_i(t))$ (output in next period) is hidden. So, besides function f_1 – the first player must define function $\varphi_1: T \times \mathcal{U}_1$ in R^n_+ so that relations

$$
(4) \quad \begin{aligned} x_1(t+1) &= \varphi_1(t-\tau_1, V_1(t-\tau_1)) - f_1(t-\tau_1, V_1(t-\tau_1)) - S_2 f_2(t-\tau_2, V_2(t-\tau_2)) \\ \varphi_1(t-\tau_1, V_1(t-\tau_1)) &\in a_1(x_1(t)) \end{aligned}
$$

are satisfied.

The problem of definition of winning strategy of the first player is emerging, i.e. the problem of searching such functions φ_1, f_1 from the fixed class so that for any strategy of the partner from this class of functions the first player can provide the victory at a step $t<+\infty$ and non-losing at steps $\tau \in (0,t)$, beginning from a state $(x_1(0), x_2(0))$. Besides, the problem definition set of initial states for which the winning strategy exists, must be solved.

In work by I.A.Poletaev (1970) a game of one-commodity models (3) with discrimination of the second player was explored. This discrimination consisted in the following: information about the state of the game was known to the second player after one step delay (i.e. $\tau_1=0, \tau_2=1$), moreover, $V_i(t)=\{(x_1(t), x_2(t))\}$ for any $t\in T$; $i=1,2$. It turned out that the whole space of initial states is divided into three sets: in the first set there is the winning strategy of the first player, in the second set the second player can win, and in the third set each player can provide himself a draw.

Different complications of this game were explored. At first, case $n>1$ was explored; from information point of view this case means more explicit knowledge by the player of the state of game (Krass I.A., Grinko L.F. (1970), Krass I.A. (1975)). Secondly, the case of continuous receiving of information (for $n=1,2$), i.e. the case of differential game was explored (Krass I.A. and Volokitin E.P. (1970); Krass I.A. and Muchsinov M.A. (1972)), and at last the case was explored when the first player knows at the initial moment only the distribution of partner state probabilities and this information is made more precise during the game development (case $n=1$, Malukov V.P. (1976)). In all these cases the existence of optimal winning strategies was proved which lay on the border of admissible control polyhedron defined by system (3), (4).

This result with some additional restrictions takes place for arbitrary n . So, changing of the information set for this play in case of infinite duration of the game has small influence on the kind of winning (i.e. optimal) strategies.

A game of the type described was explored by Malukov (1977) for $n=1$ developed only finite time T_0 which was known before to both players. If at moment $t=T_0$ both relations (1), (2) take place, then it is concluded that it is draw in this game. In this case there are no optimal strategies of the described above types, but there are only strategies in mixed expansion of the game, and they are finite mixes of pure strategies, which lay on the border of the polyhedron defined by relations (3), (4). This result may be applied for case $n>1$ with some restriction on this game.

So we may conclude that from all existing information for controlling of some economical model by player, the knowing of the time of game finishing emerges essential alteration of strategies.

REFERENCES

Poletaev I.A. (1970) Об использовании моделей производства. On using of economical models. In the book: "Исследование моделей производства", Москва, Советское Радио, 1970, 87-99.

Krass I.A., Grinko L.F. (1970) Об одной игре двух экономических моделей On one game of two economical models. In the book: "Исследование моделей производства", Москва, Советское Радио, 1970, 99-114.

Krass I.A. (1975) Конфликтное взаимодействие моделей экономической динамики и сравнение их темпов ростов. The conflict interrelations of economical dynamic models and comparing their factors of growth. In the book: Планирование и управление экономическими целенаправленными системами. Новосибирск, Наука, 1975, 72-87.

Krass I.A., Volokitin E.P. (1970) Один метод исследования дифференциальных игр качества One method of exploration of differential games of quality. In the book: Управляемые системы № 6, Новосибирск, Наука, 1970, 3-16.

Krass I.A., Muchsinov M.A. (1972) Некоторые методы исследования дифференциальных игр качества с тремя терминальными поверхностями. Some methods of exploration of,differential games of quality with third terminal surfaces. In : Управляемые системы № IO. Новосибирск, Наука, I972, IO-26.

Malukov V.P. (1976) О существовании значения динамической игры с предписанной продолжительностью. On existence value of a dynamic game with the known duration.

Malukov V.P. (1977) О конфликтном взаимодействии моделей экономической динамики. On a conflict interrelation of economical models. Transaction IV Conference on cybernetics. Moscow. 1977. 62-63.

the USSR Academy of Sciences
Siberian Branch
Institute of Mathematics

USSR
630090 Novosibirsk 90
Institute of Mathematics

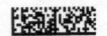